Psychoneuroimmunology, Stress, and Infection

Edited by
Herman Friedman
Thomas W. Klein
Andrea L. Friedman

CRC Press

Boca Raton   New York   London   Tokyo

Library of Congress Cataloging-in-Publication Data

Catalog record is available from the Library of Congress.

© 1996 by CRC Press, Inc.

No claim to original U.S. Government works
International Standard Book Number 0-8493-7638-6
Printed in the United States of America 1 2 3 4 5 6 7 8 9 0
Printed on acid-free paper

FOREWORD

The term "psychoneuroimmunology" defines the emerging science devoted to studying the two-way relationship between the nervous system and the immune system. Through the hypothalamus-pituitary-adrenal axis, the brain exerts profound effects on immune responses. In this regard, interaction between the nervous and immune systems entails a complex series of activities involving responses of many cells to multiple stimuli. For example, many response signals are mediated by cytokines as well as steroids and a variety of neuropeptides or neurohormones. In this regard, it is widely recognized that many types of signals to the brain (i.e., visual-tactile and emotional-) lead the hypothalamus to release corticotrophin releasing factor and, in turn, this stimulates the pituitary gland to release adrenal corticotrophic hormone into the circulation which then stimulates the adrenal cortex to produce glucocorticoid. Glucocorticoids can inhibit the response of lymphocytes to antigenic or mitogenic stimulation, depress cytokine responses by lymphocytes and also suppress natural killer cell activity as well as antibody responses to a variety of antigens, including those associated with microorganisms.

Increased glucocorticoid production is by no means the only response to stress. The levels of several neuropeptides, including vasoactive intestinal peptides, substance P, prolactin and growth hormones are also increased. The levels of several catecholamines, such as epinephrine and neuroepinephrine, are also increased. Localized increases in catecholamine levels may account for suppression of lymphocyte responses to mitogens or antigens. To close the circle, the immune system also returns signals back to the central nervous system. In this regard, immune cells are known to synthesize a number of neuropeptide hormones, including adrenal corticotrophic hormone as well as β-endorphin. Moreover, cytokines produced by lymphocytes and macrophages may act as signals on brain cells. For instance, several interleukins, as well as tumor necrosis factorα, stimulate the hypothalamus, resulting in induction of fever and release of corticosteroid hormones.

Some of the earliest studies connecting stress and infectious diseases were performed at the beginning of this century with tuberculosis and such studies continued over much of this century. More recently, with the current explosion of knowledge concerning immune response mechanisms, a relationship has been found to exist between important major histocompatibility complexes which control the immune response and microbial diseases. Products of the immune response responding to microbial infections, including not only tuberculosis, but also infections caused by other intracellular opportunistic pathogens such as viruses and parasites, have been shown to be related to altered responsiveness. The scope of this book, therefore, concerns the interaction of infectious diseases with both the nervous and the immune systems, i.e., psychoneuroimmunology. This subject has been peripheral in many reviews concerning psychoneuroimmunology as well as individual books on either stress or immunity in general. Therefore the major aim of this volume is to focus on the subject of psychoneuroimmunology as it impacts on stress and the immune system in general and describe specifically how neurological events influence susceptibility and/or resistance to infectious diseases caused by bacteria, viruses, and fungi.

The first chapter deals with an historical overview and background of psychoneuroimmunology, specifically how the brain and the immune system interacts. The author is Dr. Robert Ader, considered the founder of the modern era of investigations on psychoneuroimmunology. Dr. Ader and his colleagues showed that experimental animals primed with an inocuous substance such as saccharin, but also containing cyclophosphamide as a poison, "remembered" the deleterious effects of the cyclophosphamide on their physiology, including the immune system, and thus when exposed to saccharin alone, without cyclophosphamide, still showed profound immunosuppression. Dr. Adrian Dunn, Professor of Pharmacology at Louisiana State University in Shreveport, presents an overview of the central nervous system, hormones and the immune response mechanism.

Further chapters then review important subjects of neuropeptide hormones and infectious diseases, as well as the specific role of specific nervous system organs such as the pituitary gland, and infections. The relationship between neurohormones, aging and specific areas of psychoneuroimmunology are also covered, as well as specific subjects such as specific infectious agents like tuberculosis, Legionella pneumophila, Listeria, etc., and similar intracellular opportunistic bacteria are described. A number of reviews describe viral infections, including viruses that cause upper respiratory infections as well as Herpes viruses and stress. The role of stress and susceptibility to parasitic infections is also discussed.

It is the expectation of the editors as well as the authors of the individual chapters that this volume will provide valuable information concerning the broad aspect of stress, infection and psychoneuroimmunology, and also stimulate interest among the biomedical community, physicians and infectious disease specialists to gain a new appreciation and indeed interest in the relationship between the brain, the immune response system, stress and susceptibility to infectious diseases. Thus, it is hoped this volume will be of interest not only to immunologists, pharmacologists, and microbiologists, but also to oncologists, neuroscientists, and neuroendocrinologists. The editors take this opportunity to express gratitude to Ms. Ilona Friedman for outstanding assistance as coordinator and managing editor for this volume. The editors also wish to thank the senior editors of CRC Press, including Ms. Marsha Baker and Ms. Susan Fox, for helpful advice and assistance in the preparation of this volume.

Herman Friedman
Thomas Klein
Andrea L. Friedman

Tampa, FL
May 1995

THE EDITORS

Herman Friedman, Ph.D., is Distinguished Professor and Chairman of the Department of Medical Microbiology and Immunology at the University of South Florida College of Medicine, Tampa, FL. He received an A.B. in biology and A.M. in bacteriology from Temple University, Philadelphia, PA. He then received his Ph.D. degree in microbiology and immunology from Hahnemann University College of Medicine in 1957, in Philadelphia, PA. He served as Head of the Department of Microbiology and Immunology at the Albert Einstein Medical Center in Philadelphia for nearly 20 years and was Professor in the Departments of Microbiology and Immunology at Temple University School of Medicine and School of Dental Medicine at the same time. He relocated to Tampa in 1978, when he was appointed to the faculty of the University of South Florida. Dr. Friedman is a member of many U.S. and international biomedical societies, including the American Association of Immunologists, the American Society for Microbiology, the Association of Medical Laboratory Immunologists, the Clinical Immunology Society, etc. He is a fellow of several biomedical societies, including the New York Academy of Sciences and the American Academy of Allergy and Clinical Immunology. He serves on many advisory committees for the National Institutes of Health, including the Bacteriology and Mycology Study Section for 8 years and a study section for immunology at the National Institute on Drug Abuse. He has also been a charter member of the AIDS Basic Science Study Section at NIH for 4 years as well as a member of the Advisory Committee for Microbiology and Virology of the American Cancer Society. He received the Outstanding Alumnus Award from Hahnemann University as well as Distinguished Scientist Award from the University of South Florida College of Medicine. He also was recipient of the Becton Dickinson Award in Microbiology from the American Society for Microbiology, which has a membership of over 45,000 microbiologists. He was instrumental in founding the Diagnostic Clinical Immunology Division of the ASM and received a Distinguished Service Award from that division. He was head of the clinical immunology program for the American Academy of Microbiology for nearly a dozen years. He was also one of the organizers of the Association of Medical Laboratory Immunologists. He has served as President of the Reticuloendothelial Society, now the Leukocyte Biology Society, and was President of both the Eastern Pensylvania and the Florida branches of the American Society for Microbiology. He was a Foundation for Microbiology Visiting Lecturer several times and has been a guest lecturer at many universities in the United States and abroad, including serving as a Visiting Professor at various universities in several countries, such as Israel, Japan, China, Peru, and Germany. He has been co-editor of several editions of the Manual of Clinical Immunology, published by the ASM. He has also been on the editorial board or section editor for many national and international journals, and has been the co-editor of the Clinical Immunology Newsletters for 10 years. He has published over 500 peer reviewed journal articles and equal numbers of abstracts presented at national and international scientific meetings. He has served as chair of many scientific sessions at such meetings over the past 30 years. He has also organized and served as chairman of over a dozen international scientific symposia. He is editor or co-editor of over 55 books, including many proceedings of national and international symposia he organized and chaired.

His many research interests are in the area of immune responses to microorganisms, including bacteria, fungi and viruses, and the effects of microorganisms, especially retroviruses, on the immune response system. He has also been involved in studying the effects of environmental agents, including drugs of abuse, bacterial products, and natural as well as synthetic immunomodulators on immune response mechanisms. His research is supported by many grants from national agencies, such as the National Institutes of Health.

Thomas W. Klein, Ph.D., is Professor and Vice Chairman of the Department of Medical Microbiology and Immunology at the University of South Florida College of Medicine. He received his Ph.D. degree in microbiology from Creighton University School of Medicine, where he also received his B.S. degree, and has served as a faculty member at the University of South Florida College of Medicine since 1972. He became Professor about 10 years ago. He has served as reviewer of many scientific research papers for national and international journals as well as reviewer for research grant applications to national and federal granting agencies. He has been the co-author of numerous peer reviewed journal articles as well as symposium proceedings and book articles and has also served as co-editor for over a dozen books. He has trained many graduate students and postdoctoral fellows and has been supported by grants from the National Institutes of Health.

Dr. Klein's research interests are concerned with the effects of bacteria on the immune response and the nature and mechanism of immunity to bacterial antigens. In particular, he has been investigating the nature and mechanism of immunity to the opportunistic bacterial pathogen *Legionella pneumophila*, which infects primarily macrophages in immunocompromised individuals. He has also been investigating the mechanisms by which drugs of abuse, especially marijuana and cocaine, affect the immune response mechanism, including effects on specific cellular and humoral immune mechanisms as well as the effects of such drugs on biochemical events in the immune response system. In particular, he has been studying the role of psychoneuroimmunological events on susceptibility and resistance to infectious agents, including bacteria and viruses.

Andrea L. Friedman, Ph.D., is in the Division of Behavioral Medicine and Oncology at the Pittsburgh Cancer Institute, Pittsburgh, PA. She received her Ph.D. and M.S. degrees from the University of Miami with specialization in Clinical Health Psychology and Neuropsychology. She received her B.A. degree from Emory University. She has been active in the areas of psychoneuroimmunology and neuropsychology and has worked with several medical and neurological populations, including HIV spectrum, and Alzheimer's and Parkinson's disease patients. Her research interests include psychological interventions for chronic illness, stress and coping, and, more recently, psychooncology.

CONTRIBUTORS

ROBERT ADER, Ph.D.
University of Rochester Medical Center, Rochester, NY

MICHAEL H. ANTONI, Ph.D.
University of Miami, Coral Gables, FL

ISTVAN BERCZI, D. Mv., Ph.D.
The University of Manitoba, Winnipeg, Manitoba, Canada

J. EDWIN BLALOCK, Ph.D.
The University of Alabama at Birmingham, Birmingham, AL

BÉLA BOHUS, M.D., Ph.D.
University of Groningen, Haren, The Netherlands

DANA H. BOVBJERG, Ph.D.
Memorial Sloan-Kettering Cancer Center, New York, NY

DAVID H. BROWN, Ph.D.
The Ohio State University, Columbus, OH

YIGAL BURSTEIN, M.D.
The Weizmann Institute of Science, Rehovot, Israel

ADRIAN J. DUNN, Ph.D.
Louisiana State University Medical Center, Shreveport, LA

HERMAN FRIEDMAN, Ph.D.
University of South Florida College of Medicine, Tampa, FL

KARL GOODKIN, Ph.D.
University of Miami School of Medicine, Miami, FL

R. M. GORCZYNSKI, M.D., Ph.D.
The Toronto Hospital, University of Toronto, Toronto, Ont., Canada

SUSAN KENNEDY, Ph.D.
Denison University, Granville, OH

JAAP M. KOOLHAAS, M.D.
University of Groningen, Haren, The Netherlands

GEORGES J. M. MAESTRONI, M.D.
Center for Experimental Pathology, Istituto Cantonale di Patologia, Locarno, Switzerland

M. PECHT, Ph.D.
The Weizmann Institute of Science, Rehovot, Israel

B. RAGER-ZISMAN, Ph.D.
Ben-Gurion University of the Negev, Beer-Sheva, Israel

MICHAEL SCHLESINGER, M.D.
The Hubert H. Humphrey Center for Experimental Medicine and Cancer Research, The Hebrew University/Hadassah Medical School, Jerusalem, Israel

ARTHUR A. STONE, Ph.D.
State University of New York at Stony Brook, Stony Brook, NY

ANDOR SZENTIVANYI, M.D.
University of South Florida, Tampa, FL

N. TRAININ, M.D.
The Weizmann Institute of Science, Rehovot, Israel

DOUGLAS A. WEIGENT, Ph.D.
University of Alabama at Birmingham, Birmingham, AL

YOSHIMASA YAMAMOTO, Ph.D.
University of South Florida College of Medicine, Tampa, FL

YAIR YODFAT, M.D.
The Hebrew University/Hadassah Medical School, Jerusalem, Israel

BRUCE S. ZWILLING, Ph.D.
The Ohio State University, Columbus, OH

CONTENTS

Chapter 1

HISTORICAL PERSPECTIVES ON PSYCHONEUROIMMUNOLOGY

Robert Ader
Center for Psychoneuroimmunology Research
Department of Psychiatry
University of Rochester School of Medicine and Dentistry
Rochester, NY

Psychoneuroimmunology refers, most simply, to the study of the interactions among behavioral, neural and endocrine (or neuroendocrine), and immunological processes of adaptation. Its central premise is that homeostasis is an integrated process involving interactions among the nervous, endocrine and immune systems. The term was first used in 1980, in my presidential address to the American Psychosomatic Society.[1] Its most conspicuous use was as the title of an edited volume[2] which one reviewer referred to prophetically as "The signature volume of a new field of research." This first volume was an attempt to bring together emerging research suggesting a relationship between the brain and the immune system. Traditionally, however, the immune system has been considered an autonomous agency of defense - a system of bodily defenses regulated by cellular interactions that are independent of neural influences. Besides, there were no known connections between the brain and the immune system. To be sure, it was known that hormones or at least adrenal hormones could influence immunity; some investigators were aware that brain lesions could influence immune responses; and it was also known or, at least, suspected that emotional states were associated with the development or progression of diseases related to the immune system. Few scientists at that time, however, took such observations too seriously. After all, there were no mechanistic explanations for how such things could happen.

Considering the brief time during which multidisciplinary research has addressed brain-immune system interactions, a great deal of data has accumulated in support of the proposition that homeostatic mechanisms are the product of an integrated system of defenses of which the immune system is one component.[3] Autonomic nervous system activity and neuroendocrine outflow via the pituitary can influence immune function, and cytokines and hormones released by an activated immune system can influence neural and endocrine processes. Regulatory peptides and receptors once confined to the brain are expressed by both the nervous and immune systems and each system is thereby capable of modulating the activities of the other. It is hardly surprising, then, to find that immunologic reactivity can be modified by Pavlovian conditioning - or that the behavioral and emotional states that accompany the perception of and the effort to adapt to events in the real world can influence immune responses. Thus, psychoneuroimmunology successfully challenged the commonly held assumption of an autonomous immune system. One may, therefore, entertain the proposition that changes in immune function mediate the effects of psychosocial factors and stressful life experiences on the susceptibility to and/or the precipitation or progression of some disease processes.

It is not my intent, in this introductory chapter, to review the literature outlining the history of psychoneuroimmunology. I have, instead, taken my charge literally and chosen the more manageable task of presenting here some editorial comments and some historical perspectives on psychoneuroimmunology. These are, of necessity, brief and selected and only cover developments up until about 1980. Very much more could be written about the people and the findings described here because these are rich personal stories. Much more could also be written about what was contemporary and what came before and what came after the 1970s, but this is a chapter - not a book. The research I have chosen to highlight was not necessarily even the first of its kind; in my opinion, however, the systematic research initiated during the 1970s was "the right stuff at the right time." No one study can be said to have been (or could have been) responsible for psychoneuroimmunology. I suspect that none of the research initiatives described below would have had quite the same impact had it not been for the converging evidence of brain-immune system interactions being provided by the others at about the same time. Studies of brain-immune system relationships had been appearing in the literature for many, many years. However, it was the coalescence of research initiated during the 1970s and sustained thereafter - and the identity provided by the label, psychoneuroimmunology, itself - that reawakened long-standing interests and attracted new investigators into this "new" field.

The notion of integration is neither new nor, for the most part, can it be considered controversial. It was David Hamburg, I think, who pointed out that biochemistry, a hybrid discipline, was initially viewed as a combination of poor biology and weak chemistry. Today, it is basic and central to the study of medicine. Psychopharmacology is a recognition of the fact that drug effects depend to a large extent on the state of the organism into whom they are introduced. Neuroendocrinology reflects an appreciation of the fact that the functions of the endocrine system can not be fully understood without reference to its interactions with the nervous system. And psychoneuroendocrinology acknowledges that the feedback and feed forward pathways between these "systems" influence and are influenced by behavior. Hybrid disciplines are not always or solely attempts at integration or synthesis. Basic fields such as neurochemistry or immunopharmacology, and clinical subspecialties such as neuropsychiatry, for example, designate a focus within a parent "discipline." In fact, in keeping with the zeitgeist of the biomedical model, the latter reductionistic referent is probably the more common one.

Among other shortcomings, disciplinary boundaries tend to keep insiders in and outsiders out. Hybrid disciplines have nevertheless emerged and significantly extended our understanding of the functions of the components of interacting systems. Why is it, then, that psychoneuroimmunology precipitated - and, in some circles, continues to engender so much resistance and enmity? Certainly, the attention that psychoneuroimmunology has captured in the popular press and its exploitation by those who redefine and use psychoneuroimmunology as the scientific umbrella for their own undisciplined and untested theories and practices cannot have endeared psychoneuroimmunology or investigators who study brain-immune system interactions to the remainder of the scientific community. In my unsubstantiated view, however, the reasons lie as much within as without the biomedical community. Some scientists are willing to say they "don't believe" there's anything of substance in psychoneuroimmunology, although they are not necessarily willing to be quoted. Of course, scientists do not have recourse to "I don't believe it" as grounds for rejecting hypotheses. One can argue, "I don't believe it because..." as in: "I don't believe it because there are no connections between the brain and the immune system." Such arguments are capable of disproof and, with respect to psychoneuroimmunology, all such arguments have been contradicted by experimental data. Unfortunately for the development of the field, however, there are those in influential positions who, purportedly, believed that psychoneuroimmunology would go nowhere and acted in a consultative capacity on this "belief." There is, too, a sense of unease among some so called "hard" scientists who seem to view the scientific study of behavior as an oxymoron. In truth, the sophistication in experimental design and analysis of research by the behavioral sciences far exceeds that of the more classical biomedical sciences and even molecular biology, and is essential for addressing the quantitative questions (e.g., when, how much, under what conditions) that are raised by factoring behavioral, neural and endocrine variables into the experimental analysis of immunoregulatory processes.

Within the field, there have been some minor battles over "turf", but none has altered the defining theme of the field. The emergence of psychoneuroimmunology has actually broadened some fields of study that were more narrowly defined in the recent past (e.g., papers in psychoneuroimmunology are now solicited for publication in the *Journal of Neuroimmunology*). "Neuroimmunomodulation" and "neuroendocrinimmunology," mere mispronunciations of psychoneuroimmunology, seem to have been precipitated to disengage from the study of behavior and/or to more specifically brand the field with one's own personal or disciplinary irons. Neither label changed the substance of the interdisciplinary research it promoted. (Of course, if you can come up with still another name, you, too, can also come up with another "First Conference on....")

The first sustained program of research were the studies of Russian investigators on the classical conditioning of immune responses. This research, derived from a Pavlovian perspective, began with Metal'nikoff and Chorine[4] who were working at the Pasteur Institute in Paris. This research was reviewed in English in 1933 by no less than Clark Hull,[5] a renowned learning theorist of that era. It was also reviewed in 1933 and, again, in 1941 by Kopeloff.[6,7] The only other substantive review of this literature in English appeared in *Psychoneuroimmunology*.[8] None of these early reviews attracted much attention or had any impact on research outside the then Soviet Union, including the studies of brain lesions on immune reactions and the physiologic studies of stress derived from the work of Hans Selye. Even the research implicating the nervous system in the modulation of immune responses initiated by Rasmussen and his colleagues and others in the 1950s and 60s failed to attract much sustained attention from any but a few behavioral scientists.

Aaron Frederick Rasmussen, Jr. was certainly one of the earliest pioneers of psychoneuroimmunology. His association with Norman Brill, then Chair of the Department of Psychiatry at the UCLA School of Medicine, was probably the first collaboration between a microbiologist/immunologist and a behavioral scientist. Rasmussen died in 1984, at the age of 68, after serving as Chair of the Department of Medical Microbiology and Immunology (1962-1969) and thereafter as Associate Dean of the School of Medicine. He is remembered as a beloved and inspiring teacher and colleague and an outstanding virologist whose genetic studies laid the foundation for understanding the notorious worldwide variability in influenza viruses. Rasmussen was a meticulous experimental microbiologist, who, at the same time, never lost site of host factors in disease. He was intrigued by the unproved conventional wisdom that emotional states influence the course of infectious illness, as depicted by such great novelists as Thomas Mann and as observed by such great pre-modern clinicians as Sir William Osler. An integrative thinker not bound by disciplinary lines, Rasmussen sought out Brill to discuss his "psychomicrobiological" ideas.

In 1957, Rasmussen, Marsh and Brill demonstrated that a stressful experience, avoidance conditioning, could increase the susceptibility of mice to herpes simplex virus. In a series of landmark papers, the pathogenic effect of emotional stress on animals exposed to herpes virus,[9] Coxsackie B virus,[10] and vesicular stomatitis virus[11] was explored. He and his coworkers also found decreased susceptibility to poliomyelitis virus in stressed monkeys, an early demonstration of the variability in stress effects on disease susceptibility.[12] Anticipating modern work in psycho-oncology as well as psychoneuroimmunology, Rasmussen and his colleagues also found that stress influenced the malignancy of polyoma virus in mice,[13] and his later work on stress included measures of viral antibodies and interferon production.[14,15] I regret very much that I never met Fred Rasmussen. His research set the stage for a variety of studies dealing with stress and infection, such as those initiated by Friedman, Glasgow and Ader,[16,17] and Solomon's studies on stress and antibody responses to a novel bacterial antigen.[18] (It may not be scientifically noteworthy, but it is of personal interest that my colleague, Nicholas Cohen, was a postdoctoral student in Rasmussen's department at UCLA in the mid 1960s. Unfortunately, Rasmussen's involvement in this research was decreasing which may explain why it took Cohen so long to enter the field).

George Solomon was another of the early investigators to show that psychological or environmental stressors could influence immunity. He and his colleague, Rudolf Moos, made painstaking observations of the life histories and personality characteristics of patients, seeking a clue to the frequently observed association between emotional states and the onset or exacerbation of arthritis. Solomon described the area as "psychoimmunology" and, despite the concerns of some of his colleagues, hung a sign to that effect on his laboratory door. In retrospect, Solomon's perspective on psychoneuroimmunology derived from curiosity, serendipity, psychodynamics, and the organization of disparate observations. Equally important, he claims, is the role of tenacity, frustration tolerance, and the ability to accept the encouragement of some and to reject the negativism of others in the development of new observations and theories.

Initially, Solomon was interested in psychological factors in the onset and course of autoimmune disease.[19] This interest was instigated by his father, a psychiatrist, who was convinced that psychological factors played a role in the onset and course of rheumatoid arthritis. As a resident in psychiatry at the Langley Porter Institute, he and W. Jeffrey Fessel studied patients with systemic lupus erythematosus (SLE) who had severe psychiatric symptoms. The similarity of the symptoms in SLE to the symptoms seen in schizophrenia prompted Solomon to ask whether schizophrenia could be an autoimmune disease of the brain with genetic and psychological predisposing factors that could be influenced by stressful life experiences. After a stint in the army, Solomon returned to the University of California in San Francisco and to research on immunoglobulins and schizophrenia.[20]

He also joined forces and established a productive collaboration with Rudolf Moos, a psychologist, who was studying psychosocial factors in rheumatoid arthritis (RA). Solomon and Moos later joined the Department of Psychiatry at Stanford, but continued to do their research at UCSF; at Stanford, they encountered difficulties in obtaining access to arthritic patients for such "psychological nonsense." The most unusual study in their series of papers on rheumatoid arthritis[21] was the one comparing physically healthy relatives of RA patients (known to have a greater than average likelihood of developing autoimmune disease) with the RA patients themselves - with the additional consideration of whether or not their sera contained rheumatoid factor, an anti IgG antibody characteristic of rheumatoid arthritis. Neither subjects nor examiners knew the sera status of the study population. Those who were negative for the rheumatoid factor were like a general population: normally distributed from psychologically healthy to psychologically disturbed. However, rheumatoid factor positive relatives of RA patients were psychologically "healthy," lacking anxiety, depression, or alienation and reporting good relationships with spouses, friends, and relatives. Psychological well-being seemed to exert a protective influence in the face of a probable genetic predisposition to autoimmune disease.

The future, Solomon thought, lay in mechanistic studies - and, what's more, he saw an opening. In 1963, he read a paper by Robert Good[22] who postulated a relationship between autoimmunity and relative immunologic incompetence. Frank Dixon[23] related such incompetence to the pathogenic formation of antigen-antibody complexes which occur when the amount of antibody is low in relation to antigen. Solomon immediately strung together immunologic incompetence, adrenocorticosteroid hormones and immunosuppression with stress and corticosteroids. A naive and simplistic notion, he thought, but a heuristic one, nonetheless. These notions were presented in "Emotions, immunity, and disease: A speculative theoretical integration" published in 1964.[24] Solomon attempted to conscript Moos into developing a laboratory in which they could stress rodents. "After all, no one," he thought, "was going to believe clinical data, but they will be convinced by animal experiments." Moos, however, was not an experimentalist and chose to pursue other interests; so, Solomon was on his own. He was provided with a laboratory, but he recognized that he knew virtually no immunology. Practically all the immunologists with whom he spoke told him that the immune system was autonomous, totally self-regulatory, and, thus, not subject to neuroendocrine influences. Nevertheless, Solomon established his "psychoimmunology laboratory." Although he had the support and tutelage of good people, he was unable to develop the necessary assay procedures for this work and considered giving up this line of research. Instead, he contacted one of the most noted immunologists in the world, Sir MacFarlane Burnet, who had revolutionized immunology with his clonal-selection theory of antibody formation. In response to Solomon's letter, Sir MacFarlane Burnet replied: "I am most skeptical but your ideas are interesting. Why don't you come to Melbourne? We'll talk, and my successor, Gus (now Sir Gustav) Nossal will teach you simple techniques for stress studies." Solomon did go to Melbourne where he claims he learned something about immunology and a great deal about immunologists.

In the ensuing years, Solomon was able to enlist the collaboration of gifted colleagues with whom he conducted some of the first studies that now fall under the rubric of "psychoneuroimmunology." With Thomas Medgan, who ran the tedious bioassay, he studied the effects of stress and steroids on interferon production.[25] Using flagellin, a bacterial antigen, it was shown that handling during early life could influence subsequent primary and secondary antibody responses in rats[26] and that different stressors have different effects on antibody production.[18] He also developed a collaboration with an immunologist, Alfred Amkraut, whom Solomon describes as "brave, most competent, obsessively meticulous, and cautious." Solomon and

Amkraut studied the effects of stress on virus-induced tumors,[27] graft-vs-host reactions,[28] adjuvant-induced arthritis,[29] and other immunologic reactions. "Nobody, however, was listening." Among other things, "Alfred was not given tenure ("What does the CNS have to do with immunology?)." Thus, in the early 1970's, Solomon closed the door on this line of research. Ten years later, however, he was to return.

Solomon kept close watch on the developments in psychoneuroimmunology, "...especially after the publication of Ader and Cohen's conditioning work." Having been asked to contribute to the first edition of *Psychoneuroimmunology*, he concluded that "PNI was on the map at last," and returned to the field. AIDS was suspected of being infectious, it involved immune abnormality, and it could also affect the CNS. AIDS, then, seemed the ideal condition to study within a psychoneuroimmunologic frame of reference. In 1983, Solomon moved from Fresno to the home campus at the University of California in San Francisco to join the incipient biopsychosocial AIDS project designed to seek psychologic-immunologic (AIDS progression) correlations. There he pursued his longstanding interest in "exceptions to the rule," namely long-term survivors with AIDS from whom he felt one might learn what psychological factors and mediating mechanisms continued to health and longevity. Their informally studied group of long-term survivors were remarkable people.[30] One of these was singer Michael Callen who wrote about the study in his book, "Surviving AIDS". Callen, who died in 1994 after 12 years of asymptomatic AIDS, personified what Solomon was attempting to explain. Another personal encounter that colored Solomon's perspective on psychoneuroimmunology was his association with Norman Cousins. It was Cousins' interest in understanding the role of attitude in healing that led L.J. ("Jolly") West, then Chair of the Department of Psychiatry and Biobehavioral Sciences, to invite Solomon to join the faculty at the University of California in Los Angeles. Cousins founded a UCLA Task Force on Psychoneuroimmunology of which Solomon is still a member. In addition to continuing work on AIDS, Solomon is currently engaged in some psychologically "upper" research: research on "very healthy old people instead of sick young people."

One of the channels of communication between the neuroendocrine and immune systems is achieved through the receptors that exist on immune cells. John Hadden was prompted to ask if lymphocytes had adrenergic receptors by the emergence of adenylate cyclase as the beta receptor transduction unit in many tissues and, most specifically, by "The beta adrenergic theory of the atopic abnormality in bronchial asthma" proposed by Andor Szentivanyi.[31] Based on studies in guinea pigs of the effects of hypothalamic lesions and stimulation on anaphylactic responses,[32,33] the first studies on brain lesions and immune reactions, Szentivanyi suggested that the CNS had an impact on the immune system, at least in terms of allergic mechanisms. He further postulated a blockade of beta adrenergic receptors, with a resulting exaggeration of immune responses, as a cause of asthma. That is, it was hypothesized that beta adrenergic receptors acting via the adenylate cyclase/cyclic AMP system would down regulate allergic immune phenomena.

It was during his first year of a medical fellowship with Elliot Middleton, Jr. that Hadden learned of Szentivanyi's formulation and set out to determine if lymphocytes had adrenergic receptors that could regulate immune function in a meaningful way. Hadden and his associates showed that, in the presence of hydrocortisone, alpha-adrenergic stimulation augmented and beta adrenergic stimulation inhibited the lymphoproliferative response to the mitogen, PHA.[34] This was the first observation linking lymphocytes to the sympathetic nervous system, opening a wide door to the study of neural influences on immunity.

These findings led Hadden in several directions. The notion that beta antagonists and cyclic AMP down-regulated lymphocyte proliferation was pursued by several investigators and confirmed for a variety of lymphocyte functions.[35] Hadden and his associates elaborated on the newly detected alpha adrenergic effects as these related to glucose metabolism and transmembrane K+ transport, finally linking them to direct effects on membrane ATPases of lymphocytes.[36-38] While working in Minneapolis on transmembrane signaling, Hadden was introduced to cyclic GMP by Nelson Goldberg. Together, they found that cyclic GMP was involved in the signal induced in lymphocytes by PHA.[39] They also found that cyclic GMP was involved in lymphocyte cholinergic responses. While in the process of developing these observations, Terry Strom presented the first paper to show that T lymphocyte cytotoxicity was augmented by muscarinic cholinergic stimulation.[40] Hadden and his associates extended these results, demonstrating stimulation of RNA and DNA synthesis of lymphocytes and implicating cyclic GMP in the process.[41,42] These observations were the first to link lymphocytes to the parasympathetic nervous system, opening a door to immune regulation by the entire autonomic nervous system.

Hadden initiated some additional *in vivo* studies, but became allergic to the animals and had to abandon this line of research. Besides, he was then preoccupied with questions about signal transduction mechanisms. He was not unmindful and, indeed, was fascinated by Robert Good's stories about his personal involvement in successful demonstrations of hypnotically-induced alterations of immunity,[43] but it appeared to him that the study of the neural modulation of immunity was not yet ready to surface as a bona fide area of research. Yet, it was in 1980 that Hadden, instrumental in the organization of the new International Conferences on Immunopharmacology and starting a new journal, the *International Journal of Immunopharmacology*, invited me to present at this immunology meeting. It was important, he thought, that the immunopharmacologist be made aware of the research and the implications of the work on conditioning and immunity. "Now, 25 years later," Hadden writes, "I recognize that it has emerged and I am happy to have contributed some impetus." As evidenced by recent work on the endocrinology of the thymus,[44,45] Hadden remains committed to an understanding of neuroendocrine-immune communication.

When asked how I became involved in psychoneuroimmunology, I can not refer to a logical starting point. I say it was an accident; I was "forced" into it by my data. I was studying taste aversion learning in rats. When a novel, distinctively-flavored conditioned stimulus (CS), saccharin, is paired with the unconditioned effects of a drug, cyclophosphamide (CY), which induces a transient stomach upset, the animal learns in one such conditioning trial to avoid saccharin-flavored drinking solutions. We were conducting an experiment on the acquisition and extinction of the conditioned aversive response as a function of the strength of the CS, i.e., the volume of saccharin consumed before the animal was injected with CY. As expected, the magnitude of the conditioned response was directly related to the volume of saccharin consumed on the single conditioning trial. Also, repeated presentations of the CS in the absence of the drug resulted in extinction of the aversive response, and the rate of extinction was inversely related to the magnitude of the CS. However, in the course of these extinction trials, animals began to die. A troublesome but uninteresting observation. As more animals died, it became evident that mortality, like the magnitude of the conditioned response, varied directly with the volume of saccharin consumed on the one drug trial; a troublesome but interesting effect.

As a psychologist, I was unaware that there were no connections between the brain and the immune system. Therefore, I was free to make up any story I wanted in an attempt to explain this orderly relationship. The hypothesis was that, in the course of conditioning the avoidance behavior, we were also conditioning the immunosuppressive effects of cyclophosphamide. If, every time the conditioned animals were reexposed to the CS previously paired with the drug, the CS induced a conditioned immunosuppressive response, these animals might be more susceptible to low levels of pathogenic stimulation that may have existed in the laboratory environment. Moreover, if the strength of the conditioned response was a function of the magnitude of the CS, the greater the immunosuppressive response, the greater the likelihood of an increased susceptibility to environmental pathogens. Thus, it was the serendipitous observation of mortality in a simple conditioning study and the need to explain an orderly relationship between mortality and a conditioned aversive response that gave rise to the hypothesis that immune responses could be modified by conditioning operations.

A Letter to the Editor describing these observations and the speculation that immune responses were subject to conditioning was published in *Psychosomatic Medicine* in 1974. It was a draft of this letter that elicited the first of many unexpected and sometimes frightening responses to this work. George Engel who, having criticized me for being too conservative in the past, said that, based on my conservative reputation, people were going to believe this, just because I said it. Although meant as a compliment, I found the prospect somewhat frightening; I had not given up my right to be wrong. I was to learn, however, that if you say something unimportant, it doesn't matter whether you're right or wrong; if, however, you say something that could be important, you had better be right!

People listened politely, but I did not have much luck in generating any interest in this hypothesis - let alone the help I would need to examine it - until I met Nicholas Cohen. Cohen was the first person with sophistication in immunology who didn't think these notions were too "far out." Thus began a collaboration that is as active today as it was in 1974[46]. Still oblivious of the Russian studies of the 1920s, Cohen and I designed a study to directly examine the hypothesis that immune responses could be modified by classical conditioning. For better or worse (sometimes, we're not sure), the first experimental paradigm we adopted was successful and, with some evident trepidation on the part of the reviewers and editor, "Behaviorally conditioned immunosuppression" was published in 1975.[47] This study demonstrated that, like other physiological processes, the immune system was subject to classical (Pavlovian) conditioning, providing dramatic evidence of an inextricable relationship between the brain and the immune system. Essentially, we were forced to the conclusion that there was a relationship between the brain and the immune system. The biomedical community, however, was, to be generous, guarded and, to be precise, quite negative. Such a phenomenon simply could not occur because, as everybody knew, there were no connections between the brain and the immune system. Seminar groups were assigned the (unsuccessful) task of finding out what we had done wrong. The first replication of our findings[48] came from a study originally intended to show that, taking appropriate care and using more accurate assay procedures, the effect would not occur. The National Institutes of Health and National Institute of Mental Health, however, were not "forced" to the conclusion that one could condition alterations in immunologic reactivity, and, notwithstanding George Engel's predictions, Study Sections were loathe to use my conservative reputation as collateral. However, reading between the lines of "pink sheets" (and as confirmed by Study Section members much later), we might be right - and could they take that chance for so little money - and for only two years at a time? Our initial study was supported by a one-year grant from the Grant Foundation (where my reputation was collateral) and then, reluctantly, it seems, we were funded by the NIH. At that time, ours was the only NIH grant in this area which, on renewal, was thereafter supported by the NIMH. Today, a computer search of "psychoneuroimmunology" and "neuroimmunomodulation" lists more than 200 active research grants being supported by the U.S. Public Health Service.*

Over the next several years, there were replications and major extensions of conditioned alterations of humoral and cell-mediated immune responses.[49-51] Recent work has successfully used antigen, itself, as the unconditioned stimulus. A classically conditioned enhancement of antibody production occurred when conditioned mice were reexposed to the conditioned stimulus in the context of reexposure to a minimally immunogenic dose of that same antigen.[52] These and earlier experiments[53] documented the conditioning of immune responses, per se, in contrast to the conditioning of immunopharmacologic responses. Studies in New Zealand mice genetically susceptible to a systemic lupus erythematosus-like disease were used to demonstrate the biologic impact of conditioned alterations in immune responses. Substituting CSs for active drugs on some scheduled treatment days delayed the onset of autoimmune disease using a cumulative amount of immunosuppressive drug that was ineffective by itself in altering the progression of disease.[54] Similarly, reexposure to a CS previously paired with immunosuppressive drug treatment prolonged the survival of foreign tissue grafted onto mice.[55,56] Such results have yet to be verified in human patients. However, there has been one clinical case study describing the successful use of conditioning in reducing by one half the amount of cytoxan therapy received by a child with lupus.[57]

To date, the neural, endocrine, or neuroendocrine mechanisms underlying conditioned alterations in immune function are unknown - reason enough, apparently, for some biomedical scientists to reject the phenomenon, itself, or, in the case of *Nature*, to reject for publication a paper demonstrating conditioned enhancement of antibody production without even providing a review. One can only wonder about the implications of applying uniformly the criterion of having to identify "...the precise mechanisms involved in the phenomenon you observe..." in order to publish experimental results. Besides the fact that the precise mechanisms underlying behaviorally-induced changes in immune function are not known, it is also true that in only a few instances have the functional significance of the bidirectional communication pathways that have been identified among the nervous, endocrine and immune systems been determined.

To be sure, our studies were not always maligned. I recall, for example, the evening I met Lewis Thomas whom I have always thought of as the Montaigne of the biological sciences. After a brief exchange of pleasantries, Thomas said, "You sure are making life difficult for some people."

"Well," I answered, slowly - trying to think of an appropriate response, as I read Lewis Thomas, "that shouldn't bother you."

"It doesn't," he replied, "I love it!"

During these same years, Hugo Besedovsky was beginning his studies on endocrine-immune system interactions. Besedovsky was led into psychoneuroimmunology through a clinical route. Trained as a pediatrician at the Medical Faculty of Rosario in Argentina, he was confronted daily with patients with infectious and other diseases involving the immune system. Having been "hybridized" early in his training, Besedovsky naturally viewed the immune system as operating within the context of other physiological processes. Reflecting his pediatric training, his first studies in the early 1970s addressed endocrine influences on the immune and haematopoietic systems during ontogeny.[58] He focused on adrenocortical function which, at the time, was the only endocrine activity known to affect immunity. He discussed his interest in the possibility that neuroendocrine mechanisms could contribute to immunoregulation with Professor Bernardo Houssay who encouraged him to work with Sir Peter Medawar in London. Medawar accepted but then could not accommodate Besedovsky in his laboratory because of illness, so he went to the Swiss Research Institute in Davos, Switzerland where he was fortunate to have Ernst Sorkin as a mentor and collaborator.

Besedovsky's research on the neuroendocrine regulation of immune responses was, and still is, based on the premise that immune responses are a part of integrated homeostatic mechanisms under the control of the nervous and endocrine systems. Thus, he reasoned, it should be possible to provide evidence that: (1) antigen exposure initiates a flow of information to neuroendocrine structures about changes in the activity of immune cells; (2) as a consequence of this information, an efferent neuroendocrine response should be elicited; and (3) this efferent response should have functional significance for immunoregulatory processes and host defenses.

Besedovsky and his colleagues proceeded to demonstrate in two animal species that, independent of any "stress-induced" responses related to the procedures, immunization with different antigens was capable of inducing endocrine changes (an increase in corticosterone and a decrease in thyroxin) that were under CNS control.[59] This was followed by a collaboration with Professor Dominik Felix from the Brain Research Institute in Zurich which established that there was, in parallel with the production of antibody, an increase in the firing rate of neurons within the ventromedial hypothalamus.[60,61] This was a dramatic demonstration that the nervous system is capable of responding to signals emitted by an immune response. These results, Besedovsky recalls, were first submitted to *Nature* which rejected the paper "because it is self evident that the brain must receive information from the immune system."

Hugo Besedovsky's professional and personal relationship with Adriana del Rey, also from Argentina, began when she joined the Institute in Davos in 1977. Their first collaborative research concerned antigenic competition and the immunosuppressive role of elevations in adrenocortical steroids.[62] These studies supported their hypothesis that glucocorticoid elevations associated with antigen exposure act to prevent an abnormal expansion of the immune response which might otherwise result in a cumulative, excessive immune cell proliferation favoring the expression of autoimmune and lymphoproliferative processes and the production of potentially harmful products of activated lymphocytes.

Analogous experiments on the involvement of the sympathetic nervous system in immunoregulation included the measurement of the content and the turnover rate of splenic noradrenaline during an immune response. In highly reactive animals, there is a decrease in noradrenaline content which occurs before the peak in antibody titers;[63,64] animals that have a less active immune system show an increase in noradrenaline in lymphoid organs.[65] Also, corresponding to the increased activity of hypothalamic neurons during an immune response, Besedovsky and his associates[66] showed that there was a reduction in the noradrenaline turnover rate in the hypothalamus and brain stem. Clearly, there was a very dynamic interaction between the immune system and the sympathetic nervous system that influenced immunoregulatory processes.

The fact that there were endocrine, autonomic and neural activity changes during the course of immune responses indicated that the immune system could convey information to the CNS which led Besedovsky to suggest that the immune system acts as a "receptor sensorial organ."[60-66] This implies that the CNS can sense the activity of the peripheral immune system involved in the recognition of non-self intruders and modified self-components, as well. If so, the products of immune cells should be able to affect neuroendocrine function. His approach involved stimulating immune cells *in vitro* and transferring the supernatants obtained from such cultures into naive animals. The culture supernatants induced a pituitary-dependent increase in plasma corticosterone and a decrease in the content of noradrenaline in the brain of the rats.[67,68] Thus, Besedovsky provided the first evidence that products of activated immune cells could affect endocrine responses that were under CNS control. When purified lymphokines and monokines became available in the 1980s, the laboratory began to study the capacity of these immune system mediators (e.g., interleukin-1) to influence neuroendocrine functions.[69,70] Current research focuses on the effects of endogenously produced lymphokines and monokines.

Current research also includes a concern for the potential clinical relevance of neuroendocrine-immune system interactions. For example, some of the endocrine changes effected by the inoculation of tumor cells are mediated by cytokines rather than being a direct result of the tumor, itself, or the ensuing disease.[71] Also, the pituitary adrenal response to lipopolysaccharide is cytokine mediated[72] and IL-1 is a main factor in activation of the pituitary-adrenal axis during viral infections.[73]

The innovative research initiated by Hugo Besedovsky, Adriana del Rey and their colleagues has had a major impact on the acceptance of an integrated approach to research on homeostatic processes, in general, and on psychoneuroimmunology, in particular. It has also had a major impact on the conceptualizations and on the directions of research coming from several laboratories in the United States and in Europe. That, however, took time. Initially, the response to their work, like the response experienced by others in the field, was disheartening. On one of their several trips from Davos to Basel, Besedovsky and del Rey met with Niels Jerne, then Director of the Institute of Immunology, to discuss their ideas about the role of hormones and neurotransmitters in immunoregulation with a world famous immunologist whom they admired greatly.

Jerne listened and said, "This is too complicated. We still do not know many things about the immune system, and I think we should know, for example, whether there is a T cell receptor. Maybe you should work *in vitro*..." Needless to say, this was an unexpected and upsetting response. About five years later, Besedovsky was giving a seminar at the Hoffman-LaRoche Laboratories where the first person to arrive was Neils Jerne. Following Besedovsky's talk on the immunomodulating effects of glucocorticoids, Jerne stood up and said that "I have always believed that there is a communication between the immune and endocrine systems and ..."

Adriana del Rey, sitting near Jerne, interrupted him and shouted: "This is not true! Five years ago you told us...(and she repeated the story)."

Of course, Jerne laughed and said, "Well, what I meant to say was that I have always believed it but, after seeing these results, I think it may be true!" (Parenthetically, Nicholas Cohen, who had spent a sabbatical year at the Basel Institute in 1975, was invited to review our work on conditioning for a 1981 volume[74] honoring Niels Jerne.)

Besedovsky is now Professor of Physiology in the Medical Faculty of the University of Marburg in Germany where he has established a Department of Immunophysiology. The multidisciplinary expertise of his research group is still devoted to investigations of the complex immune-neuro-endocrine interactions that characterize the physiology of the immune system.

Similar thinking was directing the research of Edwin Blalock when, in 1979, lymphocytes were discovered to be a source of brain peptide neurotransmitters and pituitary hormones.[75] These observations were the unexpected culmination of three years of research when, as an Assistant Professor of Microbiology at the University of Texas Medical Branch in Galveston, Blalock started out to determine if the cytokine, interferon (IFN), could function as a hormone. Indeed, it appeared that, among its other endocrine activities, interferon preparations could stimulate the adrenal to synthesize glucocorticoids.[76] Since the sequences of IFN were not known at the time and IFN was functioning like ACTH, the primary regulator of the adrenal gland, Blalock and his first postdoctoral fellow, Eric Smith, wondered whether the steroidogenic activity of the cytokine might be due to the presence of a residue ACTH-like sequence within the IFN molecule. Although this appeared to be so,[75] further studies, including the cloning of IFN, showed that this was not the case.[77] But, this research led to an even more remarkable finding: supernatant fluids from human lymphocytes cultured with IFN contained ACTH and the endogenous opioid peptides, endorphins.[77] Blalock remembers quite vividly the exhilaration they felt on the day they first observed immunofluorescent pictures of lymphocytes staining positively for the production of these substances. Such observations were indeed surprising since, at the time, these peptides were thought to be the exclusive property of the brain and pituitary gland. For Blalock and for many others in the developing field of psychoneuroimmunology - this discovery suggested a molecular approach for solving the mystery of how the mind could control the immune system, e.g., how classical conditioning might modify immunity. Such a relationship could exist because the body's two principle recognition organs, the brain and the immune system, speak the same chemical language. If true, this meant that the immune system could, indeed, talk back to the brain and, perhaps, alter physiology and behavior. Research accomplished in the last several years confirms the fact that such relationships do, in fact, exist[73-79] and, because of the molecular and biochemical nature of these studies, a large measure of respectability was given to psychoneuroimmunology - but, not immediately.

As with most, if not all discoveries that challenge current dogma, Blalock's work was met with healthy as well as unhealthy skepticism and, like many pioneers in the field of psychoneuroimmunology, the messengers suffered personal and professional indignities. The NIH site visitors reviewing their first research grant proposal in this area concluded that Blalock and his colleagues were actually sane and that the work had merit, but the project was funded for only two years. According to Blalock, the study section's message was clear: you must sequence the lymphocyte's ACTH to make your point unequivocally. In retrospect, this was considered an impossible request made by Study Section members who, according to Blalock, had never, themselves, sequenced anything. However, as "green" investigators, who were also referred to as "biochemical yahoos," Blalock and his associates did not know enough to be daunted. After a year of research and encouraging results, the Study Section members were still unimpressed and Blalock's application for renewal of this research was disapproved. When later reviewed by scientists knowledgeable in the area, this very same proposal was judged to be in the top 5% of all the grants reviewed at that time. Given the time and resources, Blalock and his colleagues were able to sequence the peptides which were found to be authentic.[80] Other investigators began to pay attention and the study of neuroendocrine-immune system communication took another giant step. Today, it is accepted that brain peptides and their receptors exist within the immune system and that the products of an activated immune system function as neurotransmitters. Thus, the scientific pariahs became heroes (apparently, they have not yet experienced unreferenced descriptions of these phenomena prefaced by the phrase, "As we have long expected..."). The process, agonizing at times, was rewarding and intellectually stimulating, but, as Blalock puts it: the scientific enterprise would be more enjoyable if the scientific community recognized that "science is about unexpected discovery, not expected results."

Another critical link between the brain and the immune system was forged by David Felten. He and his colleagues brought anatomical, neurochemical, receptor binding, and *in vitro* and *in vivo* immunological techniques to bear on this relationship and provided unequivocal evidence that sympathetic noradrenergic nerve fibers signal cells of the immune system and are capable of evoking major changes in their responsiveness. Again, it was a serendipitous observation that altered the direction of Felten's research.

In 1980, Felten was examining a section of rodent spleen with fluorescence histochemistry for catecholamines to distinguish arterial and venous patterns of smooth muscle innervation. He saw and reported extensive networks of noradrenergic sympathetic nerve fibers among T cells in the white pulp, and was confused about why this had not been described in the past.[81] Felten had always looked at interactions among neuronal systems in a non-traditional fashion. From his early work at MIT as an undergraduate in Walle J.H. Nauta's laboratory, he was fascinated with integrative regulatory neuronal systems. His unexpected observation of sympathetic noradrenergic nerve fibers in apparent direct contact with lymphocytes and macrophages thus fell on fertile ground. He and his colleagues proceeded to show that these nerve fibers were localized in precise compartments of both primary (thymus, bone marrow) and secondary (spleen, lymph nodes) lymphoid organs,[81-84] and formed close, synaptic-like neuro-effector junctions with T lymphocytes and macrophages.[85]

Felten recalls that his early findings were ridiculed by many immunologists and viewed with disbelief as "minor aberrations," at best. However, with characteristic energy and persistence, he and his collaborators spent several years investigating and demonstrating that these noradrenergic nerve fibers fulfilled the criteria for neurotransmission with cells of the immune system with thymus, spleen, and lymph nodes as targets. In a detailed developmental study, it was shown that these nerve fibers formed these close contacts with lymphocytes early in ontogeny, and appeared to influence early immunological development and compartmentation.[86] At the other end of the lifespan, sympathetic nerve fibers in secondary lymphoid organs were found to diminish markedly with age.[87] Felten has proposed that this loss contributes to immunosenescence, particularly to diminished T cell functions, especially TH1 (cell mediated) responses. In other recent work, Felten's laboratory demonstrated that local denervation of adrenergic sympathetic nerves from draining lymph nodes in autoimmune disease-susceptible rats enhanced joint inflammation and bone erosion in adjuvant induced arthritis, while selective denervation of substance P nerve fibers from such draining lymph nodes protected the rats from joint pathology.[88] Such findings substantiate the functional importance of nerves supplying lymphoid organs.

In 1983, Felten was awarded a prestigious John D. and Catherine T. MacArthur Foundation Prize Fellowship at the early stage of his work in neural-immune interactions. Parenthetically, this was one of the Foundation's few ventures into psychoneuroimmunology. Several discussions and conferences in the early days held out the prospect that such new, interdisciplinary research would meet the original criteria for MacArthur Foundation support: innovative research that would face difficulties in finding support from within traditional federal funding agencies. Perhaps, however, it was too soon; purportedly, the advice received by the Foundation at that time was that psychoneuroimmunology wasn't going anywhere. David and Suzanne Felten, however, were going to the Department of Neurobiology and Anatomy at the University of Rochester School of Medicine and Dentistry to team up with Bob Ader, Nick Cohen and Sandy Livnat to develop interdisciplinary programs of research and research training.

In demonstrating a major role for sympathetic noradrenergic nerve fibers in regulating immune functions, Felten and his colleagues provided evidence for a direct, "hard-wired" connection between the CNS and the immune system. This connection has since been shown to be a major route for behavioral influences and for central cytokine influences on immune function. For Felten, the demonstration of direct neural signaling of cells of the immune system opens up several completely new directions for research. It is now possible to seek the chemical and receptor-mediated mechanisms by which behavioral and other CNS influences on immune responses are achieved. He and his colleagues are pursuing the use of neurotransmitter agonists and antagonists to specifically manipulate sites of initiation of immune responses, development and regulation of effector cell functions, and modulation of effector cell functions at diverse sites. Felten's work is a cornerstone of a mechanistic understanding of the signaling between the nervous and immune systems and provides a basis for understanding the complex systemic integration among behavioral processes, the brain, and immunophysiology.

Thus, it was during the 1970s and early 80s that independent lines of research, derived as much from the personal experiences and imagination of the investigators as from a logic dictated by different disciplinary perspectives, began converging on the theme that the immune system was part of a larger, integrated mechanism of homeostatic processes serving the survival interests of the individual. For whatever reasons - and despite overt and covert resistance - this was evidently the right stuff at the right time! A new picture of immunoregulatory processes was emerging that promised a new understanding of the functions of other narrowly conceptualized systems and a new appreciation of the multi-determined etiology of pathophysiological states. A paradigm shift was occurring and, as a result of the nearly twenty years of research precipitated by the above findings, it is no longer possible to study immunoregulatory processes as an independent function of the immune system. The research initiated by these investigators were giant steps and, despite the fact that they originated from different perspectives, they had a common effect. There were earlier, isolated studies, but most of the current research in the field derives directly or indirectly from these seminal studies. These were enabling studies in the sense that they raised questions and, further, legitimized questions that had not been asked before. And if these questions and, sometimes, the questioners were ridiculed, another almost universal experience, the evidence was, first, compelling, and then overwhelming. Thus, as Schopenhauer observed, "All truth passes through three stages. First it is ridiculed. Second it is violently opposed. Third it is accepted as being self evident." This has almost become a cliché, yet a recent textbook in immunology[89] devotes a section to neuroendocrineimmune system relationships and concludes that, "Clinical and experimental psychoneuroimmunology studies to date confirm *the long-standing belief* that the immune system does not function completely autonomously." (Italics added.)

It is the research conducted during the past 20 years that probably accounts for 95 per cent or more of what is now known about the relationships among behavioral, neural and endocrine, and immune processes of adaptation[3] and led to the general (and sometimes still begrudging) acknowledgment that, like other physiological processes operating to protect the organism, the immune system is part of an integrated system of adaptive processes and is thus subject to some regulation by the brain. Two pathways link the brain with the immune system: autonomic nervous system activity and neuroendocrine outflow via the pituitary. Both routes provide biologically active molecules which are perceived by the immune system via cell surface or internal receptors on the surface of lymphocytes, monocytes/macrophages and granulocytes. Thus, all immunoregulatory processes take place within a neuroendocrine milieu that is demonstrably sensitive to the influence of the individual's perception of and response to events occurring in the external world.

Conversely, we have learned that activation of the immune system is accompanied by changes in hypothalamic, autonomic, and endocrine processes, and by changes in behavior. For example, cytokines influence activation of the hypothalamic pituitary-adrenal (HPA) axis - and, in turn, are influenced by glucocorticoid secretion.[70] The potential interaction of neuroendocrine and immune processes is further magnified by the fact that cells of the immune system activated by immunogenic stimuli are capable of producing a variety of neuropeptides.[90] Thus, the exchange of information between the brain and the immune system is bidirectional.

Based on the above, it is hardly surprising that behavioral factors are capable of modifying immune function or that activation of the immune system would have consequences for behavior. The Pavlovian conditioning of the suppression or enhancement of immune responses[51] and, conversely, the conditioning of the physiological effects of cytokines[91] both reflect adaptive immunoregulatory processes. The majority of the behavioral research, derived in large measure from the work of Hans Selye, has addressed the immunologic effects of stressful experiences. Early studies concentrated on the immunosuppressive effects of adrenal gland activation. These pharmacologic and physiologic studies were complemented by the behavioral studies of Rasmussen's group in the 1950s, by Friedman and Ader and by Solomon in the 1960s and by a host of others, primarily physiologists, during this same time period.[92,93] There was not a lot of research of this kind from the late '60s until the publication of *Psychoneuroimmunology*.[2] In the 1980s, however, "stress and immune function" was revived and, armed with a modern technology, became the dominant theme of the behavioral component of psychoneuroimmunology.

Human studies of the immunologic changes associated with emotional states and stressful life experiences also took shape in the 1980s. Stimulated by a description of some of the immunologic effects of sudden bereavement,[94] researchers began to address the effects of losses (e.g., the death of a spouse) and of affective states, particularly, depression, on immunity. For example, Marvin Stein, then Chair of the Department of Psychiatry at the Mt. Sinai Hospital and Medical Center in New York, had, during the 1960s, been actively involved in studies of the effects of hypothalamic lesions and stimulation on anaphylactic reactions in guinea pigs.[95] Like Solomon, Stein returned to psychoneuroimmunology in the 1980s with a program of animal research on the immunologic effects of stressful experiences and a program of human studies of the immunologic changes associated with loss and with depression. In this, Stein was able to engage an interdisciplinary team of young investigators (and to stimulate the interest of several others) who now have psychoneuroimmunology laboratories of their own.

Similarly, the unique team of Janice Kiecolt-Glaser, a psychologist, and Ronald Glaser who, as Chair of a Department of Medical Microbiology and Immunology, entered the field with considerable apprehension, initiated an extensive series of studies that began with the effects of examinations in medical students on changes in immunity.[96] Glaser, like many others, became convinced of the role of behavioral factors in the modulation of immunity only when he found such relationships in his own data. Although a common event in the life of students, examination periods were found to be reliably associated with a general depression of immune function including, as a consequence, an elevation in antibodies to Epstein-Barr virus. These studies were directly and indirectly responsible for a reemergence of animal and human studies by behavioral scientists and by immunologists on the effects of stressful life experiences on immune function and susceptibility to infectious diseases.

In animals and in humans, a variety of psychosocial events interpreted as being stressful to the organism are capable of influencing a variety of immune responses. It is now clear, however, that different "stressors" have different effects on some constant outcome measure and that the same "stressor" can have different effects on different outcome measures. The direction, magnitude and duration of stress-induced alterations of immunity are influenced by: (a) the quality and quantity of stressful stimulation; (b) the capacity of the individual to cope effectively with stressful events; (c) the quality and quantity of immunogenic stimulation; (d) the temporal relationship between stressful stimulation and immunogenic stimulation; (e) sampling times and the particular aspect of immune function (or compartment) chosen for measurement; (f) the experiential history of the individual and the existing social and environmental conditions upon which stressful and immunogenic stimulation are superimposed; (g) a variety of host factors such as species, strain, age, sex, and nutritional state; and (h) interactions among these several variables. This listing of relevant variables identified in recent research paraphrases the variables identified from a much earlier analysis of the effects of stressful life experiences on behavioral and physiological responses and on susceptibility to disease.[1,98,99] Indeed, prospective as well as retrospective studies in animals and humans have also shown that, depending on interactions among the qualitative and quantitative nature of the environmental demands and the pathophysiologic process, the experimental procedures, and a variety of host factors, stressful experiences can alter the host's defense mechanisms thereby altering susceptibility to bacterial and viral infections, modifying the neuroinvasiveness of normally non-neurovirulent strains of virus, or allowing an otherwise inconsequential exposure to a pathogen to develop into clinical disease.[1,98,99]

The behavioral and emotional states that attend the perception of and the effort to adapt to environmental circumstances are accompanied by complex patterns of neuroendocrine changes. That the neural and endocrine patterns associated with behavioral and emotional states are capable of modulating immune functions lends credence to the hypothesis that changes in immune function constitute an important mediator of the pathophysiological effects of stressful life experiences. This chain of psychophysiological events may not yet have been firmly established, but, as this volume attests, the possibility is attracting renewed attention and the data are providing evidence of the relevance of psychosocially-induced alterations in immune function for differences in the susceptibility to and progression of infectious diseases.

Where is psychoneuroimmunology today? It's still working out some adolescent problems and on its way to young adulthood. It has not yet achieved the maturity of neuroendocrinology or the more recent psychoneuroendocrinology - and it has not been granted the scientific respectability it has earned and to which it is entitled. Still, psychoneuroimmunology continues to grow. *Brain, Behavior and Immunity* began publishing in 1987 and there are now two other journals devoted to the area. Research reports are now considered (and solicited) for publication in a variety of peer reviewed journals in immunology, psychology and in the neurosciences, including endocrinology. There are now two international societies and papers in the field occupy increasing blocks of time at the meetings of other scientific societies. There has been an increase in the number of students (including M.D./Ph.D. candidates) from psychology, immunology and the neurosciences interested in working in the area, and there has been an increase in the number of funded research and research training grants. There has also been a proliferation of edited volumes that address various aspects of the field (e.g., *Psychoneuroimmunology, Stress and Infection*).

Psychoneuroimmunology is, perhaps, the most recent example of an interdisciplinary field that has developed and now prospers by exploring and tilling fertile territories secreted by the arbitrary and illusory boundaries of the biomedical sciences. Disciplinary boundaries and the bureaucracies they spawned are biological fictions that can restrict imagination and the transfer and application of technologies. They lend credence to Werner Heisenberg's assertion that "What we observe is not nature itself, but nature exposed to our method of questioning." (p. 81).[100] Our own language, too, must change. The signal molecules of the nervous and immune systems are expressed and perceived by both systems. Therefore, it may no longer be appropriate to speak of "neurotransmitters" and "immunotransmitters." Also, to speak of links or channels of communication between the nervous and immune systems perpetuates the myth that these are discrete systems (or disciplines). On the contrary, the evidence indicates that relationships between so-called "systems" are as important and, perhaps, more important than relationships within "systems;" that so-called "systems" are critical components of a single, integrated network of homeostatic mechanisms. To the extent that the problems chosen for study and innovative research strategies to address these problems derive from conceptual and theoretical positions, these are important issues.

More substantively, research conducted over the past several years has resulted in a recognition and appreciation of the interactions among behavioral, neural, endocrine, and immune processes. Indeed, there has been a paradigm shift in the attempt to understand immunoregulatory function. The innervation of lymphoid organs and the availability of neurotransmitters for interactions with cells of the immune system add a new dimension to our understanding of the microenvironment in which immune responses take place. Similarly, the interaction between pituitary-, endocrine organ-, and lymphocyte-derived hormones that define the neuroendocrine milieu in which immune responses occur adds another level of complexity to the analysis of the cellular interactions that drive immune responses. Collectively, these relationships provide the foundation for previously observed behaviorally-induced alterations in immune function and for immunologically based changes in behavior. They may also provide the means by which psychosocial factors and the emotional states that accompany the perception of and response to stressful life experiences influence the development and progression of infectious, autoimmune and neoplastic disease.

ACKNOWLEDGMENTS

I am grateful to Drs. Hugo Besedovsky, J. Edwin Blalock, John Hadden, and George Solomon for providing for me a brief written description of their perspectives on psychoneuroimmunology and for their comments and corrections of an earlier draft of this paper. I am also indebted to Dr. Sherman M. Melinkoff, Professor Emeritus of Medicine and former Dean of the UCLA School of Medicine, for his personal reflections of A. Fred Rasmussen. Thanks are also due to my colleagues, Drs. David Felten and Nicholas Cohen, who contributed material for this essay. The responsibility for the selection of these particular perspectives, the editorializing, and any remaining errors are my own.

Preparation of this paper was supported by a Research Scientist Award (K05 MH06318) from the National Institute of Mental Health.

FOOTNOTE
*As I write this chapter, it seems evident that the financial support for research in psychoneuroimmunology, despite its successes, will face serious difficulties over the next several years (but that's a different chapter).

REFERENCES

1. **Ader, R.,** Presidential address: Psychosomatic and psychoimmunologic research, *Psychosom. Med.*, 42,307, 1980.
2. **Ader, R.,** Ed., *Psychoneuroimmunology*, Academic Press, New York, 1981.
3. **Ader, R., Felten, D.L., and Cohen, N.,** Eds., *Psychoneuroimmunology*, 2nd ed., Academic Press, New York, 1991.
4. **Metal'nikov, S., and Chorine, V.,** Rôle des réflèxes conditonnels dans l'imunite, *Ann. Inst. Pasteur*, 40,893, 1926.
5. **Hull, C.L.,** The factor of conditioned reflex, in *A Handbook of General Experimental Psychology*, C. Murchison, Ed., Clark Univer., Worcester, 1934.

6.   **Kopeloff, N., Kopeloff, L.M., and Raney, M.E.,** The nervous system and antibody production, *Psychiat. Quart,*, 7,84, 1933.

7.   **Kopeloff, N.,** *Bacteriology,* in *Neuropsychiatry,* Charles C. Thomas, Springfield, 1941.

8.   **Ader, R.,** A historical account of conditioned immunobiologic responses, in *Psychoneuro-immunology,* Ader, R., Ed., Academic Press, New York, 1981, p.321.

9.   **Rasmussen, A.F., Jr., Marsh, J.T., and Brill, N.Q.,** Increased susceptibility to herpes simplex in mice subjected to avoidance-learning stress or restraint, *Proc. Soc. Exp. Biol. Med.,* 96,183, 1957.

10.   **Johnson, T., Lavender, J.F., Hultin, F., and Rasmussen, A.F., Jr.,** The influence of avoidance-learning stress on resistance to Coxsackie B virus in mice, *J. Immunol.,* 91,569, 1965.

11.   **Jensen, M.M., and Rasmussen, A.F., Jr.,** Stress and susceptibility to viral infection. I. Response of adrenals, liver, thymus, spleen and peripheral leukocyte counts to sound stress, *J. Immunol.,* 90,17, 1963.

12.   **Marsh, J.T., Lavender, J.F., Chang, S.S., and Rasmussen, A.F., Jr.,** Poliomyelitis in monkeys: Decreased susceptibility after avoidance stress, *Science,* 1450,1414, 1963.

13.   **Chang, S.S., and Rasmussen, A.F., Jr.,** Effect of stress on susceptibility of mice to polyoma virus infection, *Bacteriol. Proc.,* 64,134, 1964.

14.   **Chang, S.S., and Rasmussen, A.F., Jr.,** Stress-induced suppression of interferon production in virus-infected mice, *Nature,* 205,623, 1965.

15.   **Yamada, A., Jensen, M.M., and Rasmussen, A.F., Jr.,** Stress and susceptibility to viral infections. III. Antibody response and viral retention during avoidance learning stress, *Proc. Soc. Exp. Biol. Med.,* 116,677, 1964.

16.   **Friedman, S.B., Ader, R., and Glasgow, L.A.,** Effects of psychological stress in adult mice inoculated with Coxsackie B viruses, *Psychosom. Med.,* 27,361, 1965.

17.   **Friedman, S.B., Glasgow, L.A., and Ader, R.,** Psychosocial factors modifying host resistance to experimental infections, *Ann. N. Y. Acad. Sci.,* 164,381, 1969.

18.   **Solomon, G.F.,** Stress and antibody response in rats, *Int. Arch. Allergy,* 35,97, 1969.

19.   **Solomon, G.F.,** Emotions and personality factors in the onset and course of autoimmune disease, particularly rheumatoid arthritis, in *Psychoneuroimmunology,* Ader, R., Ed., Academic Press, New York, 1981, p.159.

20.   **Solomon, G.F., Moos, R.H., Fessel, W.J., and Morgan, E.E.,** Globulins and behavior in schizophrenia, *Int. J. Neuropsychiat.,* 2,20, 1966.

21.   **Solomon, G.F., and Moos, R.H.,** The relationship of personality to the presence of rheumatoid factor in asymptomatic relatives of patients with rheumatoid arthritis, *Psychosom. Med.,* 27,350, 1965.

22.   **Good, R.A.,** Experimental models in systemic lupus erythematosus, *Arthr. Rheum.,* 6,490, 1963.

23.   **Dixon, F.J., Feldman, J, and Vasquez, J.,** Immunology and pathogenesis of experimental serum sickness, in *Cellular and Humoral Aspects of Hypersensitive States,* Lawrence, S.H., Ed., Hoeber, New York, 1959.

24.   **Solomon, G.F., and Moos, R.H.,** Emotions, immunity and disease: A speculative theoretical integration, *Arch. Gen. Psychiat.,* 11,657, 1964.

25. **Solomon, G.F., Merigan, T., and Levine, S.,** Variations in adrenal cortical hormones within physiological ranges and interferon production in mice, *Proc. Soc. Exp. Biol. Med.*, 126,74, 1967.

26. **Solomon, G.F., Levine, S., and Kraft, J.K.,** Early experience and immunity, *Nature*, 220, 821, 1968.

27. **Amkraut, A.A., and Solomon, G.F.,** Stress and murine sarcoma virus (Maloney)induced tumors, *Cancer Res.*, 32,1428, 1972.

28. **Amkraut, A.A., Solomon, G.F., Kaspar, P., and Purdue, A.,** Effect of stress and of hormonal intervention on the graft versus host response, *Adv. Exp. Med. Biol.*, 29,667, 1972.

29. **Amkraut, A.A., Solomon, G.F., and Kraemer, H.C.,** Stress, early experience, and adjuvant-induced arthritis in the rat, *Psychosom. Med.*, 33,203, 1971.

30. **Solomon, G.F., Temoshuk, L., O'Leary, A., and Zich, J.,** An intensive immunologic study of long-surviving persons with AIDS: Plot work, background studies, hypotheses, and methods, *Ann. N.Y. Acad. Sci.*, 496,647, 1987.

31. **Szentivanyi, A.,** The beta adrenergic theory of the atopic abnormality in bronchial asthma, *J. Allergy*, 42,203, 1968.

32. **Szentivanyi, A., and Filipp, G.,** Anaphylaxis and the nervous system. II., *Ann. Allergy*, 16, 143, 1958.

33. **Szentivanyi, A., and Szekely, J.,** Anaphylaxis and the nervous system. IV., *Ann. Allergy*, 16, 389, 1958.

34. **Hadden, J.W., Hadden, E.M., and Middleton, E.,** Lymphocyte blast transformation. I. Demonstration of adrenergic receptors in human peripheral lymphocytes, *J. Cell. Immunol.*, 1,583, 1970.

35. **Bourne, H., Melmon, K., and Liechtenstein, L.,** Histamine augments leukocyte adenosine 3',5'- monophosphate and blocks antigenic histamine release, *Science*, 173,743, 1971.

36. **Hadden, J.W., Hadden, E.M., and Good, R.A.,** Alpha adrenergic stimulation of glucose uptake in the human erythrocyte, lymphocyte and lymphoblast, *Exp. Cell Res.*, 68,217, 1971.

37. **Hadden, J.W., Hadden, E.M., Middleton, E., and Good, R.A.,** Lymphocyte blast transformation. II. The mechanisms of action of alpha adrenergic receptor effects, *Int. Arch. Allergy Appl. Immunol.*, 40,526, 1971.

38. **Coffey, R.G., Hadden, E.M., and Hadden, J.W.,** Norepinephrine stimulation of membrane ATPase in human lymphocytes, *Endocrinol. Res. Comm.*, 12,179, 1975.

39. **Hadden, J.W., Hadden, E.M., Haddox, M., and Goldberg N.D.,** Guanosine cyclic 3'5'monophosphate: A possible intracellular mediator of mitogenic influences in lymphocytes, *Proc. Nat. Acad. Sci.*, 69,3024, 1972.

40. **Strom, T.B., Disseroth, B., Morganroth, J., Carpenter, C, and Merrill, J.,** Alteration of the cytotoxic action of sensitized lymphocytes by cholinergic agents and activators of adenylate cyclase, *Proc. Nat. Acad. Sci.*, 69,2995, 1972.

41. **Hadden, J.W., Hadden, E.M., Meetz, G., Good, R.A., Haddox, M.K., and Goldberg, N.D.,** Cyclic GMP in cholinergic and mitogenic modulation of lymphocyte metabolism and proliferation, *Fed. Proc.*, 32,1022, 1973.

42. **Hadden, J.W., Hadden, E.M., Coffey, R.G., Johnson, E.M., and Johnson, L.D.,** Cyclic GMP and lymphocyte activation, in *Immune Recognition*, Rosenthal, A., Ed., Academic Press, New York, 1975, p.359.

43. **Good, R. A.,** Foreword: Interactions of the body's major networks, in *Psychoneuroimmunology*, Ader, R., Ed., Academic Press, New York, 1981, xvii.

44. **Hadden, J.W.,** Thymic endocrinology, *Int. J. Immunopharmac.*, 14,345, 1992.

45. **Saha, A.R. Hadden, E.M., Sosa, M., and Hadden, J.W.,** Thymus in neuroendocrine perspective, in *Nitric Oxide: Brain and Immune System*, Moncada, S., Nistico, G., and Higgs, E.A., Eds., Portland Press, England, 1993, p.43.

46. **Ader, R.,** Letter to the Editor: Behaviorally conditioned immunosuppression, *Psychosom. Med.*, 36,183, 1974.

47. **Ader, R., and Cohen, N.,** Behaviorally conditioned immunosuppression, *Psychosom. Med.*, 37,333, 1975.

48. **Rogers, M.P., Reich, P. Strom, T.B., and Carpenter, C.B.,** Behaviorally conditioned immunosuppression: Replication of a recent study, *Psychosom. Med.*, 38,447, 1976.

49. **Ader, R., and Cohen, N.,** CNS-immune system interactions: Conditioning phenomena, *Behav. Brain Sci.*, 8,379, 1985.

50. **Ader, R., and Cohen, N.,** The influence of conditioning on immune responses, in *Psychoneuroimmunology*, Second Edition, Ader, R., Felten, D.L., and Cohen, N., Eds., Academic Press, New York, 1991, p.611.

51. **Ader, R., and Cohen, N.,** Psychoneuroimmunology: Conditioning and stress, *Ann. Rev. Psychol.*, 44,53, 1993.

52. **Ader, R., Kelly, K., Moynihan, J., Grota, L.J., and Cohen, N.,** Conditioned enhancement of antibody production using antigen as the unconditioned stimulus, *Brain Behav. Immun.*, 7,334, 1993.

53. **Gorozynski, R.M., Macrae, S., and Kennedy, M.,** Conditioned immune response associated with allogeneic skin grafts in mice, *J. Immunol.*, 129,704, 1982.

54. **Ader, R., and Cohen, N.,** Behaviorally conditioned immunosuppression and murine systemic lupus erythematosus, *Science*, 214,1534, 1982.

55. **Gorozynski, R.M.,** Conditioned enhancement of skin allografts in mice, *Brain Behav. Immun.*, 4,85, 1990.

56. **Grochowicz, P., Schedlowski, M., Husband, A.J., King, M.G., Hibberd, A.D., and Bowen, K.M.,** Behavioral conditioning prolongs heart allograft survival in rats, *Brain Behav. Immun.*, 5,349, 1991.

57. **Olness, K., and Ader, R.,** Conditioning as an adjunct in the pharmacotherapy of lupus erythematosus, *J. Develop. Behav. Ped.*, 13,124, 1992.

58. **Besedovsky, H.O.,** Delay in skin allograft rejection in rats grafted with fetal adrenal glands, *Experientia*, 26,697, 1971.

59. **Besedovsky, H.O., Sorkin, E., Keller, M., and Mueller, J.,** Hormonal changes during the immune response, *Proc. Soc. Exp. Biol. Med.*, 150,466, 1975.

60. **Besedovsky, H.O., and Sorkin, E.,** Network of immune-neuroendocrine interactions, *Clin. Exp. Immunol.*, 27,1, 1977.

61. **Besedovsky, H.O., Sorkin, E., Felix, R. and Haas, H.,** Hypothalamic changes during the immune response, *Europ. J. Immunol.*, 7,323, 1977.

62. Besedovsky, H.O., del Rey, A.E., and Sorkin, E., Antigenic competition between horse and sheep red blood cells as a hormone-dependent phenomenon, *Clin. Exp. Immunol.*, 37,106, 1979.

63. Besedovsky, H.O., del Rey, A.E., Sorkin, E., DaPrada, M., and Keller, H.A., Immunoregulation mediated by the sympathetic nervous system, *Cell. Immunol.*, 48,346, 1979.

64. del Rey, A., Besedovsky, H.O., Sorkin, E., DaPrada, M., and Bondiolotti, P., Sympathetic immunoregulation: Difference between high- and low-responder animals, *Am. J. Physiol.*, 242,R30, 1982.

65. del Rey, A., Besedovsky, H.O., Sorkin, E., DaPrada, M., and Arrenbrecht, S., Immunoregulation mediated by the sympathetic nervous system. II., *Cell. Immunol.*, 63,329, 1981.

66. Besedovsky, H.O., del Rey, A.E., Sorkin, E., DaPrada, M., Burri, R., and Honegger, C., The immune response evokes changes in brain noradrenergic neurons, *Science*, 221,564, 1983.

67. Besedovsky, H.O., del Rey, A.E., and Sorkin, E., Lymphokine containing supernatants from Con A-stimulated cells increase corticosterone blood levels, *J. Immunol.*, 126,385, 1981.

68. Besedovsky, H.O., del Rey, A.E., Sorkin, E., Lotz, W., and Schwulera, U., Lymphoid cells produce an immunoregulatory glucocorticoid increasing factors (GIF) acting through the pituitary gland, *Clin Exp. Immunol.*, 59,622, 1985.

69. Besedovsky, H.O., del Rey, A., Sorkin, E., and Dinarello, C.A., Immunoregulatory feedback between interleukin-1 and glucocorticoid hormones, *Science*, 233,652, 1986.

70. Berkenbosch, J., Van Oers, J., del Rey, A., Tilders, F., and Besedovky, H.O., Corticotropin-releasing factor-producing neurons in the rat activated by interleukin-1, *Science*, 238,524, 1987.

71. Besedovsky, H.O., del Rey, A. and Normann, S., Host endocrine responses during tumor growth, in *Immunity to Cancer II*, Mitchell, M.S., Ed., Alan Liss, New York, 1989, p.203.

72. deRijk, R., van Rooujen, N., Besedovsky, H.O., del Rey, A., and Berkenbosch, F., Selective depletion of macrophages prevents pituitary-adrenal activation in response to subpyrogenic, but not to pyrogenic, doses of bacterial endotoxin in rats, *Endocrinology*, 129,330, 1991.

73. Besedovsky, H.O. and del Rey, A., Mechanism of virus-induced stimulation of the hypothalamic-pituitary-adrenal axis, *J. Steroid Biochem.*, 34,235, 1989.

74. Cohen, N. and Ader, R., Antibodies and learning: A new dimension, in *The Immune System*, Vol. 1, Steinberg, C., and Lefkovits, I., Eds., Karger, Basel, 1981, p.51.

75. Blalock, J.E. and Smith, E.M., Human leukocyte interferon: Structural and biological relatedness to adrenocorticotropic hormone and endorphins, *Proc. Nat. Acad. Sci.*, USA, 77,5972, 1980.

76. Blalock, J.E. and Harp. C., Interferon and adrenocorticotropic hormone induction of steroidogenesis, melanogenesis and antiviral activity, *Arch. Virol.*, 67,45, 1981.

77. Smith, E.M. and Blalock, J.E., Human lymphocyte production of corticotropin and endorphin-like substances: Association with leukocyte interferon, *Proc. Nat. Acad. Sci.*, USA, 78,7530, 1981.

78. Blalock, J.E., The immune system as a sensory organ, *J. Immunol.*, 132,1067, 1984.

79. Blalock, J.E., The syntax of immune-neuroendocrine communication, *Immunol. Today*, 15,504, 1994.

80. Smith, E.M., Galin, F.S., LeBoeuf, R.D., Coppenhaver, D.H., Harbour, D.V., and Blalock, J.E., Nucleotide and amino acid sequence of lymphocyte-derived corticotropin: Endotoxin induction of a truncated peptide, *Proc Nat. Acad. Sci.*, USA, 87,1057, 1990.

81. Williams, J.M., Peterson, R.G., Shea, P.A., Schmedtje, J.F., Bauer, D.C., and Felten, D.L., Sympathetic innervation of murine thymus and spleen: Evidence for a functional link between the nervous and immune system, *Brain Res. Bull.*, 6,83, 1981.

82. Livnat, S., Felten, S.Y., Carlson, S.L., Bellinger, D.L., and Felten, D.L., Involvement of peripheral and central catecholamine systems in neural-immune interactions, *J. Neuroimmunology*, 10,5, 1985.

83. Felten, D.L., Ackerman, K.D., Wiegand, S.J., and Felten, S.Y., Noradrenergic sympathetic innervation of the spleen: I. Nerve fibers associate with lymphocytes and macrophages in specific compartments of the splenic white pulp, *J. Neurosci. Res*, 18,28, 1987.

84. Felten, D.L., Felten, S.Y., Bellinger, D.L., Carlson, S.L., Ackerman, K.D., Madden, K.S., Olschowka, J.A., and Livnat, S., Noradrenergic sympathetic neural interactions with the immune system: Structure and function, *Immunol. Rev.*, 100,225, 1987.

85. Felten, S.Y., and Olschowka, J.A., Noradrenergic sympathetic innervation of the spleen: II. Tyrosine hydroxylase (TH)-positive nerve terminals form synaptic-like contacts on lymphocytes in the splenic white pulp, *J. Neurosci. Res.*, 18,37, 1987.

86. Ackerman, K.D., Felten, S.Y., Dijkstra, C.D., Livnat, S., and Felten, D.L., Parallel development of noradrenergic innervation and cellular compartmentation in the rat spleen, *Exp. Neurol.*, 103,239, 1989.

87. Bellinger, D.L., Ackerman, K.D., Felten, S.Y., Pulera, M., and Felten, D.L., A longitudinal study of age-related loss of noradrenergic nerves and lymphoid cells in the rat spleen, *Exp. Neurol.*, 116,295, 1992.

88. Felten, D.L., Felten, S.Y., Bellinger, D.L., and Lorton, D., Noradrenergic and peptidergic innervation of secondary lymphoid organs: Role in experimental rheumatoid arthritis, *Europ. J. Clin. Invest.*, 22, Suppl. 1, 37, 1992.

89. Stites, D.P., and Terr, A.L., *Basic and Clinical Immunology*, Seventh Edition, Appleton and Lange, Norwalk, 1991, p.190.

90. Weigant, A. and Blalock, J.E., Role of neuropeptides in the bidirectional communication between the immune and neuroendocrine systems, in *Neuropeptides and Immunoregulation*, Scharrer, B., Smith, E.M., and Stefano, G.B., Eds., Springer-Verlag, Berlin, 1994, p.14.

91. Dyck, D,. Janz, L., Osachuk, T.A.G., Falk, J., Labinsky, J. and Greenberg, A.H., The Pavlovian conditioning of IL-1-induced glucocorticoid secretion, *Brain Behav. Immun.*, 4,93, 1990.

92. Monjan, A.A., Stress and immunologic competence: Studies in animals, in *Psychoneuroimmunology*, Ader, R., Ed., Academic Press, New York, 1981, p.185.

93. Ader, R., and Cohen, N., Behavior and the immune system, in *Handbook of Behavioral Medicine Research*, Gentry, W.D., Ed., Guilford Press, New York, 1984, p.117.

94. Bartrop, R.W., Luckhurst, E., Lazarus, L., Kiloh, L.G., and Penny, R., Depressed lymphocyte function after bereavement, *Lancet*, i,834, 1977.

95. **Stein, M., Schleifer, S.J., and Keller, S.E.,** Hypothalamic influences on immune responses, in *Psychoneuroimmunology*, Ader, R., Ed., Academic Press, New York, 429, 1981.

96. **Glaser, R., Rice, J., Sheridan, J., Fertel, R., Stout, J., Speicher, C., Pinsky, D., Kotur, M., Post, A., Beck, M., and Kiecolt-Glaser, J.A.,** Stress-related immune suppression: Health implications, *Brain Behav. Immun.*, 1,7, 1987.

97. **Ader, R.,** The influence of psychological factors on disease susceptibility in animals, in *The Husbandry of Laboratory Animals*, Conalty, M.L., Ed., Academic Press, London, 1967, p.219.

98. **Cohen, S. and Williamson, G.,** Stress and infectious disease in humans, *Psychol. Bull.*, 109,5, 1991.

99. **Sheridan, J.F., Dobbs, C., Brown, D., and Zwilling, B.,** Psychoneuroimmunology: Stress effects on pathogenesis and immunity during infection, *Clin. Microbiol. Rev.*, 7,200, 1994.

100. **Heisenberg, W.,** *Physics and Philosophy, the Revolution in Modern Science*, Harper, New York, 1958, 81.

Chapter 2

PSYCHONEUROIMMUNOLOGY, STRESS AND INFECTION

Adrian J. Dunn

Department of Pharmacology and Therapeutics,
Louisiana State University Medical Center,
Shreveport, Louisiana 71130-3932

INTRODUCTION

Psychoneuroimmunology is concerned with the effects of the mind on the immune system. The current interest in psychoneuroimmunology was propelled by the widely held belief that the mind exerts significant effects on the health of the individual. This belief has been substantiated by a variety of research studies that rely largely on correlations between psychological states and traits and the probabilities of disease and the outcome of disease states. Physicians and others frequently exploit the use of psychological factors in the treatment of diseases (e.g., with placebos). However, it is now apparent that while the nervous system influences immune system function, the immune system also affects the nervous system. In other words, the communication between the nervous and immune systems is *bidirectional*. Thus psychoneuroimmunology (or PNI as it is often referred to) encompasses the gamut of complex interactions between the nervous and immune systems. A sound understanding of the physiological basis of the interactions between the nervous and immune systems would be both enlightening and medically useful.

STRESS

Stress has been a focal point in psychoneuroimmunology largely because of its demonstrated effects on immune function. It becomes even more intimately involved in nervous system-immune system interactions, because infections and certain other immune challenges can be regarded as stressful (see below). Thus infection induces stress, and the stress induced can alter the immune response to infection, therefore affecting the outcome of the disease. Needless to say, this complex interrelationship is presently poorly understood.

Stress has proven to be a difficult scientific concept. Although the word is frequently used and widely understood, there is not universal agreement on its definition. It is important that stress is subjective; what is stressful for one individual may not be so for another, and what is stressful for one individual at one time may not be so at another. Also, the response to stress is diminished by the extent to which the organism copes. Hans Selye, who devoted his life to the study of stress, defined stress as the nonspecific *response* of an organism to a demand made upon it.[1] The agent that induced the stress he called the stressor. Unfortunately, Selye's definition conflicted with the common use of the word stress, and its use in physics.

Nevertheless, his semantics emphasize the subjectivity of stress, and the variable nature of the the response.

THE PHYSIOLOGY OF STRESS

Walter Cannon first demonstrated increased plasma concentrations of catecholamines when cats were exposed to barking dogs. He identified the sources of the catecholamines as the sympathetic nervous system and the adrenal medulla.[2] Activation of the adrenal medulla results in secretion of epinephrine and, in most species, norepinephrine (NE), into the general circulation. The sympathetic nervous system contributes NE both locally and into the general circulation. Cannon also introduced the phrase "fight or flight" as a response to stress. He suggested that the role of the changes in peripheral catecholamines was to redistribute bloodflow and to reorient metabolism towards the production of glucose and otherwise provide energy for the musculature to enable fighting or fleeing. Later, Selye[3] demonstrated the importance in stress of the adrenal cortex and the secretion glucocorticoids. The secretion of glucocorticoids is the final stage in the activation of the hypothalamo-pituitary-adrenocortical (HPA) axis (Figure 1). The HPA axis is initiated by the secretion of corticotropin-releasing factor (CRF) from hypothalamic neurons with terminals in the median eminence. The CRF secreted is then transported by portal bloodflow to the anterior pituitary, which it stimulates to secrete adrenocorticotropin (ACTH). Circulating ACTH stimulates the secretion of glucocorticosteroids (cortisol and/or corticosterone, depending on species) from the adrenal cortex. The anterior pituitary also secretes β-lipotropin (β-LPH) and β-endorphin concomitantly with ACTH in response to CRF. Although corticosteroids are the major effectors of the HPA axis and provide negative feedback at multiple sites, it is now known that CRF, ACTH and β-endorphin each have their own independent effects.

Many have challenged the "nonspecific" aspect of Selye's definition of stress, but the two physiological responses (sympathoadrenal and HPA) occur in most situations regarded as stressful. Thus, most physiological definitions of stress cite the co-activation of the sympatho-adrenal and HPA axes, i.e. elevation of plasma concentrations of catecholamines and corticosteroids. Some ethologists do not agree with this hormonal definition of stress, and prefer a poorly defined behavioral definition. A schematic indicating the relationships within and between the sympathoadrenal and HPA systems appears in Figure 2.[4] The major known interactions are the regulation of CRF secretion by catecholamines, and the glucocorticoid regulation of a number of facets of catecholamine function, including epinephrine synthesis in the adrenal medulla, and up- and downregulation of receptors for catecholamines. There are also complex effects of the opioid peptides.

This classical physiological understanding of stress ignores the central nervous system, an organ critical in the fight-flight decision and in adaptation to stress. A large number of neurochemical changes can occur in the brain during or following some form of experimental stress. However, a critical one appears to be the activation of cerebral catecholaminergic systems. Many treatments regarded as stressful have been shown to activate cerebral noradrenergic neurons, providing a central counterpart to the peripheral sympathoadrenal activation. In addition, there is indirect evidence that CRF- and endorphin-containing neurons

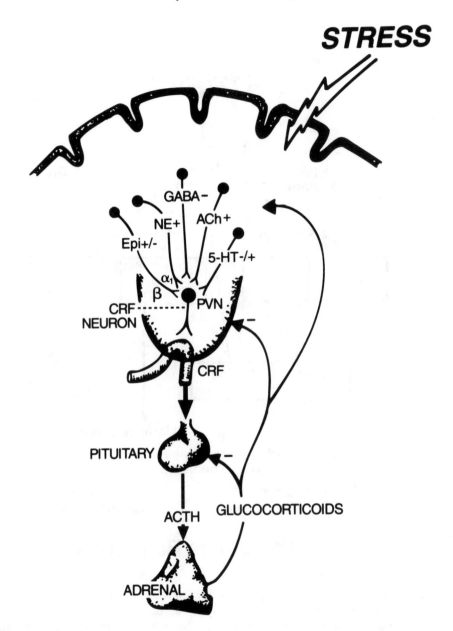

Fig 1. The hypothalamic-pituitary-adrenocortical (HPA) system. Noradrenergic (NE), cholinergic (ACh), serotonergic (5-HT), and GABAergic inputs to the hypothalamus alter the release of corticotropin-releasing factor (CRF), which stimulates release of ACTH from the anterior pituitary, which in turn stimulates release of glucocorticoids from the adrenal cortex. Glucocorticoid feedback occurs at all levels of the axis.

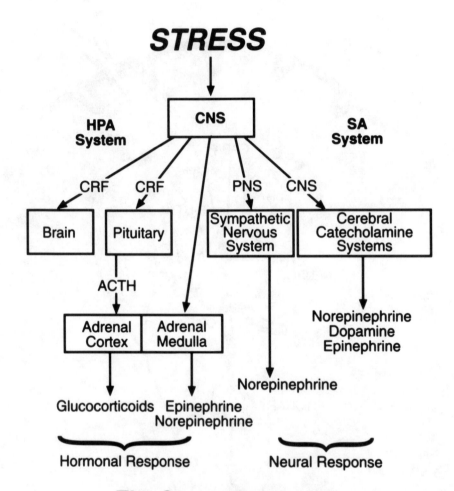

The Stress Response

Fig 2.   A schematic indicating the overall arrangement of the hypothalamic-pituitary-adrenocortical (HPA) and sympathoadrenomedullary (SA) systems. The central noradrenergic system appears to be activated in parallel with the peripheral sympathetic nervous system and can be considered its CNS counterpart. The other cerebral catecholaminergic systems (i.e., dopamine and epinephrine) may also be activated in stress. The diagram illustrates the parallel activation of both the HPA and SA systems and their interaction in the adrenal gland. (Modified from Dunn, A. J. and Kramarcy, N. R.[4] reprinted with permission of the publisher)

are also activated. The cardinal CNS responses in stress may be a co-activation of catecholaminergic systems, and those chemically related to the HPA axis. Thus the stress-related events in the CNS appear to parallel those in the periphery.

CEREBRAL MECHANISMS IN STRESS

Stressful treatments frequently decrease cerebral NE content, especially in the hypothalamus.[5] This response has been presumed to reflect increased release of NE coupled with the inability of synthetic mechanisms to keep pace, because the magnitude is greater after intense or prolonged stress. Consistent with this, neurochemical studies have indicated increased synthesis and catabolism of catecholamines during stress[4,5] and studies with *in vivo* microdialysis and *in vivo* voltammetry have indicated stress-related increases in extracellular concentrations of NE, all of which suggest that stress causes increased synaptic release of NE. This activation of cerebral NE neurons is not anatomically specific, consistent with the anatomical and biochemical data indicating that NE cell bodies in the brain stem send collaterals to widespread areas of the brain.[6]

Cerebral dopaminergic (DA) systems are also activated in response to physical and behavioral stressors. The mesocortical DA system which projects to prefrontal and cingulate cortices appears to be the most responsive, but DA projections to other regions, including the neostriatum (caudate nucleus plus putamen) and nucleus accumbens, are also affected. This DA response is readily detected by increased production of catabolites and by *in vivo* microdialysis and *in vivo* voltammetry. Other data suggest that cerebral adrenergic neurons (i.e., those secreting epinephrine) are also activated during stress. Thus it may be that neurons containing all three catecholamines in the brain contribute to the stress response.[4]

Cerebral serotonin (5-hydroxytryptamine, 5-HT) systems also appear to respond in stress. Numerous studies indicate altered 5-HT content or increased metabolism during stress (i.e., increased production of its major catabolite, 5-hydroxyindolacetic acid, 5-HIAA), and microdialysis and voltammetric studies suggest increased release. The responses are quite variable and more commonly observed following prolonged stress. The changes in cerebral 5-HT appear to depend, at least partly, on changes in cerebral free tryptophan, which are in turn dependent upon increased sympathetic nervous system activity, because ganglionic blockers and β-adrenergic antagonists prevent the CNS changes in tryptophan and 5-HIAA.[7]

Compelling data indicate an important role for CRF in the brain during stress. CRF is found in neurons in a variety of brain regions, not only in the paraventricular hypothalamic neurons concerned with activation of the HPA axis.[8] Intracerebral administration of CRF produces a wide variety of effects that resemble those observed in stress.[9,10] These include endocrine, electrophysiological, gastrointestinal, neurochemical, and behavioral effects. The endocrine effects include activation of the HPA axis and inhibition of growth hormone and gonadotropin secretion as well as sympathoadrenal activation. The electrophysiological effects include desynchronization of the cortical electroencephalogram and activation of locus coeruleus noradrenergic neurons. The complex changes in gastrointestinal motility closely resemble those observed in stress. Intracerebral CRF is a potent activator of noradrenergic and dopaminergic neurons. It also induces many behavioral effects, such as anorexia, changes in locomotor activity, increased grooming, decreased sexual behavior and exploratory behavior, anxiogenic effects on conflict behavior, and a heightened responsivity to treatments that normally induce anxiety or fear responses. Thus, CRF administration mimics most responses commonly observed in stress. That this role of CRF may be physiological is suggested because in many cases, intracerebral administration of peptide antagonists of CRF-receptors or antibodies to CRF attenuates or prevents the normal responses in stress. Thus cerebral CRF has been suggested to be a central coordinator of responses in stress.[9,11]

NERVOUS SYSTEM-DERIVED EFFECTORS OF IMMUNE FUNCTION

Currently, the major mechanism by which the central nervous system is thought to influence immune function is by chemical messengers: various hormones from the pituitary, and the secretion of catecholamines and peptides from nerve terminals of the autonomic nervous system (Figure 3). The pituitary hormones with known effects on the immune system include ACTH, β-endorphin, prolactin, growth hormone and thyroid-stimulating hormone (TSH). The adrenal medulla secretes enkephalins and dynorphins. Endorphins are also secreted by the sympathetic nervous system along with other neuropeptides (see below).

Each of these factors has been shown to influence cells of the immune system *in vitro*, and many of them may have similar effects *in vivo*. Although the actions of each of these chemical messengers are complex, for the most part, the peptides have facilitatory effects. The sympathetic innervation of organs such as the spleen, the thymus and the lymph nodes may play important regulatory functions. A potential parasympathetic innervation has not been confirmed.[12]

STRESS AND THE IMMUNE SYSTEM

Although it is regarded as common knowledge that stress impairs immune system function, the truth is significantly more complicated. Data from human and animal studies have confirmed the immunosuppressive effect of stress;[13-15] however, there is considerable evidence that stress may also enhance immune function.[13] A common human experience is that under acute stress conditions, an impending infection, such as a cold, can be held at bay, but when the pressure is relaxed, one succumbs to the cold. Some animal experiments also suggest that mild acute stressors may actually enhance measures of immunity (see, for example, References 13 and 16).

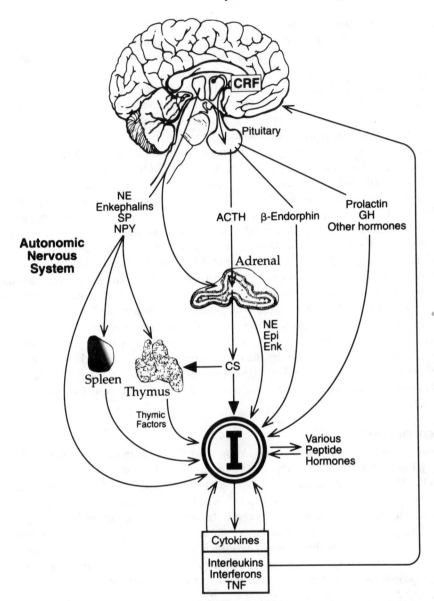

Fig 3. A schematic diagram of the interactions between the brain and components of the endocrine and immune systems. The ability of the brain to alter immune system function via a variety of endocrine pathways, and the autonomic nervous system, is emphasized, and the effects of peptides and cytokines produced by the immune system on immune cells and the brain is indicated. Abbreviations: CS, corticosteroids; Enk, enkephalins; Epi, epinephrine; GH, growth hormone; I, immunocytes. (From **Dunn, A.J.,**[99] reprinted with permission of the publisher)

THE ROLE OF THE ADRENAL GLAND

If adrenal hormones are responsible for the stress-related reductions in immunity, then adrenalectomy should prevent these effects. Because adrenalectomy removes both the medulla and the cortex, it fails to distinguish between the effects of steroids or catecholamines (or of certain neuropeptides). In some animal studies that have examined acute responses to brief stressors, adrenalectomy appears to prevent the immunosuppressive effects of stress.[14,17] However, other studies involving more prolonged stress have typically found that stress-induced changes in immunity persist in adrenalectomized animals.[18,19]

The results may also depend upon the particular immune parameters measured. The earlier studies focused on *in vitro* mitogen-stimulated proliferation assays which assess the responsivity (i.e., cell division measured by DNA synthesis) to lectin mitogens (such as concanavalin A or lipopolysaccharide, LPS). The interpretation of such assays is problematic, because the results are susceptible to a number of extraneous influences, and the assays are typically performed after several days of incubation *in vitro* separated from many normal physiological influences.[20] The data are typically highly variable. More recently, natural killer (NK) cell activity has been studied. There is evidence that NK cells are involved in the rejection of tumors.[21] Stressful treatments have been shown to suppress NK cell function in both animal and human studies.[22,23] The major effectors for the stress-induced effects on NK cell function appears to be opiates[23] and catecholamines via β-adrenergic receptors.[24]

Most of the studies of stress on immune function have used *ex vivo* procedures (i.e. stress *in vivo,* immune measures *in vitro*), so that an important factor is whether or not the population of cells sampled is altered by the *in vivo* treatment. The trafficking of lymphocytes around the body is known to be regulated by hormones and other stress-related secretions, so it is likely that the stressful treatments alter the population of cells harvested for the *in vitro* studies.

GLUCOCORTICOIDS

Glucocorticoids have long been known to have immunosuppressive effects.[25,26] Interestingly, the dogma that stress suppresses immunity has derived considerable support from these effects. The property is exploited in the medical practice of using glucocorticoids postsurgically (usually high doses of potent synthetic steroids, such as prednisolone, triamcinolone or dexamethasone) to decrease tissue inflammation and the rejection of transplanted tissues. However, the effects of supraphysiological doses of the steroids used should not be extrapolated to the normal physiological state. Also, considerable experimental data suggest that the effects of glucocorticoids are not exclusively immunosuppressive.[26,27] Moreover, glucocorticoids may not even be the major mechanism by which stress suppresses immune function.

It is often forgotten that glucocorticoids are essential for normal immune responses. Extensive experimental data from animal and human studies indicate that adrenally compromised individuals are more susceptible to infections, and that this property involves the adrenal cortex rather than the medulla.[27,28] Replacement studies have clearly implicated a role for the corticosteroids in immune defense mechanisms. Kass showed that an optimal concentration of corticosteroids was essential for normal recovery from infections in adrenalectomized animals.[29]

The glucocorticoids are immunosuppressive, but the concentrations of these compounds used clinically can cause lysis of immune cells, especially immature ones.[30] Careful studies that have used natural steroids at doses in the normal physiological range have noted stimulatory effects of steroids at lower doses.[26] Inhibitory effects occur at higher doses, typically above 10^{-6} M, the maximum concentration of free corticosterone or cortisol found in stressed animals after correcting for that bound by corticosteroid-binding globulins.[17] It should also be noted that elevations of plasma glucocorticoids following acute stress do not last long; plasma concentrations rapidly return to baseline, normally within 30 minutes of cessation of the stress.

Although glucocorticoids have direct effects on immune cells *in vitro*, there may also be indirect ones *in vivo*. A classic physiological correlate of stress is the involution of the thymus. The involution, which can decrease thymus weight by more than half, occurs largely because lymphocytes normally resident there are driven out to the periphery. Stress-induced thymic involution is prevented by adrenalectomy and can be induced by glucocorticoid administration.[25] Thus glucocorticoids can alter the bodily distribution of lymphocytes, which may in itself be an important factor marshalling the immune response to infection. The populations of lymphocytes derived by harvesting tissues from animals subjected to experimental treatments may thus be altered by the redistribution of cells caused by glucocorticoid secretion, an important consideration in interpreting the results of *ex vivo* data.

THE ROLE OF CATECHOLAMINES

Lymphocytes bear both α- and β-adrenergic receptors. Catecholamines in the circulation emanate from both the adrenal medulla (NE and epinephrine) and from sympathetic terminals (NE). In addition, lymphocytes may be directly exposed to neuronal secretions while they are resident in the thymus, spleen and lymph nodes. Anatomical studies have clearly demonstrated a sympathetic innervation of immune structures such as the bone marrow, thymus, spleen and lymph nodes.[12] Thus lymphocytes could be exposed to high local concentrations of catecholamines, as well as neuropeptides.

Studies *in vitro* have suggested separate α- and β-adrenergic effects; β-adrenergic receptors were largely inhibitory, whereas a-adrenergic receptors were stimulatory.[31] This generalization has endured to some extent, but the detailed results are very complex. There appear to be separate α- and β-adrenergic stimulatory effects on antibody production *in vitro*,[32] whereas NK activity appears to be inhibited by β-adrenergic stimulation. The results of *in vivo* studies have been bewilderingly complex. Generally, sympathectomy in adult animals depresses immune reactivity, but there are also paradoxical effects on lymphocyte proliferation and B cell differentiation. Among the confounding factors that may contribute to the complexity are: compensatory increases in adrenomedullary output, redistribution of lymphocytes, compensatory changes in the number and kind of adrenergic receptors, and the coexistence in sympathetic terminals of various neuropeptides.[33]

Several recent studies suggest that catecholamines released by the sympathetic nervous system provide a major mechanism by which NK cell activity is regulated *in vivo*. The inhibitory effect of icv CRF on NK cell activity is blocked by the ganglionic blocker, chlorisondamine,[14] and there is also direct evidence that β-adrenergic receptor blockade can prevent stress-induced effects on NK cell activity.[34]

OTHER FACTORS

Sympathetic terminals contain not only NE, but also neuropeptides which may act on the immune system. Felten *et al.* have demonstrated the presence of neuropeptide Y (NPY), substance P and vasoactive intestinal polypeptide (VIP) in the thymus spleen and lymph nodes, as well as calcitonin gene-related peptide (CGRP), in the thymus and lymph nodes, enkephalin and somatostatin in the spleen, tachykinin in the thymus, and peptide histidine isoleucine (PHI) in lymph nodes.[12]

CRF has been reported to have a variety of effects on immune cells. The direct effects of CRF are generally stimulatory. For example, CRF has been shown to stimulate B cell proliferation,[35] and NK activity,[36] as well as IL-1, IL-2 and IL-6 production.[37] Receptors for CRF have been found on immune cells,[38] providing a mechanism for these effects. Although it seems unlikely that CRF in the general circulation ever achieves concentrations high enough to stimulate these receptors, it is possible that local actions may occur, for example, in the spleen.[38]

ACTH has been shown to have some direct effects on immune function, including an inhibition of antibody production and modulation of B cell function (see reference 39), but the effects have not been striking. On the other hand, the endorphins have a plethora of effects on immune function.[39,40] Lymphocytes possess binding sites for opiates, but at least some of these are not sensitive to the opiate antagonist, naloxone.[40] Interestingly, binding sites have been found for N-acetyl-β-endorphin, which is the commonest form of β-endorphin secreted from the anterior pituitary and which has no opiate activity.[41] β-endorphin and other opioid peptides can exert effects on lymphocytes *in vitro*,[40] which by and large, are facilitatory. Such effects have been observed on NK activity, as well as on proliferative responses.[39] Opioid peptides are also chemoattractants for lymphocytes. By contrast with the enhancing effects *in vitro, in vivo* opiates are largely inhibitory, especially on NK activity. This apparent contradiction can be explained, because, at least in the case of morphine, the site of opiate action appears to be in the CNS.[42] Moreover, the effects appear to be mediated via the adrenal gland, most probably by catecholamines.[21]

Perhaps the most interesting stress-related hormone that affects the immune system is prolactin. Its effects appear to be largely stimulatory.[43] Reduction of pituitary prolactin secretion (e.g, by dopaminergic agonists or opiate antagonists) impairs immune function and increases susceptibility to infections, such as *Listeria monocytogenes,* whereas stimulation of prolactin secretion (e.g., by D2 dopaminergic antagonists or opiates) can enhance it. Bernton *et al.*[43] have postulated that prolactin may be the counter-regulatory hormone to glucocorticoids. Opposing interactions between these two hormones on immune function can be demonstrated *in vivo.* Direct effects of prolactin on lymphocyte function have been difficult to demonstrate, but antibodies to prolactin impair proliferative responses *in vitro.* Lymphocytes can produce a prolactin-like protein, although its identity with prolactin has not been demonstrated.

INFECTION AS A STRESSOR

We all recognize that sickness is stressful. In his autobiography, Selye[44] indicated that it was the common characteristics of sickness, regardless of the underlying disease, "the syndrome of just being sick" that first interested him in research on stress and led him to advance his proposal of the *nonspecificity* of stress. During World War I, it was noted that fatalities from infections were associated with striking morphological changes in the adrenal cortex,[27] suggesting that the HPA axis is activated following infections. In the 1950's, it was discovered that endotoxin (LPS) stimulated the HPA axis.[45,46] Later, it was shown that *E. coli* infection of rats increased the secretion of ACTH.[47] More recently, it has been shown that the responses to viral infections and immune challenges resemble those to physical or psychological stressors.[48] Infections activate the HPA axis and the sympathoadrenal system. They also appear to increase the synaptic release of NE and 5-HT in the brain.[49] Thus by the physiological criteria discussed above, infections can be regarded as stressful. Infections also induce behavioral responses that resemble those to other stressors.

From a neurochemical perspective, there are two important differences between the responses to infections and immune challenges and those to physical and behavioral stressors. Firstly, the noradrenergic responses to infections and immune challenges are substantially greater in the hypothalamus than in other brain regions, suggesting a selectivity for the A1 and A2 noradrenergic nuclei in the brainstem relative to the locus coeruleus (A6). Secondly, the catecholaminergic response is most often selective for NE; dopaminergic systems are not normally affected. Also, when DA systems are activated (as, for example, following administration of LPS), the responses do not show the selective enhancement of the prefrontal cortical systems normally observed following electric footshock or restraint.[50]

MECHANISMS OF IMMUNE SYSTEM EFFECTS ON THE NERVOUS SYSTEM

Besedovsky et al.[51] first showed that increases of plasma corticosterone accompanied the appearance of cells producing antibodies to such commonly used antigens as sheep red blood cells (SRBCs). The observation complemented earlier Russian observations of the electrophysiological activation of cells in the medial hypothalamus accompanying an immune response, replicated by Besedovsky *et al.*[52] and Saphier.[53] Besedovsky *et al.*,[54] also showed that SRBC inoculation changed the apparent turnover of NE in the hypothalamus.

The HPA activations related to treatment with SRBC and other antigens occur 5-8 days following treatment and coincided with the peak production of antibody. Similar observations have been noted by others with myelin basic protein[55] or keyhole limpet hemocyanin.[56] However, there is also an acute HPA response to immune stimulation that occurs within the first few hours. This acute response is observed following LPS and NDV administration.[56-58]

The mechanism of the activation of the HPA axis by immune stimuli has been studied extensively. Blalock and Smith[59] suggested that peptides synthesized by lymphocytes may be secreted in sufficient quantities to have systemic actions. Specifically, they suggested that ACTH derived from lymphocytes might activate the adrenal cortex to secrete glucocorticoids. Many reports suggest that lymphocytes can synthesize and secrete peptides. The spectrum of peptides reported is large, and includes many of the known peptide hormones, as well as the hypophysiotropic factors. They include: ACTH, CRF, growth hormone, thyrotropin, prolactin, human chorionic gonadotropin, the endorphins, enkephalins, substance P, somatostatin and VIP.[60,61] The quantities of the peptides produced are very small, and often their presence can only be inferred by the ultrasensitive assays used to detect their mRNAs.[62] Thus the physiological significance of this phenomenon is not clear. Also, there is considerable variability in the ability of different populations of lymphocytes to produce a specific peptide, an issue which has received very little serious attention. Because in many cases lymphocytes display receptors for these same peptides, they may function as chemical messengers within the immune system. It is also possible that there may be a local bidirectional communication between lymphocytes and other cells. An example of this may be in the spleen, where CRF appears to be present in the innervating neurons, and CRF-receptors are present on resident macrophages.[38] Also, β-endorphin produced locally in an area of inflammation by lymphocytes may exert an analgesic action directly on sensory nerve terminals.[63] Such a mechanism is appealing, because the concentrations of the peptides produced locally may be adequate to exert such effects, and the metabolic lability of peptides would ensure that the effect was localized.

In support of the notion that lymphocyte-derived ACTH may be responsible for the Newcastle disease virus (NDV)-induced HPA response, Blalock reported that administration of NDV to hypophysectomized mice elicited an increase in plasma corticosterone.[64] Unfortunately, no control (i.e., non-hypophysectomized) animals were included in the study to compare the effectiveness of NDV in intact and hypophysectomized mice. Moreover, several other groups have failed to replicate an NDV-induced increase of plasma corticosterone in hypophysectomized mice, when the completeness of the hypophysectomy was verified adequately.[65,66] Hypophysectomy also prevented the increases in plasma ACTH and corticosterone in response to LPS.[49,67] Thus it seems most likely that the mice used in the single experiment of Blalock and Smith[64] were not completely hypophysectomized, as has been found with other batches of "hypophysectomized" mice from the same supplier (Ref. 66 and Besedovsky, personal communication).

Most researchers favor an important role for cytokines, and especially interleukin-1 (IL-1), as mediators of the HPA response to immune stimulation. Besedovsky *et al.*[68] initially reported that supernatants of immune cells challenged *in vitro* with concanavalin A (i.e., conditioned media) had the ability to activate the HPA axis when injected into rats. The active factor synthesized and secreted in response to the mitogens was suggested to be IL-1 because supernatants of NDV-treated lymphocytes could be neutralized with an antibody to IL-1, and injection of purified recombinant human IL-1,β was found to be a potent stimulator of the HPA axis.[69] A scheme for this arrangement is depicted in Figure 4.

There is evidence that IL-1 can activate the HPA axis at multiple levels: the hypothalamus, the pituitary and the adrenal. However, the bulk of the evidence strongly favors the need for hypothalamic CRF. Deafferentation of the hypothalamus and lesions of the paraventricular nucleus block the ACTH response to IL-1.[70] Moreover, antibodies to CRF attenuate or block the ACTH and corticosterone responses to IL-1.[71,72] The pituitary appears to be essential because hypophysectomy prevents the response to IL-1.[49,72] The conflicting reports rely on *in vitro* studies which are notoriously subject to artifact. Although several authors have reported that IL-1 can stimulate hormone release from the pituitary and the adrenal *in vitro*, the reports are remarkably inconsistent and prolonged incubation *in vitro* is necessary to observe such responses.[72]

IL-1 administration mimics the neurochemical responses to LPS and NDV very closely. The increased NE metabolism is focused on the hypothalamus, with a regionally nonspecific increase in tryptophan and in 5-HT metabolism.[57,58,73] The only distinction is that LPS appears to affect DA metabolism. Thus the neurochemical data complement the HPA data implicating IL-1 as a mediator of the responses.

Because NE is considered to be a major regulator of CRF secretion, it is reasonable to ask whether NE is involved in the response to IL-1. Studies in rats lesioned with 6-

hydroxydopamine suggest that the ventral noradrenergic bundle input to the hypothalamus is indeed necessary for HPA responses to IL-1,[74] but studies with adrenergic blockers have been less successful in demonstrating adrenergic control of the HPA response to IL-1,[75] although we observe a partial blockade of the HPA response to IL-1 by the α_1-adrenergic antagonist, prazosin, but not by α_2- or β-adrenergic receptor antagonists.

THE SIGNIFICANCE OF HPA EFFECTS ON THE IMMUNE SYSTEM

The activation of the HPA axis associated with immune responses has been interpreted to indicate that the immune system can act as a sensory system, signalling the brain to indicate the presence of a threat from the external environment and triggering a classical stress response.[59] The HPA activation has been suggested to provide a negative feedback mechanism via the immunosuppressive activity of the glucocorticoids. The inhibitory activity of the glucocorticoids has been proposed to limit the inflammatory responses and prevents the immune system over-reacting resulting in autoimmune problems.[69,76]

Consistent with this, adrenal corticosteroids appear to play a critical role in recovery from experimental allergic encephalomyelitis (EAE). Rats treated with myelin basic protein produce antibodies to this protein 11-14 days after immunization, concurrently with paralysis of the tail and hind limbs and elevation of plasma corticosterone.[55] Intact rats normally recover within a few days; the spontaneous recovery does not occur in adrenalectomized rats, unless glucocorticoid replacement therapy is instituted. Also, in Lewis rats, a decreased responsivity of the HPA axis is associated with the susceptibility to arthritis . Lewis rats show an arthritic response which mimics human rheumatoid arthritis in response to administration of a streptococcal cell wall peptidoglycan polysaccharide, whereas the histocompatible Fischer rats do not. The arthritic response in Lewis rats can be prevented by dexamethasone, and can be induced in Fischer rats by the glucocorticoid-receptor antagonist RU 486.[77] The deficit in Lewis rats appears to be associated with a deficient activation of the HPA streptococcal cell wall peptidoglycan polysaccharide, IL-1, and CRF (compared to Fischer rats), and may be associated with an inappropriate regulation of the CRF gene. These two examples indicate a potential physiological significance for the glucocorticoid feedback on the immune response.

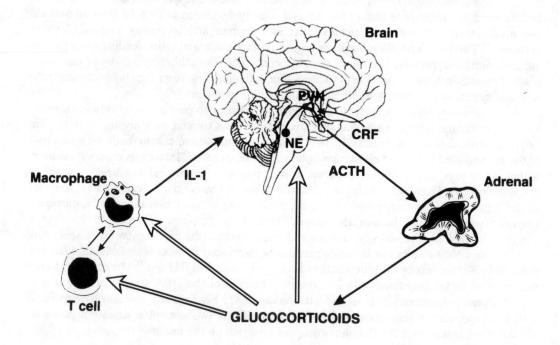

Fig 4.   A schematic diagram of the relationship between the brain, the axis, and immune cells. Interleukin-1 (IL-1) produced by lymphocytes during the immune response activates noradrenergic (NE) projections for the brain to the hypothalamic paraventricular nucleus (PVN). This input activates the hypothalamic-pituitary-adrenocortical (HPA) axis, stimulating the release of corticotropin-releasing factor (CRF) in the median eminence region of the hypothalamus, which in turn stimulates the secretion of ACTH from the anterior lobe of the pituitary, which then activates the adrenal cortex to synthesize and secrete glucocorticoid hormones. The glucocorticoids in turn provide a negative feedback on cytokine production by lymphocytes. (From   **Dunn, A.J.,**[99] reprinted with the permission of the publisher)

CYTOKINES AS IMMUNOTRANSMITTERS

The best candidates for messengers from the immune system to the nervous system are cytokines. Although cytokines play a major role in coordinating the immune response, they also have substantial effects on other tissues including the nervous system. The cytokines currently thought to have most relevance for the nervous system are IL-1, IL-6, tumor necrosis factor α (TNFα) and the interferons, but many others may soon be recognized. Shortly following most immunologic challenges, macrophages produce TNFα, followed closely by IL-1, and then IL-6.[78] Surprisingly, each of these three cytokines is capable of inducing the others, so that IL-1 administration induces TNFα synthesis and *vice versa;* IL-1 and TNFα induce IL-6, and so on.[79]

These three cytokines have a plethora of effects, many of which are related to the physiology of infections. IL-1, IL-6 and TNFα are each pyrogenic.[79] IL-1 is not the only cytokine that can affect the HPA axis; other cytokines, such as IL-6 and TNFα, can have similar effects,[80] although they are significantly less potent.[81,82] Interestingly, the production of TNFα and IL-1 is inhibited by glucocorticoids, so that the HPA activation elicited by the cytokines provides feedback regulation of cytokine synthesis.[78] IFN-α also affects the HPA axis, its effects being largely inhibitory, and apparently mediated through opiate receptors. IL-1 and IL-6 have a variety of other endocrine effects.[72]

These cytokines are also behaviorally active. IL-1 and TNFα are somnogenic.[83] IL-1, TNFα and IFNα decrease feeding.[84,85] IL-1[86] and IFNα[85] decrease locomotor activity, and IL-1 decreases exploratory activity.[87] Sexual activity is also diminished. These responses can in most cases be elicited by both central and peripheral administration, and some may be mediated by CRF.[87] It can be recognized that each of these responses is characteristic of sickness, and it has been argued that IL-1, perhaps in combination with other cytokines, accounts for many if not most of the physiological and behavioral responses to infections. These behavioral responses have been rationalized in terms of what has been termed "sickness behavior."[88,89] The fever may be instrumental in fighting invading organisms and the sleep, hypomotility, anorexia and reduced libido may cause the animal to hide and thus escape predators.[88]

The potential "immunotransmitter activity" of cytokines has been tested in some experiments with antagonists. The IL-1-receptor antagonist (IRAP or IL-1ra, a naturally occurring polypeptide synthesized and secreted along with IL-1) can prevent the HPA-activating and neurochemical effects of some immune challenges, such as NDV.[58] IL-1ra also blocked the effects of LPS on social behavior in rats, but not the anorexia,[89] and variable results have been obtained with HPA activation.[81,90,91] Thus IL-1 may not be the only factor mediating responses to LPS. A TNFα antibody was also ineffective against LPS, even in combination with IL-1ra.[81] Because immune challenges produce a "cocktail" of cytokines, it seems likely that differences in these cocktails account for the different patterns of responses to different immune stimuli.

It is not yet clear how IL-1 or other cytokines affect the brain. The molecule (M.Wt. 17,500) appears too large to readily cross the blood-brain barrier,[92] but it may act directly on the organum vasculosum laminae terminalis (OVLT)[93] or the median eminence.[80] It has been suggested that there is a specific brain uptake system for IL-1,[94] but the function of such a system has not been clearly established. Curiously, binding sites for IL-1 have been demonstrated in the brains of mice, but not rats.[95,96] IL-1 does not appear to be present in the normal healthy brain.[72,97] It does appear, however, when the brain is infected or lesioned. The blood-brain barrier is then breached allowing invasion of macrophages from the periphery, which then proliferate in the CNS as microglia which can synthesize IL-1. The IL-1 acts as potent growth factor for astroglia, causing them to proliferate, sealing off the lesion and restoring the damaged blood-brain barrier.[98] Thus IL-1 plays an important role in a classical pathological mechanism for protection of the brain.

CONCLUSIONS

There are a variety of mechanisms by which the nervous system can affect the immune system and *vice versa*. Many of the factors released by the nervous system during stress have profound effects on immune system function.

However, infections which activate the immune system appear to act as stressors, probably via the production of cytokines. Thus stress, infection and immune system function are intimately involved in complex relationships critical for the survival of the organism.

REFERENCES

1. **Selye, H.,** The Physiology and Pathology of Exposure to Stress, *Acta Inc.*, Montreal, 1950.
2. **Cannon, W. B.,** The emergency function of the adrenal medulla in pain and the major emotions, *Amer. J. Physiol.*, 33,356, 1914.
3. **Selye, H.,** Thymus and adrenals in the response of the organism to injuries and intoxications, *Brit. J. Exptl. Pathol.*, 17,234, 1936.
4. **Dunn, A. J. and Kramarcy, N. R.,** Neurochemical responses in stress: relationships between the hypothalamic-pituitary-adrenal and catecholamine systems, in *Handbook of Psychopharmacology*, L. L. Iversen, S. D. Iversen and S. H. Snyder, Eds., Vol. 18, Plenum Press, New York, 1984, p.455.
5. **Stone, E. A.,** Stress and catecholamines, in *Catecholamines and Behavior. Vol.* 2 *Neuropsychopharmacology*, A. J. Friedhoff, Ed., Vol. 2, Plenum Press, New York, 1975, p.31.
6. **Moore, R. Y. and Bloom, F. E.,** Cerebral catecholamine neuron systems: anatomy and physiology of the norepinephrine and epinephrine systems, *Ann. Rev. Neurosci.*, 2,113, 1979.
7. **Dunn, A. J. and Welch, J.,** Stress- and endotoxin-induced increases in brain tryptophan and serotonin metabolism depend on sympathetic nervous system activation, *J. Neurochem.*, 57,1615, 1991.
8. **Swanson, L. W., Sawchenko, P. E., Rivier, J. and Vale, W. W.,** Organization of ovine corticotropin-releasing factor immunoreactive cells and fibers in the rat brain: an immunohistochemical study, *Neuroendocrinol.*, 36,165, 1983.
9. **Dunn, A. J. and Berridge, C. W.,** Physiological and behavioral responses to corticotropin-releasing factor administration: is CRF a mediator of anxiety or stress response?, *Brain Res. Rev.*, 15,71, 1990.
10. **Owens, M. J. and Nemeroff, C. B.,** Physiology and pharmacology of corticotropin-releasing factor, *Pharmacol. Rev.*, 43,425, 1991.

11. **Koob, G. F. and Bloom, F. E.**, Corticotropin-releasing factor and behavior, *Fed. Proc.*, 44,259, 1985.
12. **Felten, S. Y. and Felten, D. L.**, Innervation of lymphoid tissue, in *Psychoneuroimmunology*, R. Ader, D. L. Felten and N. Cohen, Eds., Academic Press, San Diego, 1991, p.27.
13. **Ader, R. and Cohen, N.**, Psychoneuroimmunology: conditioning and stress, *Ann. Rev. Psychol.*, 44,53, 1993.
14. **Irwin, J. and Livnat, S.**, Behavioral influences on the immune system: stress and conditioning, *Progr. Neuro-Psychopharmacol. Biol. Psychiat.*, 11,137, 1987.
15. **Kiecolt-Glaser, J. K. and Glaser, R.**, Stress and immune function in humans, in *Psychoneuroimmunology*, R. Ader, D. L. Felten and N. Cohen, Eds., Academic Press, San Diego, 1991, p.849.
16. **Wood, P. G., Karol, M. H., Kusnecov, A. W. and Rabin, B. S.**, Enhancement of antigen-specific humoral and cell-mediated immunity by electric footshock stress in rats, *Brain Behav. Immun.*, 7,121, 1993.
17. **Dunn, A. J.**, Psychoneuroimmunology for the psychoneuroendocrinologist: a review of animal studies of nervous system-immune system interactions, *Psychoneuroendocrinol.*, 14,251, 1989.
18. **Keller, S. E., Weiss, J. M., Schleifer, S. J., Miller, N. E. and Stein, M.**, Stress-induced suppression of immunity in adrenalectomized rats, *Science*, 221,1301, 1983.
19. **Bonneau, R. H., Sheridan, J. F., Feng, N. G. and Glaser, R.**, Stress-induced modulation of the primary cellular immune response to herpes simplex virus infection is mediated by both adrenal-dependent and independent mechanisms, *J. Neuroimmunol.*, 42,167, 1993.
20. **Maier, S. F. and Laudenslager, M. L.**, Inescapable shock, shock controllability and mitogen stimulated lymphocyte proliferation, *Brain Behav. Immun.*, 2,87, 1988.
21. **Ben-Eliyahu, S. and Page, G. G.**, *In vivo* assessment of natural killer cell activity in rats, *Prog. NeuroEndocrinImmunol.*, 5,199, 1992.
22. **Irwin, M., Patterson, T., Smith, T. L., Caldwell, C., Brown, S. A., Gillin, C. J. and Grant, I.**, Reduction of immune function in life stress and depression, *Biol. Psychiat.*, 27,22, 1990.
23. **Shavit, Y., Lewis, J. W., Terman, G. W., Gale, R. P. and Liebeskind, J. C.**, Opioid peptides mediate the suppressive effect of stress on natural killer cell cytotoxicity, *Science*, 223,188, 1984.
24. **Takamoto, T., Hori, Y., Koga, Y., Toshima, H., Hara, A. and Yokoyama, M. M.**, Norepinephrine inhibits human natural killer cell activity *in vitro*, *Intl. J. Neurosci.*, 58,127, 1991.
25. **Claman, H. N.**, How corticosteroids work, *J. Allergy Clin. Immunol.*, 55,145, 1975.
26. **Tsokos, G. C. and Balow, J. E.**, Regulation of human cellular immune responses by glucocorticosteroids, in *Enkephalins and Endorphins: Stress and the Immune system*, N. P. Plotnikoff, R. E. Faith, A. J. Murgo and R. A. Good, Eds., Plenum Press, New York, 1986, p.159.
27. **Jefferies, W. M.**, Cortisol and immunity, *Med. Hypoth.*, 34,198, 1991.
28. **Beisel, W. R. and Rapoport, M. I.**, Inter-relations between adrenocortical functions and infectious illness, *N. Eng. J. Med.*, 280,541, 1969.
29. **Kass, E. H. and Finland, M.**, Corticosteroids and infections, *Adv. Intl. Med.*, 9,45, 1958.

30. **Ritter, M. A.,** Embryonic mouse thymocyte development: enhancing effect of corticosterone at physiological levels, *Immunol.,* 33,241, 1977.
31. **Hadden, J. W., Hadden, E. M. and Middleton, E.,** Lymphocyte blast transformation - 1. Demonstration of adrenergic receptors in human peripheral lymphocytes, *Cell. Immunol.,* 1,583, 1970.
32. **Sanders, V. M. and Munson, A. E.,** Norepinephrine and the antibody response, *Pharmacol. Rev.,* 37,229, 1985.
33. **Madden, K. S. and Livnat, S.,** Catecholamine action and immunologic reactivity, in *Psychoneuroimmunology,* R. Ader, D. L. Felten and N. Cohen, Eds., Academic Press, San Diego, 1991 p.283.
34. **Cunnick, J. E., Lysle, D. T., Kucinski, B. J. and Rabin, B. S.,** Evidence that shock-induced immune suppression is mediated by adrenal hormones and peripheral β-adrenergic receptors, *Pharmacol. Biochem. Behav.,* 36,645, 1990.
35. **McGillis, J. P., Park, A., Rubin-Fletter, P., Turok, C., Dallman, M. F. and Payan, D. G.,** Stimulation of rat B-lymphocyte proliferation by corticotropin-releasing factor, *J. Neurosci. Res.,* 23,346, 1989.
36. **Leu, S.-J. C. and Singh, V. K.,** Modulation of natural killer cell-mediated lysis by corticotropin-releasing neurohormone, *J. Neuroimmunol.,* 33,253, 1991.
37. **Leu, S.-J. C. and Singh, V. K.,** Stimulation of interleukin-6 production by corticotropin-releasing factor, *Cell. Immunol.,* 143,220, 1992.
38. **Webster, E. L., Tracey, D. E., Jutila, M. A., Wolfe, S. A. and De Souza, E. B.,** Corticotropin-releasing factor receptors in mouse spleen: identification of receptor-bearing cells as resident macrophages, *Endocrinol.,* 127,440, 1990.
39. **Heijnen, C. J., Kavelaars, A. and Ballieux, R. E.,** Corticotropin-releasing hormone and pro-opiomelanocortin-derived peptides in the modulation of immune function, in *Psychoneuroimmunology,* R. Ader, D. L. Felten and N. Cohen, Eds., Academic Press, San Diego, 1991, p.429.
40. **Carr, D. J. J.,** The role of endogenous opioids and their receptors in the immune system, *Proc. Soc. Exptl. Biol. Med.,* 198,710, 1991.
41. **Shahabi, N. A., Peterson, P. K. and Sharp, B. M.,** β-Endorphin binding to naloxone-insensitive sites on a human mononuclear cell line (U937): effects of cations and guanosine triphosphate, *Endocrinol.,* 126,3006, 1990.
42. **Weber, R. J. and Pert, A.,** The periaqueductal gray matter mediates opiate induced immunosuppression, *Science,* 245,188, 1989.
43. **Bernton, E. W., Bryant, H. U. and Holaday, J. W.,** Prolactin and immune function, in *Psychoneuroimmunology,* R. Ader, D. L. Felten and N. Cohen, Eds., Academic Press, San Diego, 1991 p.403.
44. **Selye, H.,** *The Stress of My Life,* 2nd edition ed., Van Nostrand Reinold, New York, 1979, p.1
45. **Bliss, E. L., Migeon, C. J., Eik-Nes, K., Sandberg, A. A. and Samuels, L. T.,** The effects of insulin, histamine, bacterial pyrogen, and the antabuse-alcohol reaction upon the levels of 17-hydroxycorticosteroids in the peripheral blood of man, *Metabolism,* 3,493, 1954.

46. **Conn, J. W., Fajans, S. S., Louis, L. H., Seltzer, H. S. and Kaine, H. D.,** A comparison of steroidal excretion and metabolic effects induced in man by stress and by ACTH, *Rec. Progr. Horm. Res.,* 10,471, 1954.

47. **Ando, S., Guze, L. B. and Gold, E. M.,** ACTH release *in vivo* and *in vitro:* extrapituitary mediation during *Esch. coli* bacteremia, *Endocrinol.,* 74,894, 1964.

48. **Dunn, A. J., Powell, M. L., Meitin, C. and Small, P. A.,** Virus infection as a stressor: influenza virus elevates plasma concentrations of corticosterone, and brain concentrations of MHPG and tryptophan, *Physiol. Behav.,* 45,591, 1989.

49. **Dunn, A. J.,** Role of cytokines in infection-induced stress, *Ann. N. Y. Sci.,* 697,189, 1993.

50. **Dunn, A. J.,** Stress-related changes in cerebral catecholamine and indoleamine metabolism: lack of effect of adrenalectomy and corticosterone, *J. Neurochem.,* 51,406, 1988.

51. **Besedovsky, H. O., Sorkin, E., Keller, M. and Muller, J.,** Changes in blood hormone levels during the immune response, *Proc. Soc. Exptl. Biol. Med.,* 150,466, 1975.

52. **Besedovsky, H., Sorkin, E., Felix, D. and Haas, H.,** Hypothalamic changes during the immune response, *Europ. J. Immunol.,* 7,323, 1977.

53. **Saphier, D.,** Neurophysiological and endocrine consequences of immune activity, *Psychoneuroendocrinol.,* 14,63, 1989.

54. **Besedovsky, H. O., del Rey, A. E., Sorkin, E., Da Prada, M., Burri, R. and Honegger, C.,** The immune response evokes changes in brain noradrenergic neurons, *Science,* 221,564, 1983.

55. **MacPhee, I. A. M., Antoni, F. A. and Mason, D. W.,** Spontaneous recovery of rats from experimental allergic encephalomyelitis is dependent on regulation of the immune system by endogenous adrenal corticosteroids, *J. Exptl. Med.,* 169,431, 1989.

56. **Stenzel-Poore, M., Vale, W. W. and Rivier, C.,** Relationship between antigen-induced immune stimulation and activation of the hypothalamic pituitary-adrenal axis in the rat, *Endocrinol.,* 132,1313, 1993.

57. **Dunn, A. J.,** Endotoxin-induced activation of cerebral catecholamine and serotonin metabolism: comparison with interleukin-1, *J. Pharmacol. Exptl. Therapeut.,* 261,964, 1992.

58. **Dunn, A. J. and Vickers, S. L.,** Neurochemical and neuroendocrine responses to Newcastle disease virus administration in mice, *Brain Res.,* 645,103, 1994.

59. **Blalock, J. E. and Smith, E. M.,** A complete regulatory loop between the immune and neuroendocrine systems, *Fed. Proc.,* 44,108, 1985.

60. **Carr, D. J. J. and Blalock, J. E.,** Neuropeptide hormones and receptors common to the immune and neuroendocrine systems: bidirectional pathway of intersystem communication, in *Psychoneuroimmunology,* R. Ader, D. L. Felten and N. Cohen, Eds., Academic Press, San Diego, 1991, p.573.

61. **Goetzl, E. J., Turck, C. W. and Speedharan, S. P.,** Production and recognition of neuropeptides by cells of the immune system, in *Psychoneuroimmunology,* R. Ader, D. L. Felten and N. Cohen, Eds., Academic Press, San Diego, 1991, p.263.

62. **Lolait, S. J., Clements, J. A., Markwick, A. J., Cheng, C., McNally, M., Smith, A. I. and Funder, J. W.,** Pro-opiocortin messenger ribonucleic acid and post-translational processing of beta endorphin in spleen macrophages, *J. Clin. Invest.,* 77,1776, 1986.

63. Stein, C., Hassan, A. H. S., Przewlocki, R., Gramsch, C., Peter, K. and Herz, A., Opioids from immunocytes interact with receptors on sensory nerves to inhibit nociception in inflammation, *Proc. Natl. Acad. Sci. USA*, 87,5935, 1990.

64. Smith, E. M., Meyer, W. J. and Blalock, J. E., Virus-induced corticosterone in hypophysectomized mice: a possible lymphoid adrenal axis, *Science*, 218,1311, 1982.

65. Olsen, N. J., Nicholson, W. E., DeBold, C. R. and Orth, D. N., Lymphocyte-derived adrenocorticotropin is insufficient to stimulate adrenal steroidogenesis in hypophysectomized rats, *Endocrinol.*, 130,2113, 1992.

66. Dunn, A. J., Powell, M. L., Moreshead, W. V., Gaskin, J. M. and Hall, N. R., Effects of Newcastle disease virus administration to mice on the metabolism of cerebral biogenic amines, plasma corticosterone, and lymphocyte proliferation, *Brain Behav. Immun.*, 1, 216, 1987.

67. Wexler, B. C., Dolgin, A. E. and Tryczynski, E. W., Effects of a bacterial polysaccharide (piromen) on the pituitary-adrenal axis: further aspects of hypophyseal-mediated control of response, *Endocrinol.*, 61,488, 1951.

68. Besedovsky, H. O., del Rey, A. and Sorkin, E., Lymphokine-containing supernatants from Con A-stimulated cells increase corticosterone blood levels, *J. Immunol.*, 126,385, 1981.

69. Besedovsky, H. O., del Rey, A., Sorkin, E. and Dinarello, C. A., Immunoregulatory feedback between interleukin-l and glucocorticoid hormones, *Science*, 233,652, 1986.

70. Rivest, S. and Rivier, C., Influence of the paraventricular nucleus of the hypothalamus in the alteration of neuroendocrine functions induced by intermittent footshock or interleukin, *Endocrinol.*, 129,2049, 1991.

71. Uehara, A., Gottschall, P. E., Dahl, R. R. and Arimura, A., Interleukin-1 stimulates ACTH release by an indirect action which requires endogenous corticotropin releasing factor, *Endocrinol.*, 121,1580, 1987.

72. Dunn, A. J., Interleukin-1 as a stimulator of hormone secretion, *Prog. NeuroEndocrinImmunol.*, 3,26, 1990.

73. Dunn, A. J., Systemic interleukin-1 administration stimulates hypothalamic norepinephrine metabolism paralleling the increased plasma corticosterone, *Life Sci.*, 43,429, 1988.

74. Chuluyan, H., Saphier, D., Rohn, W. and Dunn, A. J., Noradrenergic innervation of the hypothalamus participates in the adrenocortical responses to interleukin-1, *Neuroendocrinol.*, 56,106, 1992.

75. Rivier, C., Vale, W. and Brown, M., In the rat, interleukin-1α and -1β stimulate adrenocorticotropin and catecholamine release, *Endocrinol.*, 125,3096, 1989.

76. Munck, A. and Guyre, P., Glucocorticoid physiology, pharmacology and stress, *Adv. Exptl. Med. Biol.*, 196,81, 1986.

77. Sternberg, E. M., Young, W. S., Bernardini, R., Calogero, A. E., Chrousos, G. P., Gold, P. W. and Wilder, R. L., A central nervous system defect in biosynthesis of corticotropin-releasing hormone is associated with susceptibility to streptococcal cell wall-induced arthritis in Lewis rats, *Proc. Natl. Acad. Sci. USA*, 86,4771, 1989.

78. Zuckerman, S. H., Shellhaas, J. and Butler, L. D., Differential regulation of lipopolysaccharide-induced interleukin-l and tumor necrosis factor synthesis: effects of endogenous and exogenous glucocorticoids and the role of the pituitary-adrenal axis, *Europ. J. Immunol.*, 19,301, 1989.

79. **Dinarello, C. A.,** Interleukin-1 and its related cytokines, in *Macrophage Derived Cell Regulatory Factors: Cytokines,* C. Sorg, Ed., Vol. 1, Karger, Basel, 1989, p.105.

80. **Matta, S. G., Singh, J., Newton, R. and Sharp, B. M.,** The adrenocorticotropin response to interleukin-1,β instilled into the rat median eminence depends on the local release of catecholamines, *Endocrinol.,* 127,2175, 1990.

81. **Dunn, A. J.,** The role of interleukin-1 and tumor necrosis factor α in the neurochemical and neuroendocrine responses to endotoxin, *Brain Res; Bull.,* 29,807, 1992.

82. **Besedovsky, H. O., del Rey, A., Klusman, I., Furukawa, H., Monge Arditi, G. and Kabiersch, A.,** Cytokines as modulators of the hypothalamus pituitary-adrenal axis, *J. Steroid Biochem. Molec. Biol.,* 40,613, 1991.

83. **Opp, M. R., Kapas, L. and Toth, L. A.,** Cytokine involvement in the regulation of sleep, *Proc. Soc. Exptl Biol. Med.,* 201,16, 1992.

84. **McCarthy, D. O., Kluger, M. J. and Vander, A. J.,** Effect of centrally administered interleukin-1 and endotoxin on food intake of fasted rats, *Physiol. Behav.,* 36,745, 1986.

85. **Crnic, L. S. and Segall, M. A.,** Behavioral effects of mouse interferon-α and interferon-γ and human interferon-α in mice, *Brain Res.,* 590,277, 1992.

86. **Otterness, I. G., Seymour, P. A., Golden, H. W., Reynolds, J. A. and Daumy, G. O.,** The effects of continuous administration of murine interleukin-1α in the rat, *Physiol. Behav.,* 43,797, 1988.

87. **Dunn, A. J., Antoon, M. and Chapman, Y.,** Reduction of exploratory behavior by intraperitoneal injection of interleukin-1 involves brain corticotropin-releasing factor, *Brain Res. Bull.,* 26,539, 1991.

88. **Hart, B. L.,** Biological basis of the behavior of sick animals, *Neurosci. Biobehav. Rev.,* 12, 123, 1988.

89. **Kent, S., Bluthe, R.-M., Kelley, K. W. and Dantzer, R.,** Sickness behavior as a new target for drug development, *TIPS,* 13,24, 1992.

90. **Ebisui, O., Fukata, J., Murakami, N., Kobayashi, H., Segawa, H., Muro, S., Hanaoka, I., Naito, Y., Masui, Y., Ohmoto, Y., Imura, H. and Nakao, K.,** Effect of IL-1 receptor antagonist and antiserum to TNF-α on LPS-induced plasma ACTH and corticosterone rise in rats., *Amer. J. Physiol.,* 266,E986, 1994.

91. **Schotanus, K., Tilders, F. J. H. and Berkenbosch, F.,** Human recombinant interleukin-1 receptor antagonist prevents adrenocorticotropin, but not interleukin-6 responses to bacterial endotoxin in rats, *Endocrinol.,* 133,2461, 1993.

92. **Coceani, F., Lees, J. and Dinarello, C. A.,** Occurrence of interleukin-1 in cerebrospinal fluid of the conscious cat, *Brain Res.,* 446,245, 1988.

93. **Katsuura, G., Arimura, A., Koves, K. and Gottschall, P. E.,** Involvement of organum vasculosum of lamina terminalis and preoptic area in interleukin 1β-induced ACTH release, *Amer. J. Physiol.,* 258,E163, 1990.

94. **Banks, W. A., Ortiz, L., Plotkin, S. R. and Kastin, A. J.,** Human interleuken (IL)1α, murine IL-1α and murine IL-1β are transported from blood to brain in the mouse by a shared saturable mechanism, *J. Pharmacol. Exptl. Therapeut.,* 259,988, 1991.

95. **Haour, F., Ban, E., Marquette, C., Milon, G. and Fillion, G.,** Brain Interleukin-1 receptors: mapping, characterization and modulation, in *Interleukin-1 in the Brain,* N. J. Rothwell and R. D. Dantzer, Eds., Pergamon Press, Oxford, 1992, p.13.

96.   **Takao, T., Newton, R. C. and De Souza, E. B.**, Species differences in [^{125}I]interleukin-1 binding in brain, endocrine and immune tissues, *Brain Res.*, 623,172, 1993.

97.   **Tchelingerian, J. L., Quinonero, J., Booss, J. and Jacque, C.**, Localization of TNFα and IL-1α immunoreactivities in striatal neurons after surgical injury to the hippocampus, *Neuron,* 10,213, 1993.

98.   **Giulian, D. and Lachman, L. B.**, Interleukin-1 stimulation of astroglial proliferation after brain injury, *Science,* 228,497, 1985.

99.   **Dunn, A.J.**, Interactions between the nervous system and the immune system: implications for psychopharmacology, in *Psychopharmacology: The Fourth Generation of Progress*, F. E. Bloom and D. J. Kupfer, Eds., Raven Press, New York, 1995, p. 719.

Chapter 3

NEUROENDOCRINE PEPTIDE HORMONES AND RECEPTORS IN THE IMMUNE RESPONSE AND INFECTIOUS DISEASES

Douglas A. Weigent
Department of Physiology and Biophysics
University of Alabama at Birmingham
Birmingham, AL 35294-0005

J. Edwin Blalock
Department of Physiology and Biophysics
University of Alabama at Birmingham
Birmingham, AL 35294-0005

INTRODUCTION

The analysis of the immune system over the past 14 years in terms of the production of neuropeptide hormones and their receptors is beginning to reveal an important dimension of immunity. The idea that cells of the immune system produce neuropeptides was established as early as 1980.[1] Since this time, the number of neuropeptides produced by cells of the immune system is over 20. In addition, neuropeptides have been shown to have a substantial influence on a variety of immune functions.[2,3] Evidence is emerging that the hormones synthesized by cells of the immune system act mostly as autocrine and paracrine factors controlling basic aspects of the local control of immune function. This review will describe the neuropeptides produced by immune cells, the presence of neuropeptide receptors, and the evidence these molecules are important in the immune system.

A.   NEUROPEPTIDE HORMONES IN THE IMMUNE SYSTEM
1.   ACTH

The generation of neuropeptides by cells of the immune system is well documented (Table 1). In the case of hormones originally identified in the pituitary, ACTH was the first described in the immune system.[4] Many laboratories have reported that human and mouse lymphocytes produce POMC messenger RNA and POMC-derived peptides.[4-15] The most compelling evidence that cells of the immune system can be a source of POMC products is the finding of a POMC RNA in lymphocytes, including the ACTH and the β-lipotropin coding regions.[16] In addition, the data show that lymphocytes basally transcribe at least two POMC transcripts that lack exons 1 and 2 but contain either part or all of exon 3 that can be upregulated by corticotrophin-releasing hormone (CRH).[17,18] The complete identity between pituitary and leukocyte ACTH has been established by showing that the amino acid and nucleotide sequences of mouse spleen and pituitary ACTH are identical.[18]

Table 1.
Stimuli for leukocyte-derived peptide hormones

Hormone	Stimuli	References
Corticotropin (ACTH) and endorphin	Newcastle disease virus Lymphotropic viruses Bacterial lipopolysaccharide Corticotropin releasing factor (CRF) Arginine vasopressin Interleukin-1	(1-22)
[Met]enkephalin	Concanavalin A	(23-25)
Thyrotropin (TSH)	Staphylococcal enterotoxin A Thyrotropin releasing hormone (TRH)	(26-29)
Luteinizing hormone (LH)	Luteinizing hormone releasing hormone	(30-32)
Follicle-stimulating hormone (FSH)	Concanavalin A	(33)
Prolactin	Concanavalin A Unknown	(35-45)
Growth hormone (GH)	Growth hormone releasing hormone (GHRH) Growth hormone	(46-57)
Arginine vasopressin	Unknown	(75)
Oxytocin	Unknown	(75)
Neuropeptide Y	Unknown	(72,73)
Vasoactive intestinal peptide	Histamine liberators Hypersensitivity Unknown	(69,70)
Somatostatin	Hypersensitivity Unknown	(71)
Substance P	Unknown	(74)
Parathyroid hormone-related protein	Human T cell lymphotropic virus I	(76)

Table 1 continued

Calcitonin gene-related peptide	Hypersensitivity	(70)
Corticotropin releasing factor	PHA TPA	(58-60)
Growth hormone releasing hormone	Unknown	(62,64)
Luteinizing hormone releasing hormone	Unknown	(66,67)

Another report has demonstrated by in situ hybridization and immunohistochemistry that full-length POMC transcripts and β-endorphin-immunoreactivity are detectable in a very small population of rat lung and spleen cells.[19] ACTH production by lymphocytes is affected by viruses and circulating hormones. ACTH can be detected in lymphocytes treated with CRF and AVP and production blocked by glucocorticoids.[17] Upon closer examination, it was found that CRH causes IL-1 production by macrophages which then elicits POMC production by B cells.[5] The discovery of POMC production by lymphocytes has also led to the identification of unique processing pathways. Our studies have found that CRF or virus elicits the production of ACTH[1-39] and β-endorphin whereas LPS induces the production of ACTH (1-25) and α- or γ-endorphin from B lymphocytes.[20] An enzyme has been identified in B cells induced or activated by LPS and at pH 5 cleaves ACTH (1-39) to ACTH (1-24).[21] The origin of ACTH depends on the stimulus for production. Thus, peripheral blood cells harboring NDV all produce POMC peptides whereas LPS induces ACTH peptides in B cells. In other work with PBL, some investigators were unable to detect POMC while another report suggests that POMC transcripts were expressed by a small subset of macrophages.[19] It is unfortunate that in this latter work only spleens from control nonstimulated animals were evaluated by *in situ* hybridization. At the present time, there appears to be little doubt that cells of the immune system may produce POMC peptides. However, the selective expression of low levels of POMC according to the state of activation, and/or location of the cells, reflects the complexity of the system as well as the variable findings. In addition, the details and significance of the coexpression of a ligand and its receptor in the same cell and specific sorting and secretion of gene products lacking signal sequences have yet to be worked out. Also, the structure of intracellular ACTH and potential additional cellular binding sites are yet to be determined. It very recently has been shown that although small amounts of ACTH are secreted by activated lymphocytes, it nevertheless is bioactive. This was shown in a coculture system of lymphocytes and adrenal cells where the lymphocyte release of ACTH stimulated corticosterone secretion by the adrenal cells and the effect could be blocked by antibodies to ACTH.[22] Overall, the data clearly support the notion that POMC is expressed in activated lymphoid cells. In addition, cells of the immune system produce preproenkephalin mRNA[23] that appears to be translated and the product detected in culture supernatant fluids.[24] Recently, by in situ hybridization and immunohistochemistry on lymph nodes an increase in the expression of proenkephalin was seen *in vivo* after treatment of rats with lipopolysaccharide or adrenaline.[25] These data indicate at least a facilitating role for mediators of the adrenergic system in the LPS-induced proenkephalin response.

2.   TSH

Shortly after the discovery that human lymphocytes produce ACTH, it was observed they also could generate TSH. Stimulation of human PBL with staphylococcal enterotoxin A (SEA), a T cell mitogen, increased TSH beta subunit-specific immunofluorescence.[26] In this study, it was also shown that lymphocyte TSH was de novo synthesized and had a similar subunit structure to pituitary TSH. The TSH-β-related RNA was detected in human and mouse lymphocytes.[27] The production of TSH has also been studied in cell lines where a transcript the same size as pituitary TSH was detected in the T cell lymphoma line Hut-78.[28] The production of TSH in this cell line was increased by thyrotropin-releasing hormone and reduced by triiodothyronine. In a more recent report, the TSH β-subunit has been amplified from a human lymphoma cell line (WIL2) and the nucleotide sequence analysis suggests it is not identical to pituitary TSH.[29] Interestingly, only a 3% difference was found which had no effect on the amino acid sequence. These mostly silent changes were not observed at the exon-intron boundaries. The results from this latter study confirmed the effect of thyromimetic compounds on the expression of TSH-β-RNA. Besides TSH, two other glycoprotein hormones once thought to be restricted to the pituitary have been identified in the immune system. In response to luteinizing hormone releasing hormone (LHRH), lymphocytes from humans and mice produced and secreted an immunoreactive LH.[30,31] The lymphocyte-derived LH was similar to the pituitary hormone in subunit structure and antigenicity.[32] Two biologically active immunoreactive follicle-stimulating hormones (FSH) have been detected with different molecular weights after treatment of rat lymphocytes with the T cell mitogen, concanavalin A.[33] One form was similar to pituitary FSH while the other was larger. Both bone marrow and the spleen produce inhibin and activin and it remains to be determined whether these hormones play a role in the lymphoid levels of FSH.[34]

3.   Prolactin

The possibility that lymphocytes could serve as a source of prolactin (PRL) was suggested when it was reported that activated lymphocytes contained a prolactin-related RNA species[35] and PRL-like molecules.[36] Immunocytochemical evidence that lymphocytes produce PRL was also provided; however, PRL-like molecules were identified with a larger molecular weight.[36] Additional studies in a B cell line revealed the PRL RNA transcript was about 150 nucleotides longer than pituitary PRL RNA while the size of the two proteins was identical.[37] Additional controversy occurred when it was indicated that the PRL found in IL-2-stimulated T cells resulted from internalization of the hormone from the medium.[38] Several groups since these earlier observations have published their findings from human and rat cells showing that cells of the immune system produce PRL RNA similar to pituitary PRL RNA by restriction[39] and sequence analysis.[40,41] There appears, however, to be considerable size heterogeneity with protein products ranging from 11 kDa to 36 kDa and 46 kDa and 60 kDa.[38,41-44] The size variations observed in these proteins may be due to posttranslational modification including proteolysis, glycosylation, and aggregation. PRL gene regulation in lymphoid cells may involve glucocorticoids as well as molecules involved in activation of lymphocytes and as yet undefined factors.[45]

4. Growth Hormone

The production of GH by rodent and human lymphocytes as well as in cell lines has been reported in a number of studies.[35,46-50] Our own studies mostly in rats have shown that lymphocytes produce and secrete an immunoreactive GH that is similar to pituitary GH by a variety of immunologic and biologic techniques.[46] In a more recent study, we have cloned and sequenced a cDNA from rat lymphocytes corresponding in sequence to that obtained from the pituitary (Rohn and Weigent, in preparation). Similar findings have been obtained in the human Burkitt lymphoma Ramos cell line.[50] Although the studies are just beginning, it appears that the synthesis and secretion of GH by immune cells, for the most part, is similar to that described for pituitary GH.[46-50] Thus, GHRH appears to stimulate GH synthesis whereas IGF-I and SOM seem to inhibit lymphocyte GH synthesis. However, one important difference with the pituitary appears to be that the gene in lymphocytes is under tonic suppression and the removal of spleen or thymic tissues from animals results in rapid, spontaneous gene transcription. In another report, exogenous GH has been suggested to augment endogenous GH secretion from nonstimulated and phytohemagglutinin (PHA) stimulated peripheral blood mononuclear cells.[48] Dwarf mice which lack an important transcription factor for GH expression in the pituitary called Pit-1 have near normal expression of GH in lymphocytes. This finding raises the possibility that another protein may substitute for Pit-1 in cells of the immune system.[51] In addition to the hormones common to the neuroendocrine system influencing GH production by lymphocytes, it seems likely that cytokines and GH-axis hormones endogenous to the immune system will also play an important role. Several laboratories have reported that mitogens increase the levels of GH produced by lymphocytes.[46,48] In addition, we and others have identified leukocytes positive for IGF-I production by direct immunofluorescence. Our results using immunoaffinity purification, HPLC and a fibroblast proliferation bioassay suggests that the de novo synthesized leukocyte-derived irIGF-I' is similar in molecular weight, antigenicity, and bioactivity to serum IGF-I. The levels of leukocyte-derived IGF-I increase after treatment of lymphoid cells with GH.[52] In addition, the levels of leukocyte-derived IGF-I decreased after treatment of lymphoid cells with antibodies to GH.[52] In another study, we have blocked the production of lymphocyte-derived GHRH with antisense oligonucleotides and reduced the levels of GH produced.[53]

The idea that GH produced by lymphocytes may function in a mostly paracrine or autocrine rather than endocrine role is supported by the low levels of GH produced by immune cells. In the case of GH, it appears that approximately 5% of lymphoid cells are positive by immunofluorescence whereas only 0.1% of cells are secreting GH as determined by the reverse hemolytic plaque assay.[54] Thus much more of the lymphocyte hormone appears to remain intracellular and may function in an intracellular fashion like prolactin or in an autocrine manner.[52,55] In other studies, we have analyzed the production of GH mRNA and secretion of GH by purified subpopulations of rat lymphoid cells.[56] The results show that mononuclear leukocytes from various tissues, including spleen, thymus, bone marrow, Peyer's patches and peripheral blood cells, all have the ability to produce GH mRNA and secrete GH. Data obtained from cells separated by adherence, nylon wool columns, and positive and negative sorting with monoclonal antibodies that define B-, monocyte, T-helper and T-cytotoxic cells show that several different cell types have the ability to produce GH mRNA. These results suggest that B-cells and macrophages produce more GH mRNA than do T-cells. Natural killer cells also produce detectable levels of GH mRNA. Most interesting was the finding of a remarkably low level of GH mRNA expression in T-suppressor cells relative to T-helper cells. This supports the idea that GH generally promotes, or has a positive effect on, immune cell functions. In addition to these findings, we have also shown that the same cells that produce GH also produce IGF-I.[57] Taken together, the results suggest that an autocrine regulatory circuit may be important for the production of leukocyte-derived GH and IGF-I within the immune system.

5.   Hypothalamic Releasing Hormones

Since it has been shown that cells of the immune system manufacture neuroendocrine hormones and that hypothalamic releasing hormones have direct effects upon the immune system, we and others hypothesized that immune cells might produce releasing factors. The first report in this area described the presence of corticotropin releasing factor (CRF) mRNA and irCRF in nonstimulated human peripheral blood lymphocytes (PBL) and neutrophils.[58] In this report, no biological function was studied and the immunoreactive CRF was not parallel to bona fide hypothalamic CRF in an RIA dilution curve. More recent work suggests the molecule produced in the rat thymus and spleen and human T lymphocytes is structurally similar to hypothalamic CRF but that the regulation of lymphocyte CRF is different.[59,60] Synthesis and secretion of lymphocyte CRF increased after phytohemagglutinin and TPA treatment[59] and nordihydroguaiaretic acid treatment, respectively, while IL1 had no effect.[60] The lymphocyte CRF was shown to be bioactive on pituitary cells whereas its function in the immune system is yet to be determined. However, the demonstration that lymphocytes can produce POMC peptides after stimulation with CRF-41[17] and other studies showing potent direct immunosuppressive effects of CRF-41 on lymphocytes[61] suggests that the lymphocyte product could exert direct immunomodulatory effects or perhaps indirectly influence immune function through the induction of POMC intermediates.

The second hypothalamic releasing factor identified as being produced by cells of the immune system was GHRH.[62] Our studies in the rat showed that lymphocyte GHRH-related RNA had the same molecular mass as hypothalamic GHRH RNA. Antibody affinity chromatography followed by size separation columns demonstrates a peak (5 kda) that was de novo synthesized, could bind to the GHRH receptor, and increased GH mRNA synthesis in lymphocytes and pituitary cells. In another study, we have shown that hypothalamic GHRH is active on leukocytes and that cells of the immune system have receptors for GHRH similar to those described on pituitary cells.[63] These data strongly support a functional basis for bidirectional communication. Our work in the rat has been confirmed with human PBL with one important difference.[64] Our results suggest that irGHRH is similar in size to hypothalamic GHRH in the rat, whereas in the human the data suggest a molecular weight of approximately 50 kD. Although parallel displacement curves were obtained by RIA, the human lymphocyte molecule behaves differently on reverse-phase high performance liquid chromatography than hypothalamic GHRH.

In another series of studies, rat spleen lymphocytes have been shown to contain an immunoreactive and bioactive luteinizing hormone-releasing hormone.[65] Increasing amounts of lymphocyte irLHRH displaced ^{125}I-LHRH antibody in an RIA in a manner parallel to that produced by synthetic hypothalamic LHRH. The lymphocyte product stimulated the release of LH in a dose-dependent fashion and could be significantly inhibited by an LHRH antagonist. These same authors using reverse transcription and the polymerase chain reaction synthesized a 375 base pair cDNA from lymphocytes and the hypothalamus that hybridized to a specific LHRH cDNA probe.[66] A more recent report has confirmed this finding and in addition shown that the cDNA sequence of hypothalamic and lymphocyte LHRH are identical.[67]

6.   Neuropeptides

In addition to hormones commonly found in the neuroendocrine system, a number of other peptides have also been identified as being produced by cells of the immune system (Table 1). Some of these peptides were initially identified in lymphoid tissues innervated with noradrenergic fibers, but their origin was not known. These peptides include neuropeptide Y, somatostatin, vasoactive intestinal peptide, parathyroid hormone, and calcitonin gene-related peptides and substance P.[68] Although most of the studies done to date have examined the effects of these substances on selected immune functions, several reports have appeared documenting cells of the immune system as a source of these peptides. Thus, vasoactive intestinal peptide (VIP) and somatostatin (SOM) have been immunologically detected in platelets, mononuclear leukocytes, mast cells, and polymorphonuclear (PMN) leukocytes.[69] VIPs from rat basophilic leukemia cells consist of VIP_{10-28} free acid and a mixture of amino terminally extended "large" BIPs, which are apparently obtained from a novel prepro-VIP encoded by an alternatively spliced mRNA.[70] One species of SOM mRNA similar to what has been seen in the hypothalamus was identified in the rat spleen and thymus, and the bursa of Fabricius of the chicken by an SI nuclease protection assay using a SOM complementary RNA probe.[71] In the same study, the protein was identified in cells by immunocytochemistry and RIA. High levels of neuropeptide Y and its mRNA have been found in rat spleen, bone marrow, and peripheral blood lymphocytes.[72] Recently, neuropeptide Y has been shown by reverse transcription and PCR to be inducible in human lymphocytes and monocytes and expressed at sites where these cells are activated *in vivo*.[73] The results demonstrated that NPY mRNA was constitutively expressed at low levels and inducible by stimulation with mitogens and phorbol myristate acetate. A number of other neuropeptides, including substance P,[74] arginine vasopressin, oxytocin,[75] parathyroid hormone related protein,[76] and calcitonin gene-related peptide[70] have all also been shown to be synthesized by or associated with cells of the immune system. When one considers the increasing number of neuropeptides so far identified, it seems that virtually all known neuropeptides will be found in the immune system and many produced by cells of the immune system. It thus seems likely that cells of the immune system may contain neuropeptides by several mechanisms including uptake from extracellular fluids where hormones have been secreted from neighboring cells, including neurons or from distant cells in the neuroendocrine system. On the other hand, there seems to be a low constitutive level of synthesis of peptides by immune cells which can be elevated by a number of ways but usually by mitogen activation. These possibilities stress the importance of combining immunohistochemistry along with cellular localization of the mRNA positive cells.

B.   NEUROPEPTIDE HORMONE RECEPTORS IN THE IMMUNE SYSTEM

An increasing body of evidence strongly supports the idea that neuropeptides modulate the immune response. While these numerous reports suggest the presence of specific peptide binding sites, the actual work characterizing the receptors on lymphocytes is still in the beginning stages (Table 2). Despite this, however, at least ten neuropeptide receptors have been studied and the data show these receptors on cells of the immune system have characteristics which are similar to those receptors found on neuropeptide tissues.[77]

Table 2.
Receptors for neuropeptide hormones on cells of the immune system

Receptor	Binding Cell Type	References
Adrenocorticotropin	Rat spleen T and B cells	(3,78)
β-endorphin	Spleen	(79)
Thyrotropin	Neutrophils, monocytes, B cells	(80-82)
Prolactin	Human and mouse T and B cells	(40,85-90)
Growth hormone	Rat, mouse, human PBL, spleen, thymus	(93-97)
Growth hormone releasing hormone	Rat spleen, human PBL	(63)
Corticotropin releasing factor	Human PBL	(100)
Thyrotropin releasing hormone	T cell line	(82,101)
Luteinizing hormone releasing hormone	Rat thymocytes	(30)
Arginine vasopressin		(3,99)
Somatostatin	Human PBL	(102,103)
Substance P	Mouse T and B cells	(105,106)
Calcitonin gene-related peptide	Rat T and B cells	(110,111)

Table 2 continued

Vasoactive intestinal peptide	T cells	(77,- 107-109)

1. ACTH

Previous work has described specific, high affinity receptors for ACTH on T and B lymphocytes from rats.[78] The cells possess both high and low affinity receptor binding sites while nonstimulated thymocytes display few receptors. B lymphocytes possess three times the number of high affinity binding sites compared to T lymphocytes while the treatment of lymphocytes with the mitogen ConA increases the number of high affinity ACTH receptors on T and B cells by 2- to 3-fold.[78] ConA stimulation of thymocytes results in a 100-fold increase in the high affinity ACTH receptor binding site. The binding of ACTH to its receptor initiates a signal transduction pathway that involves both cyclic AMP and the mobilization of Ca^{++}.[3] Activation of the high affinity site increases calcium influx while ACTH binding to the low affinity receptor is thought to affect regulatory elements in the adenylate system. The mechanism of stimulation of cells of the immune system by ACTH is similar to the one advanced for the actions of ACTH in the stimulation of steroidogenesis in adrenal cells. The receptors for other POMC peptides, including the opioids, have also been studied. Opioid receptors were first identified in neuronal tissue and are designated into four distinct classes termed μ, δ, ε, and K based on ligand binding characteristics. The characterization of lymphocyte opioid receptors suggests that the same four classes defined by neuronal tissue binding have been identified on cells of the immune system.[79] The lymphocyte receptors for the opioid peptides appear to share many of the unique features, including molecular size, immunogenicity, and the use of specific intracellular signalling pathways with those described for neuronal tissue.[79]

2. Thyrotropin

Thyrotropin, a glycoprotein hormone, was shown to bind to neutrophils, and monocytes and more recently B cells.[80,81] The murine spleen TSH receptor is specific to B lymphocytes and B cell lines including a pre-B cell and IgM and IgG secreting cell lines.[81] Murine splenocytes enriched for T and B cells display no measurable TSH receptors; however, LPS stimulated B cells express high affinity receptors while staphylococcal enterotoxin A (SEA) stimulated T cells do not.[81] In one report, TSH stimulated Ig production and increased intracellular cAMP levels in mouse lymphocytes, while in another study done on human immune cells, no change was seen in adenylate cyclase activity after TSH treatment.[81,82]

3. Growth Hormone

It has now been clearly demonstrated that lymphoid cells also contain receptors for prolactin and GH and that both hormones are potent immunomodulators of the immune response.[83] The prolactin and GH receptors have been shown to be members of the superfamily of cytokine receptors involved in the growth and differentiation of lympho-hematopoietic cells.[84] In the case for PRL receptors, they have been found on normal human and mouse lymphocytes including all the major immune cell types.[85,86] Specific mRNA for the PRL receptor has been demonstrated in the NB2 cell line and ConA-stimulated mouse spleen and thymus cells.[40,87] cDNA cloning of the Nb2-PRL receptor permitted its characterization as an intermediate form lacking 198 amino acids in the cytoplasmic domain.[88] A systematic survey of PRL receptor expression on distinct lymphocyte subsets in primary and secondary organs of the immune system has recently been accomplished.[85,89,90] The findings revealed by flow cytometry show that PRL receptors are universally expressed in normal hematopoietic tissues with some differences in density which could be promoted by ConA treatment. The frequencies of PRL-R-bearing T cells increased with age in N2B mice suggesting that some imbalance might occur in autoimmune situations.[89] A recent interesting report in the GH3 tumor cell line showed that PRL could serve as an autocrine growth factor by binding to intracellular rather than cell surface receptors.[91] They also determined that treatment of the cells with IL1 caused the receptors to migrate to the cell surface. These findings support the idea that cytokines may play a role in the actions of neuropeptides and the expression of their receptors.

The past few years have seen a dramatic increase in knowledge in relation to the molecular structure of the GH receptor. GH receptors from a number of species have been cloned and sequenced and very recently the co-crystallization of hGH with the GH receptor has been achieved.[92] One molecule of GH binds in an asymmetric fashion to two molecules of GH binding protein (GHBP). There is some evidence for GH receptor heterogeneity and hGH can bind to both the GH and PRL receptors. GH binding and cellular processing of the GH receptor have been studied in the IM-9 cell line which has the characteristics of a B lymphocyte.[93] GH can stimulate the proliferation of the IM-9 cell line and the GH receptor can be down regulated by phorbol esters.[94] Several lines of evidence indicate that activation of the GH receptor increases tyrosine kinase activity and that the GH receptor is associated with a tyrosine kinase in several cell types, including the IM-9 cell line.[95] More recently, the GH receptor associated kinase has been identified as a member of the JAK family.[95] It appears that the binding of GH by GH receptor results in the formation of a ligand bound GH receptor dimer capable of binding a JAK kinase. This leads to the activation of JAK tyrosine kinase activity which presumably phosphorylates other proteins. In IM-9 cells, another member of the JAK kinase family termed TYK2 has been identified that binds to the GH receptor.[95] After the binding of GH to its receptor, the cell membrane becomes refractory and GH binding decreases markedly until a new supply of receptors arrives from de novo synthesis or from recycling of processed receptors. The data suggest exocytosis of intact hormone via recycling endosomes as well as degradation in the lysosomes. Overall, very little receptor recycling occurs and a rapid receptor synthetic process is required to maintain the cell surface complement of GH receptors.[96] Future studies on the intracellular fate of lymphocyte GH and its receptor will show whether these pathways play important roles in initiating, maintaining, or catalyzing biological responses to GH.

The recent discovery of circulating GHBP in human and rabbit plasma has offered new perspectives on GH action.[97] GH circulates in the blood bound to two different binding proteins (BP). One is a high affinity GHBP which complexes almost 90% of the 22 kd GH protein and a second low affinity GHBP which complexes the minor 20 kd GH variant. There is a single gene encoding the GH receptor and no gene for the GHBP has been identified. A recent report in the human spleen shows that both GHR mRNA transcripts containing and lacking exon 3 are present in equal proportions.[98] The significance of the lack of 22 amino acids near the amino terminus remains to be elucidated. In humans, the GHBP is thought to arise from receptor cleavage while in rodents two mRNA species have been identified consistent with alternative splicing. Very little is known about the mechanism and secretion of GHBP. The high affinity GHBP inhibits binding of GH to receptors *in vitro* while GHBP enhances the growth-promoting actions of GH *in vivo*. This paradox can be explained by the ability of the GHBP to prolong the half-life of GH. Although the physiologic roles of GHBP remain to be defined, it seems very likely that local effects of GHBP, including sites of inflammation, will be identified that influence GH binding and action.

4.   Other Receptors

The receptors for other neuroendocrine hormones have also been identified on cells of the immune system although in some instances they are not as well characterized. AVP receptors have been categorized into V1 receptors causing calcium mobilization or V2 receptors which stimulate cAMP production.[99] Cells of the immune system appear to have novel V1-like receptors as determined by the blocking ability of different V1 and V2 antagonists.[3] CRF mediates a variety of effects on cells of the immune system that take place through specific cell receptors. Radioligand binding studies in membrane homogenates and autoradiographic studies in tissue sections have identified, characterized, and localized CRF receptors in the brain, pituitary, and spleen.[100] The relative density of CRF binding was highest in the anterior lobe of the pituitary and low but detectable binding on spleen macrophages, and B cells. A number of similarities have been identified between the pituitary and spleen binding of CRF, including affinity, binding in the presence of cations and guanine nucleotides, stimulation of adenylate cyclase, and apparent subunit molecular weight.[100] The GHRH receptor has also been identified on cells of the immune system. The GHRH receptor binding sites are saturable and are found on both thymocytes and splenic lymphocytes. Following GHRH binding to its receptor, there is a rapid increase in intracellular Ca^{2+} which is associated with the stimulation of lymphocyte proliferation.[63] Additionally, luteinizing hormone-releasing hormone receptors have been functionally identified on thymocytes.[30] Finally, leukocytes have been shown to respond to TRH treatment as well by producing TSH mRNA and protein. More recent work has shown the presence of two receptor types for TRH present on T cells.[81] One of these sites satisfies the criteria for a classical TRH receptor and is involved in the release of IFN1-γ from T cells.[101]

Several groups have demonstrated that cells of the immune system also express receptors for neuropeptides.[102-110] Recently, one group has described the existence of distinct subsets of SOM receptors on the Jurkat line of human leukemic T cells and U266 IgG-producing human myeloma cells.[103] They showed that these cells have both high and low affinity receptors with kd values in the pM and nM range respectively. The authors speculate that two subsets of receptors may account for the biphasic concentration-dependent nature of the effects of SOM in some systems. Substance P is an 11 amino acid transmitter that acts through a receptor, denoted NK-1, coupled to phosphoinositide turnover in astrocytes.[104] Substance P receptors with high affinity have also been identified on human and mouse T and B cells with kd values in the nM range and 200 to 1,000 sites per cell.[105] Studies using crosslinking with ^{125}I-SP suggest the receptor is a single polypeptide chain with a molecular weight of 58 kDa.[105] Antipeptide receptor antibodies have been used to recognize a similar sized molecule in lymphocytes.[106] High affinity sites for VIP receptors functionally distinct from those on epithelial cells have been identified on T cells in the human.[107,108] The findings suggest a 200-500 pM value for the kd and 500-3,000 sites per cell.[77] Studies have also been done on cell lines where the receptor binding site has a molecular weight of 47 kDa. Receptor activation stimulates the guanine nucleotide binding protein complex and the catalytic subunit of adenylate cyclase, resulting in the generation of cAMP and the activation of a cAMP-dependent protein kinase.[109] Other receptors have been identified on immune cells, including α-MSH, nerve growth factor, and calcitonin-gene related peptide (CGRP).[77] In the case of CGRP, high affinity adenylyl cyclase linked receptors have been identified on a murine macrophage cell line[110] and T and B cells of the rat. CGRP binding proteins on rat lymphocytes had a molecular mass of 74 and 220 kDa.[111]

C.   IMMUNOREGULATION OF NEUROPEPTIDE HORMONES
1.   Neuroendocrine Hormones

A large amount of data has been published that when taken together demonstrate dramatic effects of neuropeptides on the immune system (Table 3). Both *in vitro* and *in vivo* studies suggest that exogenous neuropeptides influence numerous immune cell functions mediated by every major immune cell type. Although neuropeptides produced by cells of the immune system have been less well studied, it appears they also have dramatic effects on immune functions.

The evidence that ACTH and endorphin-like activities were made by lymphocytes prompted a study of the effect of these hormones on antibody production.[112] It was discovered that ACTH profoundly inhibited antibody production. ACTH's ability to suppress the antibody response to T dependent antigens (Ag) was more effective than T independent Ag suggesting that ACTH may interfere with the production or action of helper T cell signals. ACTH 1-24, like ACTH 1-39, has full steroidogenic activity, yet had no effect on antibody production suggesting that the immunoregulatory effects of ACTH may be associated with different structural parts of ACTH than those involved in steroidogenesis. Since this, ACTH has been shown to: suppress MHC class II expression by murine peritoneal macrophages;[113] stimulate natural killer cell activity;[114] modulate the rise in intracellular-free calcium concentration after T cell activation;[115] suppress the production of IFN-γ;[116] and function as a late-acting B cell growth factor that can synergize with IL5; and stimulate the growth and differentiation of human tonsillar B cells.[117] The endogenous opiates β-, γ-, and α-endorphin are also contained in the polyprotein proopiomelanocortin (POMC) and have been shown to modulate the activity of cells in the immune system.[77,112] α-endorphin is a potent inhibitor of the anti-SRBC plaque-forming cell (PFC) response while β- and γ-endorphin are mild inhibitors.[112] Many other aspects of immunity are modulated by the opiate peptides including: (a) enhancement of the natural cytotoxicity of lymphocytes and macrophages toward tumor cells; (b) enhancement or inhibition of T cell mitogenesis; (c) enhancement of T cell rosetting; (d) stimulation of human peripheral blood mononuclear cells; and (e) inhibition of major histocompatibility (MHC) class II antigen expression.[77] Several additional reports have appeared that strongly point to the presence of POMC peptides in lymphocytes and suggest a role for POMC peptides in immunoregulation.

Table 3.
Modulation of immune responses by neuropeptides

Hormone	Modulation	References
Corticotropin	Antibody synthesis IFNγ B-lymphocyte growth	(112) (116) (117)
Endorphins	Antibody synthesis Mitogenesis NK activity	(112) (77)
Thyrotropin	Antibody synthesis Comitogenic with ConA	(3,27) (127)
GH	Cytotoxic T cells Mitogenesis	(130)
LH and FSH	Proliferation Cytokine	(127) (129)
PRL	Comitogenic with ConA Induces IL2 receptors	(55,136-138,140-146,148)
CRF	IL1 production Enhance NK Immunosuppressive	(177-180)
TRH	Antibody synthesis	(27,81,82)
GHRH	Stimulate proliferation Inhibit NK activity Inhibit chemotactic response	(63,133)
Substance P	Stimulate chemotaxis Stimulate proliferation Modulate cytokine levels	(151-160)
VIP	Inhibit proliferation	(161-166)
AVP	T cell helper Functions for IFNγ production	(175)
SOM	Inhibit proliferation Reduces IFNγ production	(155,167-170)

Table 3 continued

CGRP	B cell differentiation T cell chemotaxis	(181,182)

Early on, it was shown that virus infection of hypophysectomized mice caused a time-dependent increase in corticosterone production in animals whose spleens were positive for ACTH by immunofluorescence.[118] In another report, hypophysectomized chickens were shown to produce ACTH and corticosterone response to *Brucella abortus*. This ACTH and corticosterone in response was ablated if B lymphocytes were deleted by bursectomy prior to hypophysectomy.[119] In certain instances, similar results have been observed in humans. For example, when children who were pituitary ACTH-deficient were pyrogen tested, they showed an increase in the percentage of ACTH-positive mononuclear leukocytes.[2] Two laboratories have been unable to reproduce a pituitary-independent ACTH response in mice and rats.

More recently, lymphocyte POMC expression *in vivo* has been confirmed and documented in an arthritis animal model,[120] a diabetes animal model,[112] in rats injected with CRF,[122] and in chickens following antigen challenge.[123] In the study on chickens, the authors concluded that the increase in corticosterone and redistribution of lymphocytes might be important in the initiation of antibody production in chickens. The proof that lymphocytes secrete biologically active corticotropin with steroidogenic activity has been shown *in vitro* when lymphocytes are cocultured with adrenal cells.[22]

Gram-negative bacterial infections and endotoxin shock may also represent a situation in which leukocyte hormones influence the immune system. For instance, endorphins have been implicated in the pathophysiology associated with these maladies since the opiate antagonist, naloxone, improved survival rates and blocked a number of cardiopulmonary changes associated with these conditions.[124] Considering the potent immunological effects of endotoxin, as well as its *in vitro* ability to induce leukocyte production of endorphins, cells of the immune system seem the most likely source of endogenous opiates that are observed during gram-negative sepsis and endotoxin shock. Consistent with this idea is the observation that lymphocyte depletion, like naloxone treatment, blocked a number of endotoxin-induced cardiopulmonary changes.[125] In a different approach, LPS-resistant inbred mice which have essentially no pathophysiologic response to LPS were shown to have a defect in leukocyte processing of POMC to endorphins. If leukocyte-derived endorphins were administered to such LPS-resistant mice, they then showed much of the pathophysiology associated with LPS administration to sensitive mice.[20]

A relationship between the thyroid gland and the immune system was suggested when hypothyroidism was observed in athymic mice.[126] Since this time, only a few reports have appeared documenting effects of TSH on immune cell functions. First it was shown that TSH could augment both T-dependent and T-independent antibody production.[3] TSH had to be present in the medium during the first 24-48 hr of culture for enhancement of the antibody response to occur. In additional studies, it was shown that thyrotropin releasing hormone (TRH) also enhanced the antibody plaque forming cell response and induced splenocyte production of TSH.[27] This enhancement by TRH was specifically blocked by antibodies to the TSH β-subunit, which demonstrated that the action of TRH was through its ability to induce

TSH production by lymphocytes. This was the first demonstration that a pituitary hormone could function as an autocrine or paracrine regulator of the immune system.[27] Subsequently, studies have suggested that TSH may elevate cAMP levels and influence differentiation in B cell lines.[81] In another report, it was shown that while TSH increased DNA synthesis and intracellular cAMP levels of FRTL-5 rat thyroid cells, it did not have much stimulatory effects on lymphocytes.[82] In this report, however, it was confirmed that TSH caused a moderate increase in immunoglobulin production by activated B cells. A more recent study reported that TSH at various concentrations significantly increased the proliferative response of mouse lymphocytes to ConA and PHA and stimulated IL2-induced NK cell activity without modifying the basal levels of cytotoxicity.[127] The results support the immunoregulatory role of TSH on both T and B cells.

Very little work has been done examining gonadotropic hormones (LH and FSH) and immunity. As discussed earlier, LH receptors have been documented on lymphoid T cells[128] and LH has been shown to modulate cytokine and gamma globulin secretion in mice.[129] In addition, LH at various concentrations increased the proliferative response to mitogens. Overall, it appears that LH modulates both humoral and cellular immunity. It has been shown that inhibin and activin, whose major function is the control of FSH release from the pituitary, can modulate lymphocyte function.[34] A significant dose-related increase in monocyte chemotaxis was induced by inhibin. While activin increased the migrational activity of monocytes, inhibin significantly decreased interferon-γ production, and its effect was reversed by activin. Inhibin and/or activin had no significant effect on either phytohemagglutinin-induced lymphocyte proliferation or lymphocyte cytotoxic capability. The present findings show that inhibin and activin may affect some immune parameters and suggest a possible involvement of these hormones along with LH and possibly FSH in regulating cell-mediated immune function.

Numerous studies over the past years have clearly demonstrated that GH exerts important effects on the immune system.[83,130] The potential role of GH in immunoregulation has been demonstrated *in vitro* for numerous immune functions, including stimulation of DNA and RNA synthesis in the spleen and thymus of normal and hypophysectomized rats. GH also affects hematopoiesis by stimulating neutrophil differentiation, augments erythropoiesis, increases proliferation of bone marrow cells, and influences thymic development. Proliferative responses of lymphoid cells are greater when treated with GH *in vitro*. GH affects the functional activity of cytolytic cells including T lymphocytes and NK cells. GH was necessary for T lymphocytes to develop cytolytic activity against an allogeneic stimulus in serum-free media. The cytolytic activity of NK cells is reduced by hypophysectomy, and this effect can be partially reversed by administration of GH *in vivo*. GH has been shown to stimulate the production of superoxide anion formation from macrophages.[130] It appears the binding of GH to neutrophils to stimulate superoxide anion occurs through the PRL receptor.[131] Also, it is not clear whether GH directly influences intra- or extrathymic development or acts indirectly by augmenting the synthesis of thymulin or IGF-I. Recent work has shown that stimulation of T cells with IL2 induced expression of IGF-I receptors consistent with the idea that IGF-I is important in the activation of T cells.[132] These observations suggest that GH stimulates local production of IGF-I which acts to promote tissue growth and action in a paracrine/autocrine fashion. Many of the leukocyte functions stimulated by GH can also be augmented by IGF-I treatment. Thus, it appears that the action of GH on cells of the immune system may in part be mediated by IGF-I. The hypothalamic releasing hormone for GH has also been shown to modulate the immune system. In the case of GHRH, it has been shown to stimulate an increase in GH mRNA in leukocytes and cause a two-fold increase in thymidine and uridine incorporation of nonstimulated lymphocytes.[63] By others, GHRH has been shown to stimulate lymphocyte proliferation, inhibit NK activity, and inhibit the chemotactic response.[133]

The idea that GH influences cells of the immune system *in vivo* was initially shown when mice injected with antibody to GH developed thymic atrophy. In addition, dwarf mice which lack GH showed a reduced ability to synthesize antibodies. Although a number of immune events are abnormal in GH-deficient humans, they are generally not considered to be immunodeficient. The changes in humans include thymic hypoplasia, reduced activity of NK cells, and reduced ability of lymphoid cells to respond or stimulate cells in an allogeneic mixed lymphocyte reaction.[130] In contrast to humans, many studies have been done in both normal and GH-deficient animals that demonstrate the ability of GH to modulate immune function.[98,130] Overall, injections of GH increase thymic size, stimulate thymocyte proliferation, and induce expression of the DNA-binding protein c-myc in hypophysectomized rats. GH augments antibody synthesis and skin-graft rejection when injected into hypopituitary animals. It can reverse the leukopenia caused by stress. GH injected *in vivo* increases both basal and lectin-induced proliferative responses from spleen cells of dwarf and aged rats.[134] In other *in vivo* studies, it has been shown that GH can stimulate the production of IL1, IL2, TNFα, and thymulin; induce the cytotoxic activity of NK cells and restore the normal architecture of the thymus in aged animals.[130] More recently, it has been shown that GH promotes lymphocyte engraftment in immunodeficient mice.[134] Overall, it appears that GH has a pleiotropic effect upon the thymus, functionally altering the lymphoid compartment as well as thymulin production by thymic epithelial cells.

Most of the studies done to date have examined the effect of exogenously added GH on selected immune responses. The presence of lymphocyte-derived GH produced by cells of the immune system makes it more difficult to study. Nevertheless, we have employed antisense oligodeoxynucleotides (ODNs) and antibodies to block the endogenous activity of GH.[135] These studies showed that treatment of rat lymphocytes with a specific GH antisense ODN decreased the amount of leukocyte GH synthesized and lymphocyte proliferation. The antisense ODN growth inhibition could be prevented by complementary GH sense ODN and reversed by the exogenous addition of rat GH. In studies with antibodies to GH, we were able to measure a two-fold decrease in the number of cells positive for IGF-I strongly supporting an important role for endogenously produced GH in the induction of leukocyte-derived IGF-I.[52] Taken together, our findings support the idea that GH-axis hormones function in a paracrine/autocrine loop.

The role of lymphocyte-derived GH *in vivo* has not yet been well investigated. Most of the data point toward a paracrine/autocrine role within the immune system but an endocrine role cannot be ruled out. We have conducted studies to examine whether syngeneic transfer of GH-producing leukocytes could stimulate the growth of dwarf mice. The results showed that normal spleen cells alone or spleen cells treated with GHRH did not appear to significantly stimulate the growth of dwarf mice although there was a trend toward growth.[51] The immune system of dwarf mice receiving spleen cells, however, was significantly altered. Spleen cells from dwarf animals showed enhanced immunoglobulin production, IL6, IL2, and IGFγ production whereas no significant change was apparent in natural killer cell activity. Overall, the evidence to date indicates that lymphocyte-derived GH is an important regulator of the immune system. This is supported by a number of *in vitro* and *in vivo* studies reporting the effects of GH on immunity. Also, it has been found that GHRH, somatostatin, and IGF-I and their receptors are produced in the immune system. A great deal more research will be necessary to dissect the role(s) of the GH-axis hormones in lymphocytes as well as their potential role in health, disease, and aging.

The evidence implicating PRL in the regulation of the immune system has been accumulating over several years now.[83] Just ten years ago, it was reported that the drug bromocriptine which inhibits PRL secretion suppressed antibody formation and the delayed hypersensitivity response in rats.[136] More recent data show that suppression of prolactin secretion in mice with bromocriptine: 1) increases the lethality of a *Listeria* challenge, 2) abrogates T-lymphocyte-dependent activation of macrophages as well as the production of lymphocyte interferon following inoculation with Listeria or mycobacteria, and 3) suppresses T-lymphocyte proliferation without affecting the production of IL2. All of these changes could be prevented *in vivo* by administration of prolactin.[138] In other studies, the administration of prolactin was seen to result in increases in lymphocyte proliferative responses to mitogens.[137] Overall, it appears that both humoral and cell-mediated immune responses are dependent on PRL. Some work has also been done *in vitro* examining the proliferation of Nb2 node lymphoma cells in response to PRL.[137] Nb2 cells are of T-cell origin and can be stimulated to grow by both PRL and IL2.[139] The effects of the two hormones were additive and appeared to be mediated by different receptors. In splenocytes from ovariectomized rats, PRL induced IL2 receptor expression and increased the mitogenic effect of this cytokine.[140] In cloned T-cells and ConA-stimulated splenocytes, IL2 in turn stimulated PRL uptake and nuclear translocation of PRL via an endosomal/lysosomal pathway.[55] IL2 stimulated [^{3}H]thymidine incorporation into splenocytes, but the IL2-stimulated increase in [^{3}H]thymidine incorporation was blocked in the presence of antibodies to PRL. This inhibition was reversible by the presence of PRL. Thus, Nb2 cell proliferation was shown to be a PRL receptor-mediated event, and antibodies to the PRL receptor were able to abolish PRL-induced proliferation of Nb2 cells. One group has transfected Nb2 cells with plasmids coding for mouse PRL with and without a signal peptide. In the absence of a signal peptide, PRL accumulated in the intracellular compartment but failed to stimulate proliferation, suggesting that in this model intracellular PRL could not trigger cell growth. In cells bearing a plasmid coding for PRL including a signal peptide, PRL was secreted into the medium and triggered cell growth by an autocrine mechanism. Several PRL receptors have been cloned which show heterogeneity to each other and extensive homology with the GH receptor and other cytokine receptors.[84] Guanine nucleotides have been implicated in the signal transduction mechanism mediated by PRL in Nb2 cells,[141] and another early event following the interaction of PRL with its receptor is activation of the Na^+/H^+ exchanger.[142] This PRL-mediated event may involve activation of protein kinase C and was not restricted to Nb2 cells, but also was induced in lymphoid cells of the spleen.[143] PRL stimulates ornithine decarboxylase and activates protein kinase C which are important enzymes in the differentiation, proliferation, and function of lymphocytes.[144] There are recent data showing in dwarf mice that PRL and GH can correct the defective induction of IL2-receptor expression after ConA stimulation.[134]

Stimulation of a cloned T-cell with PRL was shown to induce the expression of interferon regulatory factor-1 suggesting that PRL may regulate cell proliferation by enhancing expression of some genes required for entry into S-phase.[145] In addition to triggering resting lymphocytes to cell division, the hormone can also control the magnitude of their response to polyclonal stimuli depending on the concentration.[146] Translocation of PRL into the nucleus appears necessary for IL2-stimulated proliferation.[147] Finally, it has been shown that PRL appears to exert counterregulatory actions which may modify glucocorticoid actions on immune tissues.[148]

It is clear from all the studies to date that PRL and GH have important and sometimes similar effects on the immune system. Both animal studies and the available clinical data in humans suggest that absence of either PRL or GH can lead to deficiencies in both cell-mediated and humoral immunological functions. The deficiencies can be corrected by replacement therapy with PRL or GH. Numerous studies point toward the presence of functional receptors for PRL[89,90] and GH[93] on cells of the immune system, the physiological significance of which will continue to be explored.

2.   Neuropeptides

A number of neuropeptides commonly found in the peripheral nervous system and now known to be produced by cells of the immune system have profound effects on the immune system (Table 3). Several have been selected and will be discussed below, including substance P, CGRP, SOM, VIP, AVP, and neuropeptide Y to show these peptides can be powerful immunomodulators.

Substance P is an 11-amino acid neuropeptide widely distributed throughout the central and peripheral nervous system.[149] In addition to its role on epithelial tissue where substance P receptors have been cloned,[150] specific receptors for substance P have also been identified on immune cells[106] consistent with an important immunoregulatory role. Substance P can modulate cytokine production in monocytes by inducing the production of IL1, TNFα, and IL6.[151] Recently, substance P was shown to increase TNFα mRNA in murine mast cells.[152] It also was shown that substance P could act as a cofactor for the mitogen-induced production of IL2 in human systems by normal T cells and T cell lines.[153] Substance P is thought to act posttrans-criptionally by stabilizing the IL2 mRNA induced by PHA + PMA in Jurkat cells and not to alter the IL2 gene transcription.[154] In addition to being a mediator of pain, substance P stimulates smooth muscle contraction, vasodilation, and plasma extravasation, all of which contribute to local inflammation. Substance P may regulate other aspects of peripheral inflammatory/immune responses by stimulating the differentiation and antibody-secreting potential of B cells[155,156] and macrophage respiratory burst.[157] In fact, three neuropeptides (VIP, substance P, and SOM) have been found to be stimulatory for a cloned T cell line responding to specific antigen.[158] More recently, it has been shown that substance P can enhance antigen-induced IFNγ secretion by splenic and granuloma T cells[159] whereas substance P does not influence IL1 secretion or gene expression.[160]

Another neurotransmitter found in neurons, polymorphonuclear leukocytes, eosinophils and mast cells is vasoactive intestinal peptide.[69] VIP has been reported to have an inhibitory effect on a number of immune responses. It inhibits mitogen-stimulated proliferation of mouse lymphocytes isolated from spleen, Peyer's patches, and mesenteric lymph nodes.[108] T lymphocytes seem to be the primary target of VIP-mediated inhibition since proliferative responses to T cell mitogens such as ConA and PHA were affected, but proliferation induced by the B cell mitogen LPS was not inhibited. The findings are consistent with the observation that T cells, but not B cells, have high affinity receptors for VIP.[161] VIP has been shown to inhibit the generation of IL2 and IFNγ from mitogen-stimulated murine and human lymphocytes,[162] as well as modulate immunoglobulin production,[155] and lymphocyte traffic.[163] It has been proposed that these inhibitory effects may be the result of activation of adenyl cyclase.[164] In addition to these findings, it has been suggested that VIP suppresses mitogen and antigen-induced T cell proliferation by inhibiting IL2 production in mice infected with *Schistosoma mansoni*.[165] Another report by this same group suggests that VIP stimulates T cells to release IL5 in murine *Schistosoma mansoni* infection.[166] It is thought that substance P secretion is dominant in the early phases of granuloma development whereas VIP and IL5 come into action later depending on the granulomatous response.[159]

SOM is a small neuropeptide found in the hypothalamus, central and peripheral nervous systems, gastrointestinal tract, and in cells of the immune system.[71] Whereas GH and PRL have immunoenhancing capabilities, SOM has potent inhibitory effects on immune responses. SOM has been shown to significantly inhibit Molt-4 lymphoblast proliferation and PHA stimulation of human T lymphocytes.[167] Also, nanomolar concentrations were able to inhibit the proliferation of both spleen- and Peyer's patches-derived lymphocytes.[155] Other immune responses, such as SEA-stimulated IFNγ secretion,[168] endotoxin-induced leukocytosis,[169] and colony-stimulating activity release,[170] are also inhibited by SOM. In another study, it appears that SOM, a product of granuloma macrophages[171] may participate in the modulation of lymphokine secretion in murine *Schistosomiasis mansoni* infection. Granuloma T cells have SOM receptors and respond to SOM with decreased IFNγ secretion.[172] The observations with VIP, SOM, and substance P indicate how complex the interactions are between neuropeptides and T cells in the granulomata response.[173]

Regulatory control of the immune system by the central nervous system has been postulated for several other neuropeptides.[174] The posterior pituitary releases arginine vasopressin (AVP) which is produced in the hypothalamus and subsequently transported to the pituitary where it stimulates the release of ACTH and enhances the corticotropin-releasing hormone (CRF)-mediated ACTH release. This hormone, in addition to its antidiuretic and vasopressor activities, also influences the immune response. AVP plays an important role in enhancing IFN-γ production by providing a costimulatory signal to lymphocytes that is able to replace IL2.[175] Overall, the effects of AVP appear to be positive and opposite to that of ACTH on T cell function.

The influence of hypothalamic hormones on immunity is being extensively investigated and includes studies of CRF.[176] The most important function of CRF may be to stimulate production of lymphocyte POMC peptides after stimulation with CRF. Thus, the effects of ACTH and endorphins discussed above may be initiated in the immune system via hypothalamic releasing hormones. Along another line, some studies suggest that CRF modulates the immune response to stress in the rat by inhibiting lymphocyte proliferation and natural killer cell activity.[177] The neuropeptide α-MSH has powerful antipyretic and anti-inflammatory properties.[178] In addition, it appears CRF mimics the immunosuppressive effects of α-melanocyte stimulating hormone (MSH) but has a longer duration of action.[179] The activation of granulocytes can be suppressed by ACTH and MSH.[180] In the same study, it was observed that human immunodeficiency virus induces high levels of ACTH and MSH and suggested that these hormones may be part of the immunosuppression observed during virus infection.[180] Another peptide that has been studied is CGRP where it has been shown to modulate B lymphocyte differentiation[181] and T cell chemotaxis.[182] Thus, the release of CGRP from nerve endings could influence T cell traffic. Overall, the data strongly support the view that neuropeptides play a major role in the immune response.

CONCLUSION

The studies to date clearly show that cells of the immune system contain receptors for neuropeptides and can also be considered a source of pituitary, hypothalamic, and neural peptides. For example, the inflammatory cells found in the granulomas of murine *Schistosomiasis mansoni* produce substance P, VIP, and SOM peptides and have their receptors which serve to modulate lymphokine secretion by regulatory T lymphocytes and macrophage activation. The production and regulation of these neuropeptides by leukocytes in some instances is similar to that observed by neural tissue as well as there appear to be noteworthy differences. The plasma hormone concentrations contributed by lymphocytes usually do not reach the levels required when the pituitary gland is the source, but because immune cells are mobile, they can locally deposit the hormone at the target site. Once secreted, these peptide hormones seem to function in at least two capacities. They are endogenous regulators of the immune system as well as conveyors of information from the immune to the neuroendocrine system. It is our hypothesis that the relay of information to the neuroendocrine system represents a sensory function for the immune system wherein leukocytes recognize stimuli that are not recognizable by the central and peripheral nervous systems (i.e., bacteria, tumors, viruses, and antigens). The recognition of such noncognitive stimuli by immunocytes is then converted into information, in the form of peptide hormones and neurotransmitters and cytokines, which is conveyed to the neuroendocrine system and a physiologic change occurs.

ACKNOWLEDGMENTS

The research was supported in large part by grants from the National Institute of Neurology and Communicative Disorders (RO1 NS24636) and the National Institute of Arthritis, Diabetes, Digestion and Kidney Disease (RO1 DK38024). The authors are indebted to former graduate students, postdoctoral fellows, and faculty colleagues for their contributions to the work reviewed here. We also thank Diane Weigent for excellent editorial assistance and typing the manuscript.

REFERENCES

1.   **Blalock, J. E., and Smith, E. M.,** Human leukocyte interferon: Structural and biological relatedness to adrenocorticotropic hormone and endorphins, *Proc. Natl. Acad. Sci. USA*, 77,5972, 1980.

2.   **Blalock, J.E.,** Production of peptide hormones and neurotransmitters by the immune system, in *Neuroimmunoendocrinology,* Blalock, J. E., Ed., 2nd rev ed, *Chem Immunol.,* Karger, Basel, 52,1, 1992.

3.   **Johnson, H..M., Downs, M. O., and Pontzer, C. H.** Neuroendocrine peptide hormone regulation of immunity, in *Neuroimmunoendocrinology,* Blalock, J. E., Ed., Karger, Basel, 2nd rev ed, *Chem. Immunol.,* 1992, 52,49.

4.   **Smith, E. M., and Blalock, J. E.,** Human lymphocyte production of ACTH and endorphin-like substances: Association with leukocyte interferon, *Proc. Natl. Acad. Sci. USA,* 78,7530, 1981.

5.   **Kavelaars, A., Ballieux, R.E., and Heijnen, C.J.,** The role of IL-1 in the corticotropin-releasing factor and arginine-vasopressin-induced secretion of immunoreactive beta-endorphin by human peripheral blood mononuclear cells, *J. Immunol.,* 142,2338, 1989.

6.   **Lolait, S.J., Clements, J. A., Markwick, A. J., Cheng, C., McNally, M., Smith, A. I., and Funder, J. W.,** Pro-opiomelanocortin messenger ribonucleic acid and posttranslational processing of beta endorphin in spleen macrophages, *J. Clin. Invest.,* 77,1776, 1989.

7.   **Buzzetti, R., McLoughlin, L., Lavender, P. M., Clark, A. J. L., and Rees, L. H.,** Expression of pro-opiomelanocortin gene and quantification of adrenocorticotropic hormone-like immunoreactivity in human normal peripheral mononuclear cells and lymphoid and myeloid malignancies, *J. Clin. Invest.,* 83,733, 1989.

8.   **Westley, H. J., Kleiss, A. J., Kelley, K. W., Wong, P. K.Y., and Yuen, P. H.,** Newcastle disease virus-infected splenocytes express the pro-opiomelanocortin gene, *J. Exp. Med.,* 163,1589, 1986.

9.   **Meyer, W. J. I., Smith, E. M., Richards, G. E., Cavallo, A., Morrill, A. C., and Blalock,** *In vivo* immunoreactive ACTH production by human leukocytes from normal and ACTH-deficient individuals, *J. Clin. Endocrinol. Metabol.,* 64,98, 1987.

10.   **Przewlocki, R., Hassan, A. H. S., Lasson, W., Epplen, C., Herz, A., and Stein, C.,** Gene expression and localization of opioid peptides in immune cells of inflamed tissue: Functional role in antinociception, *Neuroscience,* 48,491, 1992.

11.   **Reder, A. T.,** Regulation of production of adrenocorticotropin-like proteins in human mononuclear cells, *Immunology,* 77,491, 1992.

12.   **Westley, H. F., Goldstien, D., and Borden, E. C.,** Beta-interferon augments pro-opiomelanocortin gene expression in human peripheral blood cells, *FASEB J.,* 2, A1640, 1988.

13.   **Bayle, J. E., Guellati, M., Ibos, F., and Roux, J.,** *Brucella abortus* antigen stimulates the pituitary-adrenal axis through the extra-pituitary B-lymphoid system, *Prog. NeuroEndocrinImmunol.,* 4,99, 1991.

14.   **Oates, E. L., Allaway, G. P., Armstrong, G. R., Boyajian, R. A., Kehr, J. R., and Prabhakar, B. S.,** Human lymphocytes produce pro-opiomelanocortin gene related transcripts. Effects of lymphotropic viruses, *J. Biol. Chem.,* 263,10041, 1988.

15.   **Galin, F. S., LeBoeuf, R. D., and Blalock, J. E.,** A lymphocyte mRNA encodes the adrenocorticotropin/β-lipotropin region of the pro-opiomelanocortin gene, *Prog. NeuroEndocrinImmunol.,* 3,205, 1990.

16. **Galin, F. S., LeBoeuf, R. D., and Blalock, J. E.,** Corticotropin-releasing factor upregulates expression of two truncated pro-opiomelanocortin transcripts in murine lymphocytes, *J. Neuroimmunology,* 31,51, 1991.

17. **Smith, E. M., Morrill, A. C., Meyer, W.J., and Blalock, J. E.,** Corticotropin releasing factor induction of leukocyte-derived immunoreactive ACTH and endorphins, *Nature,* 322,881, 1986.

18. **Smith, E. M., Galin, F. S., LeBoeuf, R. D., Coppenhaver, D. H., Harbour, D. V., and Blalock, J. E.,** Nucleotide and amino acid sequence of lymphocyte-derived corticotropin, Endotoxin induction of a truncated peptide, *Proc. Natl. Acad. Sci. USA,* 87,1057, 1990.

19. **Mechanick, J. I., Levin, N., Roberts, J. L., and Autelitano, D. J.,** Proopiomelanocortin gene expression in a distinct population of rat spleen and lung leukocytes, *Endocrinology,* 131,518, 1992.

20. **Harbour, D. V., Galin, F. S., Hughes, T. K., Smith, E. M., and Blalock, J. E.,** Role of leukocyte-derived pro-opiomelanocortin peptides in endotoxic shock, *Circ. Shock,* 35,181, 1991.

21. **Harbour, D. V., Smith, E. M., and Blalock, J. E.,** A novel processing pathway for proopiomelanocortin in lymphocytes, Endotoxin induction of a new pro-hormone cleaving enzyme, *J. Neurosci. Res.,* 18,95, 1987.

22. **Clarke, B. L., Gebhardt, B. M., and Blalock, J. E.,** Mitogen-stimulated lymphocytes release biologically active corticotropin, *Endocrinology,* 132,983, 1993.

23. **Zurawski, G., Benedik, M., Kamp, B. J., Abrams, J. S., Zurawski, S. M., and Lee, F.D.,** Activation of mouse T-helper cells induces abundant pre-proenkephalin mRNA synthesis, *Science,* 232,772, 1986.

24. **Kuis, W., Villiger, P. M., Leser, H.G., and Lotz, M.,** Differential processing of proenkephalin A by human peripheral blood monocytes and T lymphocytes, *J. Clin. Invest.,* 88,817, 1991.

25. **Behar, O., Ovadia, H., Polakiewicz, R. D., and Rosen, H.,** Lipopolysaccharide induces proenkephalin gene expression in rat lymph nodes and adrenal glands, *Endocrinology,* 134,475, 1994.

26. **Smith, E. M., Phan, M., Coppenhaver, D., Kruger, T. E., and Blalock, J. E.,** Human lymphocyte production of immunoreactive thyrotropin, *Proc. Natl. Acad. Sci. USA,* 80,6010, 1983.

27. **Kruger, T. E., Smith, L.R., Harbour, D.V., and Blalock, J. E.,** Thyrotropin, An endogenous regulator of the *in vitro* immune response, *J. Immunol.,* 142,744, 1989.

28. **Harbour, D. V., Kruger, T. E., Coppenhaver, D., Smith, E. M., and Meyer, W. J.,** Differential expression and regulation of thyrotropin (TSH) in T cell lines, *Mol. Cell. Endocrin.,* 64,229, 1989.

29. **Peele, M. E., Carr, F. E., Baker, Jr., J. R., Wartofsky, L., and Burman, K. D.,** TSH beta subunit gene expression in human lymphocytes, *Amer. J. Med. Sci.,* 305,1, 1993.

30. **Blalock, J. E., and Costa, O.,** Immune neuroendocrine interactions, Implications for reproductive physiology, *Ann. NY Acad. Sci.,* 564,261, 1990.

31. **Standaert, F. E., Chew, B. P., Wong, T. S., and Michal, J. J.,** Porcine lymphocytes secrete factors in response to LHRH to stimulate progesterone production by granulosa cells *in vitro* (Abstract)., *Biol. Reproduction*, 42,75, 1990.

32. **Ebaugh, M. J., and Smith, E. M.,** Human lymphocyte production of immunoreactive luteinizing hormone (Abstract), *FASEB J.*, 2,1642, 1988.

33. **Gorospe, W. C., and Kasson, B. G.,** Lymphokines from concanavalin-A-stimulated lymphocytes regulate rat granulosa cell steroidogenesis *in vitro*, *Endocrinology*, 123, 2462, 1989.

34. **Petraglia, F., Sacerdote, P., Cossarizza, A., Angioni, S., Genazzani, A.D., Franceschi, C., Muscettola, M., and Grasso, G.,** Inhibin and activin modulate human monocyte chemotaxis and human lymphocyte interferon-gamma production, *J. Clin. Endocrin. Met.*, 72,496, 1991.

35. **Hiestand, P. C., Mekler, P., Nordmann, R., Grieder, A., and Permmongko, I. C.,** Prolactin as a modulator of lymphocyte responsiveness provides a possible mechanism of action for cyclosporin, *Proc. Natl. Acad. Sci. USA*, 83,2599, 1986.

36. **Montgomery, D. W., Zukowski, C. F., Shah, N. G., Buckley, A. R., Pacholczyk, T., and Russell, D. H.,** Concanavalin A-stimulated murine leukocytes produce a factor with prolactin-like bioactivity and immunoreactivity, *Biochem. Biophys. Res. Commun.*, 145, 692, 1987.

37. **DiMattia, G. E., Gellersen, B., Bohnet, H. G., and Friesen, H. G.,** A human B-lymphoblastoid cell line produces prolactin, *Endocrinology*, 122,2508, 1988.

38. **Clevenger, C. V., Russell, D. H., Appasamy, P. M., and Prystowsky, M. B.,** Regulation of interleukin-2 driven T-lymphocytes, *Proc. Natl. Acad. Sci., USA*, 87,6460, 1990.

39. **Jurcovicova, J., Day, R. N., and MacLeod, R. M.,** Expression of prolactin in rat lymphocytes, *Prog. NeuroendocrinImmunology*, 5,256, 1992.

40. **Pellegrini, I,, Lebrun, J.-J., Ali, S., and Kelly, P. A.,** Expression of prolactin and its receptor in human lymphoid cells, *Mol. Endocrin.*, 6,1023, 1992.

41. **O'Neal, K. D., Montgomery, D. W., Truong, T. M., and Yu-Lee, L-y.,** Prolactin gene expression in human thymocytes, *Mol. Cell. Endocrin.*, 87,R19, 1992.

42. **Kenner, J. R., Holaday, J. W., Bernton, E. W., and Smith, P. F.,** Prolactin-like protein im murine lymphocytes, morphological and biochemical evidence, *Prog. NeuroendocrinImmun.*, 3,188, 1990.

43. **Montgomery, D. W., Shen, G. K., Ulrich, E. D., Steiner, L. L., Parrish, P. R., and Zukoski, C. F.,** Human thymocytes express a prolactin-like messenger ribonucleic acid and synthesize bioactive prolactin-like proteins, *Endocrinology*, 131,3019, 1992.

44. **Sabharwal, P., Glaser, R., Lafuse, W., Varma, S., Liu, Q., Arkins, S., Kooijman, R., Kutz, L., Kelley, K. W., and Malarkey, W. B.,** Prolactin synthesized and secreted by human peripheral blood mononuclear cells, An autocrine growth factor for lymphoproliferation, *Proc. Natl. Acad. Sci. USA*, 89,7713, 1992.

45. **Friesen, H. G., DiMattia, G. E., and Too, C. K. L.,** Lymphoid tumor cells as models for studies of prolactin gene regulation and action, *Prog. NeuroEndocrinImmunology*, 4,1, 1991.

46.   **Weigent, D. A., Baxter, J. B., Wear, W. E., Smith, L. R., Bost, K. L., and Blalock, J. E.**, Production of immunoreactive growth hormone by mononuclear leukocytes, *FASEB Journal*, 2,2812, 1988.

47.   **Kao, T. L., Harbour, D. V., Smith, E. M., and Meyer, W. J.**, Immunoreactive growth hormone production by cultured lymphocytes, *Endocrine Society, 71st Annual Meeting*, Seattle, WA, Abstract 343, 1989.

48.   **Hattori, N., Shimatsu, A., Sugita, M., Kumagai, S., and Imura, H.**, Immunoreactive growth hormone (GH) secretion by human lymphocytes augmented exogenous GH, *Biochem. Biophys. Res. Commun.*, 168,396, 1990.

49.   **Varma, S., Sabharwal, P., Sheridan, J. F., and Malarkey, W. B.**, Growth hormone secretion by human peripheral blood mononuclear cells detected by an enzyme-linked immunoplaque assay, *J. Clin. Endocrin. Met.*, 76,49, 1993.

50.   **Lytras, A., Quan, M. E., Vrontakis, M. E., Shaw, J. E., Cattini, P. A., and Friesen, H. G.**, Growth hormone expression in human Burkitt lymphoma serum-free Ramos cell line, *Endocrinology*, 132,620, 1993.

51.   **Weigent, D. A., and Blalock, J. E.**, Effect of the administration of growth-hormone-producing lymphocytes on weight gain and immune function in dwarf mice, *Neuroimmunomodulation*, 1,50, 1994.

52.   **Baxter, J. B., Blalock, J. E., and Weigent, D. A.**, Characterization of immunoreactive insulin-like growth factor-I from leukocytes and its regulation by growth hormone, *Endocrinology*, 129,1727, 1991.

53.   **Payne, L. C., Rohn, W., and Weigent, D. A.**, Lymphocyte-derived growth hormone releasing hormone is an autocrine modulator of lymphocyte-derived growth hormone, *Endocrine Society*, Anaheim, CA, 1994.

54.   **Kao,T-L, Supowit, S. C., Thompson, E. A., and Meyer, III,W. J.**, Immunoreactive growth hormone production by human lymphocyte cell lines, *Cell. Mol. Neurobiol.*, 12, 483, 1992.

55.   **Clevenger, C. V., Sillman, A. L., and Prystowsky, M. B.**, Interleukin-2 driven nuclear translocation of prolactin in cloned T-lymphocytes, *Endocrinology*, 127,3151, 1990.

56.   **Weigent, D. A., and Blalock, J. E.**, The production of growth hormone by subpopulations of rat mononuclear leukocytes, *Cellular Immunol.*, 135,55, 1991.

57.   **Weigent, D. A., Baxter, J. B., and Blalock, J. E.**, The production of growth hormone and insulin-like growth factor-I by the same subpopulation of rat mononuclear leukocytes, *Brain Behav. Immun.*, 6,365, 1992.

58.   **Stephanou, A., Jessop, D. S., Knight, R. A., and Lightman, S. L.**, Corticotropin-releasing factor-like immunoreactivity and mRNA in human leukocytes, *Brain Behav. Immun.*, 4,67, 1990.

59.   **Ekman, R., Servenius, B., Castro, M. G., Lowry, P. J., Cederlund, A. S., Bergman, O., and Sjogren, H. O.**, Biosynthesis of corticotropin-releasing hormone in human T lymphocytes, *J. Neuroimmunology*, 44,7, 1993.

60.   **Aird, F., Clevenger, C. V., Prystowsky, M. B., and Rebei, E.**, Corticotropin-releasing factor mRNA in rat thymus and spleen, *Proc. Natl. Acad. Sci. USA*, 90,7104, 1993.

61. Rajeev, J., Zwickler, D., Hollander, C. S., Brand, H., Saperstein, A., Hutchinson, B., Brown, C., and Audhya, T., Corticotropin-releasing factor modulates the immune response to stress in rats, *Endocrinology*, 128,1329, 1991.

62. Weigent, D. A., and Blalock, J. E., Immunoreactive growth hormone releasing hormone in rat leukocytes, *J. Neuroimmunology*, 29,1, 1990.

63. Guarcello, V., Weigent, D. A., and Blalock, J. E., Growth hormone releasing factor receptors on lymphocytes, *Cellular Immunology*, 136,291, 1991.

64. Stephanou, A., Knight, R. A., and Lightman, S. L., Production of a growth hormone-releasing hormone-like peptide and its mRNA by human lymphocytes, *Neuroendocrinology*, 53,628, 1991.

65. Emanuele, N. V., Emanuele, M. A., Tentler, J., Kirsteins, L., Azad, N., and Lawrence, A. M., Rat spleen lymphocytes contain an immunoreactive and bioactive luteinizing hormone-releasing hormone, *Endocrinology*, 126,2482, 1990.

66. Azad, N., Emanuelle, N. V., Halloran, M. M., Tentler, J., and Kelley, M. R., Presence of luteinizing hormone-releasing hormone (LHRH) mRNA in spleen lymphocytes, *Endocrinology*, 128,1679, 1992.

67. Maier, C. C., Marchetti, B., LeBoeuf, R. D., and Blalock, J. E., Thymocytes express a mRNA that is identical to hypothalamic luteinizing hormone-releasing hormone mRNA, *Cell. Mol. Neurosci.*, 1992,447, 1992.

68. Felten, D. L., Felten, S. Y., Carlson, S. L., Olschowka, J. A., and Livnat, S., Noradrenergic and peptidergic innervation of lymphoid tissue, *J. Immunol.*, 135,755s, 1985.

69. Gomariz, R. P., Lorenzo, M. J., Cacicedo, L., Vicente, A., and Zapata, A. G., Demonstration of immunoreactive vasoactive intestinal peptide (IR-VIP) and somatostatin (IR-SOM) in rat thymus, *Brain Behav. Immun.*, 4,151, 1990.

70. Goetzl, E. J., Grotmol, T., VanDyke, R. W., Turck, C. W., Wershil, B., Galli, S. J., and Sreedharan, S. P., Generation and recognition of vasoactive intestinal peptide by cells of the immune system, *Ann. NY Acad. Sci.*, 594,34, 1990.

71. Aguila, M. C., Dees, W. L., Haensly, W. E., and McCann, S. M., Evidence that somatostatin is localized and synthesized in lymphoid organs, *Proc. Natl. Acad. Sci. USA*, 88,11485, 1991.

72. Ericsson, A., Schalling, M., McIntyre, K. R., Lundberg, J. M., Larhammar, D., Seroogy, K., Hokfelt, T., and Persson, H., Detection of neuropeptide Y and its mRNA in megakaryocytes, Enhanced levels in certain autoimmune mice, *Proc. Natl. Acad. Sci. USA*, 84,5585, 1987.

73. Schwarz, H., Villiger, P. M., von Kempis, J., and Lotz, M., Neuropeptide Y is an inducible gene in the human immune system, *J. Neuroimmunology*, 51,53, 1994.

74. Pascual, D. W., and Bost, K. L., Substance P production by macrophage cell lines, a possible autocrine function for this neuropeptide, *Immunology*, 71,52, 1990.

75. Geenen, V., Robert, F., Martens, H., Benhida, A., DeGiovanni, G., Defresne, M-P., Boniver, J., Legios, J-J, Martial, J., and Franchimont, P., Biosynthesis and paracrine (cryptocrine actions of 'self' neurohypophysial-related peptides in the thymus), *Mol. Cell. Endocrin.*, 76,C27, 1991.

76. **Adachi, N., Yamaguchi, K., Miyake, Y., Honda, S., Hagasaki, K., Akiyama, Y., Adachi, I., and Abe, K.,** Parathyroid hormone-related peptide is a possible autocrine growth inhibitor for lymphocytes, *Biochem. Biophys. Res. Commun.*, 166,1088, 1990.

77. **Carr, D. J. J.,** Neuroendocrine peptide receptors on cells of the immune system, in *Neuroimmunoendocrinology*, 2nd rev ed, Blalock, J. E., Ed., *Chem. Immunol.*, 52,49, 1992, Karger, Basel.

78. **Clarke, B.L., and Bost, K. L.,** Differential expression of functional adrenocorticotropic hormone receptors by subpopulations of lymphocytes, *J. Immunol.*, 143,464, 1991.

79. **Carr, D. J. J.,** The role of endogenous opioids and their receptors in the immune system., *Soc. Exp. Biol. Med.*, 198,710, 1991.

80. **Habaud, O., and Lissitzky, S.,** Thyrotropin-specific binding to human peripheral blood monocytes and polymorphonuclear leukocytes, *Mol. Cell. Endocrinol.*, 7,79, 1977.

81. **Harbour, D. V., Leon, S., Keating, C., and Hughes, T. K.,** Thyrotropin modulates B-cell function through specific bioactive receptors, *Prog. NeurEndocrinImmunol.*, 3, 266, 1990.

82. **Coutelier, J-P, Kehrl, J. H., Bellur, S. S., Kohn, L. D., Notkins, A. L., and Prabhakar, B.S.,** Binding and functional effects of thyroid stimulating hormone on human immune cells, *J. Clin. Immunol.*, 10,204, 1990.

83. **Gala, R. R.,** Prolactin and growth hormone in the regulation of the immune system, *Proc. Soc. Exp. Biol. Med.*, 198,513, 1991.

84. **Bazan, J. F.,** Structural design and molecular evolution of a cytokine receptor superfamily, *Proc. Natl. Acad. Sci., USA*, 87,6934, 1990.

85. **Dardenne, M., de Moraes, M. d. C. L., Kelly, P. A., and Gagnerault, M.-C.,** Prolactin receptor expression in human hematopoietic tissues analyzed by flow cytofluorometry, *Endocrinology*, 134,2108, 1994.

86. **Matera, L., Mucciolli, G., Cesano, A., Bellussi, G., and Genazzani, E.,** Prolactin receptors on large granular lymphocytes, Dual regulation by cyclosporin A, *Brain Behav. Immun.*, 2,1, 1988.

87. **O'Neal, K. D., Schwarz, L. A., and Yu-Lee, L. Y.,** Prolactin receptor gene expression in lymphoid cells, *Mol. Cell. Endocrinol.*, 82,127, 1991.

88. **Ali, S., Pellegrini, I., and Kelly, P. A.,** A prolactin-dependent immune cell line (Nb2) expresses a mutant form of prolactin receptor, *J. Biol. Chem.*, 266,20110, 1991.

89. **Gagnerault, M.-C., Touraine, P., Savino, W., Kelly, P. A., and Dardenne, M.,** Expression of prolactin receptors in murine lymphoid cells in normal and autoimmune situations, *J. Immunol.*, 150,5673, 1993.

90. **Gala, R. R., and Shevach, E. M.,** Identification by analytical flow cytometry of prolactin receptors on immunocompetent cell populations in the mouse, *Endocrinology*, 133,1617, 1993.

91. **Krown, K. A., Wang, Y-F., and Walker, A. M.,** Autocrine interaction between prolactin and its receptor occurs intracellularly in the 235-1 mammotroph cell line, *Endocrinology*, 134,1546, 1994.

92. **De Vos, A. M., Ultsch, M., and Kossiakoff, A. A.,** Human growth hormone and extracellular domain of its receptor, Crystal structure of the complex, *Science*, 255, 306, 1992.

93. **Silva, C. M., Weber, M. J., and Thorner, M. O.,** Stimulation of tyrosine phosphorylation in human cells by activation of the growth hormone receptor, *Endocrinology*, 132, 101,1993.

94. **Suzuki, K., Suzuki, S., Saito, Y., Ikebuchi, H., and Terao, T.,** Human growth hormone-stimulated growth of human cultured lymphocytes (IM-9) and its inhibition by phorbol diesters through down-regulation of the hormone receptors, *J. Biol. Chem.*, 265,11320, 1990.

95. **Argetsinger, L. S., Campbell, G. S., Yang, X., Witthuhn, B. A., Silvennoinen, O., Ihle, J. N., and Carter-Su, C.,** Identification of JAK2 as a growth hormone receptor-associated tyrosine kinase, *Cell*, 74,237, 1993.

96. **Roupas, P., and Herington, A. C.,** Cellular mechanisms in the processing of growth hormone and its receptor, *Mol. Cell. Endocrin.*, 61,1, 1989.

97. **Baumann, G.,** Growth hormone-binding proteins, *Proc. Soc. Exp. Biol. Med.*, 202,392, 1993.

98. **Mercado, M., Davila, N., McLeod, J. F., and Baumann, G.,** Distribution of growth hormone receptor messenger ribonucleic acid containing and lacking exon 3 in human tissues, *J. Clin. Endocrin. Met.*, 78,731, 1994.

99. **Fahrenholz, F., Kajro, E., Muller, M., Boer, R., Lohr, R., and Grzonka, Z.,** Iodinated photoreactive vasopressin antagonist, Labelling of hepatic vasopressin receptor subunits, *Eur. J. Biochem.*, 161,321, 1986.

100. **De Souza, E. B.,** Corticotropin-releasing factor and interleukin-1 receptors in the brain-endocrine-immune axis, *Ann. NY Acad. Sci.*, 697,9, 1993.

101. **Harbour, D. V., and Hughes, T. K.,** Thyrotropin releasing (TRH) induces gamma interferon release, *FASEB J*, A5884, 1991.

102. **Hiruma, K., Nakamura, K. H., Sumida, T., Maeda, T., Tomioka, H., Yoshida, S., and Fujita, T.,** Somatostatin receptors on human lymphocytes and leukaemia cells, *Immunol.*, 71,480, 1990.

103. **Fais, S., Annibale, B., Boirivant, M., Santoro, A., Pallone, F., and Delle Fave, G.,** Effects of somatostatin on human intestinal lamina propria lymphocytes, Modulation of lymphocyte activation, *J. Neuroimmunol.*, 31,211, 1991.

104. **Marriott, D. R., Wilkin, G. P., and Wood, J. N.,** Substance P-induced release of prostaglandins from astrocytes, Regional specialisation and correlation with phosphoinositol metabolism, *J. Neurochem.*, 56,259, 1991.

105. **McGillis, J. P., Mitsuhashi, M., and Payan, D. G.,** Immunomodulation by tachykinin neuropeptides, *Ann. NY Acad. Sci.*, 594,85, 1990.

106. **Pascual, D. W., Blalock, J. E., and Bost, K. L.,** Antipeptide antibodies that recognize a lymphocyte substance P receptor, *J. Immunol.*, 143,3697, 1989.

107. **Blum, A.M., Mathew, R., Cook, G. A., Metwali, A., Felman, R., and Weinstock, J. V.,** Murine mucosal T cells have VIP receptors functionally distinct from those on intestinal epithelial cells, *J. Neuroimmunol.*, 39,101, 1992.

108. **Ottaway, C., and Greenberg, G.,** Interaction of vasoactive intestinal peptide with mouse lymphocytes, Specific binding and modulation of mitogen responses, *J. Immunol.*, 132,417, 1984.

109.   Ottaway, C., Vasoactive intestinal peptide and immune function, in *Psychoneuroimmunology*, 2nd ed., Ader R., Felten, D. L., and Cohen, N., Eds., Academic Press, San Diego, 1991, pp 225.

110.   Abello, J., Kaiserlian-Nicolas, D., Cuber, J. C., Revillard, J. P., and Chayvialle, J. A., Identification of high affinity calcitonin gene-related peptide receptors on a murine macrophage-like cell line, *Ann. NY Acad. Sci.*, 594,364, 1990.

111.   McGillis, J. P., Humphreys, S., and Reid, S., Characterization of functional calcitonin gene-related peptide receptors on rat lymphocytes, *J. Immunol.*, 147,3482, 1991.

112.   Johnson, H. M., Smith, E. M., Torres, B. A., and Blalock, J. E., Neuroendocrine hormone regulation of *in vitro* antibody production, *Proc. Natl. Acad. Sci. USA*, 79, 4171, 1982.

113.   Zwilling, B. S., Lafuse, W. P., Brown, D., and Pearl, D., Characterization of ACTH mediated suppression of MHC class II expression by murine peritoneal macrophages, *J. Neuroimmunol.*, 39,133, 1992.

114.   McGlone, J. J., Lumpkin, E. A., and Norman, R. L., Adrenocorticotropin stimulates natural killer cell activity, *Endocrinology*, 129,1653, 1991.

115.   Kavelaars, A., Ballieux, R. E., and Heijnen, C., Modulation of the immune response by proopiomelanocortin-derived peptides, *Brain Behav. Immun.*, 2,57, 1988.

116.   Johnson, H. M., Torres, B. A., Smith, E. M., Dion, L. D., and Blalock, J. E., Regulation of lymphokine (c-interferon) production by corticotropin, *J. Immunol.*, 132, 246, 1984.

117.   Brooks, W. H., and Walmann, M., Adrenocorticotropin functions as a late-acting B cell growth factor and synergizes with IL5, *FASEB J.*, A479, 1989.

118.   Smith, E. M., Meyer, W. J., and Blalock, J. E., Virus-induced increases in corticosterone in hypophysectomized mice, A possible lymphoid adrenal axis, *Science*, 218,1311, 1982.

119.   Bayle, J. E., Guellati, M., Ibos, F., and Roux, J., *Brucella abortus* antigen stimulates the pituitary-adrenal axis through the extrapituitary B lymphoid system, *Prog. Neuro. Endocrin. Immunol.*, 4,99, 1991.

120.   Stephanou, A., Sarlis, N. J., Knight, R. A., Chowdrey, H. S., and Lightman, S. L., Response of pituitary and spleen pro-opiomelanocortin mRNA, and spleen and thymus interleukin-1B mRNA to adjuvant arthritis in the rat, *J. Neuroimmunol.*, 37,59, 1992.

121.   Law, V., Payne, L. C., and Weigent, D. A., Effects of streptozotocin-induced diabetes on lymphocyte POMC and growth hormone gene expression in the rat, *J. Neuroimmunol.*, 49,35, 1994.

122.   Kavelaars, A., Berkenbosch, F., Croiset, G., Ballieux, R. E., and Heijnen, C. J., Induction of beta-endorphin secretion by lymphocytes after subcutaneous administration of corticotropin-releasing factor, *Endocrinology*, 126,759, 1990.

123.   Mashaly, M. M., Trout, J. M., and Hendricks, G. L., The endocrine function of the immune cells in the initiation of humoral immunity, *Poultry Sci.*, 72,1289, 1993.

124.   Reynolds, D. G., Guill, N. V., Vargish, T., Hechner, R. B., Fader, A. I., and Holaday, J. W., Blockage of opiate receptors for naloxone improves survival and cardiac performance in canine endotoxic shock, *Circ. Shock*, 7,39, 1980.

125. **Bohs, C. T., Fish, J. C., Miller, T. H., and Traber, D. L.**, Pulmonary vascular response to endotoxin in normal and lymphocyte depleted sheep, *Circ. Shock*, 6,13, 1979.

126. **Pierpaoli, W., Baroni, C., Fabris, N., and Sorkin, E.**, Hormones and immunological capacity. II. Reconstruction of antibody production in hormonally deficient mice by somatotropic hormone, thyrotropic hormone and thyroxine, *Immunology*, 16,217, 1969.

127. **Provinciali, M., DiStefano, G., and Fabris, N.**, Improvement in the proliferative capacity and natural killer cell activity of murine spleen lymphocytes by thyrotropin, *Int. J. Immunopharm.*, 14,865, 1992.

128. **Rouabhia, M., Ghanmi, Z., and Deschaux, P.**, Luteotrophic hormone, receptor and inhibition of T cytotoxicity *in vitro*, *Per. Biol.*, 89,119, 1987.

129. **Rouabhia, M., Chakir, J., and Deschaux, P.**, Interaction between the immune and endocrine systems, Immunomodulatory effects of luteinizing hormone, *Prog. Neuro. Endocrin. Immunol.*, 4,86, 1991.

130. **Kelley, K. W.**, Growth hormone, lymphocytes, and macrophages, *Biochem. Pharmacol.*, 35,705, 1989.

131. **Fu, Y-K, Arkins, S., Fuh, G., Cunningham, B. C., Well, J. A., Fong, S., Cronin, M. J, Dantzer, R., and Kelley, K.W.**, Growth hormone augments superoxide anion secretion of human neutrophils by binding to the prolactin receptor, *J. Clin. Invest.*, 89,451, 1992.

132. **Johnson, E. W., Jones, L. A., and Kozak, R. W.**, Expression and function of insulin-like growth factor receptors on anti-CD3-activated human T lymphocytes, *J. Immunol.*, 148,63, 1992.

133. **Zelazowski, P., Dohler, K. D., Stepien, H., and Pawlikowski, M.**, Effect of growth hormone releasing hormone on human peripheral blood leukocyte chemotaxis and migration in normal subjects, *Neuroendocrinology*, 50,236, 1989.

134. **Gala, R. R., and Shevach, E. M.**, Influence of prolactin and growth hormone on the activation of dwarf mouse lymphocytes *in vivo*, *Proc. Soc. Exp. Biol. Med.*, 204,224, 1993.

135. **Weigent, D. A., LeBoeuf, R. D., and Blalock, J. E.**, An antisense oligonucleotide to growth hormone mRNA inhibits lymphocyte proliferation, *Endocrinology*, 128,2053, 1991.

136. **Nagy, E., Berczi, I., and Freisen, H. G.**, Regulation of immunity in rats by lactogenic and growth hormone, *Acta Endocrinology*, 102,351, 1983.

137. **Bernton, E. W., Bryant, H. U., and Holaday, J. W.**, Prolactin and immune function, in *Psychoneuroimmunology, 2nd ed.*, Ader, R., Felten, D. L., and Cohen, N., Eds, Academic Press, San Diego, 1991, pp 403-428.

138. **Tanaka, T., Shiu, R. P. C., Gout, P. W., Beer, C. T., Noble, R. L., and Freisen, H. G.**, A new sensitive and specific bioassay for lactogenic hormones, Measurement of prolactin and growth hormone in human serum, *J. Clin. Endocrin. Met.*, 51,1058, 1980.

139. **Croze, F., Walker, A., and Friesen, H. G.**, Stimulation of growth of Nb2 lymphoma cells by interleukin-2 in serum-free and serum-containing media, *Mol. Cell. Endocrin.*, 55,253, 1988.

140. **Mukherjee, P., Mastro, A. M., and Hymer, W. C.**, Prolactin induction of interleukin-2 receptors on rat splenic lymphocytes, *Endocrinology*, 126,88, 1990.

141. **Too, C. K. L., Murphy, P. R., and Friesen, H. G.,** G-proteins modulate prolactin- and interleukin-2-stimulated mitogenesis in rat Nb2 lymphoma cells, *Endocrinology*,124, 2185, 1989.

142. **Too, C. K. L., Cragoe, Jr. E. J., and Friesen, H. G.,** Amiloride-sensitive Na+/H+ exchange in rat Nb2 node lymphoma cells. Stimulation by prolactin and other mitogens, *Endocrinology*, 121,1512, 1987.

143. **Berczi, I., Nagy, E., de Toledo, S. M., Matusik, R. J., and Friesen, H. G.,** Pituitary hormones regulate c-myc and DNA synthesis in lymphoid tissue,*J. Immunol.*, 146,2201, 1991.

144. **Russell, D. H., and Larson, D. F.,** Prolactin-induced polyamine biosynthesis in spleen and thymus. Specific inhibition by cyclosporine, *Immunopharm.*, 9,165, 1985.

145. **Clevenger, C. V., Sillman, A.. L., Hanley-Hyde, J.,** and Prystowsky, M. B., Requirement for prolactin during cell cycle regulated gene expression in cloned T-lymphocytes, *Endocrinology*, 130,3216, 1992.

146. **Matera, L., Cesano, A., Bellone, G., and Oberholtzer, E.,** Modulatory effect of prolactin on the resting and mitogen-induced activity of T, B, and NK lymphocytes, *Int. J. Immunopharm.*, 14,1235, 1992.

147. **Clevenger, C. V., Altmann, S. W., and Prystowsky, M. B.,** Requirement of nuclear prolactin for interleukin-2-stimulated proliferation of T lymphocytes, *Science*, 253,77, 1991.

148. **Bernton, E., Bryant, H., Holaday, J., and Dave, J.,** Prolactin and prolactin secretagogues reverse immunosuppression in mice treated with cysteamine, glucocorticoids, or cyclosporin-A, *Brain Behav. Immun.*, 6,394, 1992.

149. **Pernow, B.,** Substance P., *Pharmacol. Rev.*, 35,85, 1983.

150. **Gerard, N. P., Garraway, L. A., Eddy, R. .L.J., Shows, T. B., Iijima, H., Paquet, J.-L., and Gerard, C.,** Human substance P receptor (NK-1), organization of the gene, chromosome localization and functional expression of cDNA clones, *Biochem.*, 30,10640, 1991.

151. **Lotz, M., Vaughan, J. H., and Carson, D. A.,** Effect of neuropeptides on production of inflammatory cytokines by human monocytes, *Science*, 241,1218, 1988.

152. **Ansel, J. C., Brown, J. R., Payan, D. G., and Brown, M. A.,** Substance P selectively activates TNF-a gene expression in murine mast cells, *J. Immunol.*, 150,4478, 1993.

153. **Calvo, C.-F., Chavanel, G., and Senik, A.,** Substance P enhances IL-2 expression in activated human T cells, *J. Immunol.*, 148,3498, 1992.

154. **Calvo, C.-F.,** Substance P stabilizes interleukin-2 mRNA in activated Jurkat cells, *J. Neuroimmunol.*, 51,85, 1994.

155. **Stanisz, A. M., Befus, D., and Bienenstock, J.,** Differential effects of vasoactive intestinal peptide, substance P and somatostatin on immunoglobulin synthesis and proliferation by lymphocytes from Peyer's patches, mesenteric lymph nodes and spleen, *J. Immunol.*, 136,152, 1986.

156. **Pascual, D. W., Xu-Amano, J., Kiyono, H., McGhee, J. R., and Bost, K. L.,** Substance P acts directly upon cloned B lymphoma cells to enhance IgA and IgM production, *J. Immunol.*, 146,2130, 1991.

157. **Hartung, H.-P., Wolters, K., and Toyka, K. V.,** Substance P, Binding properties and studies on cellular responses in guinea pig macrophages, *J. Immunol.*, 136,3856, 1986.
158. **Nio, D. A., Moylan, R. N., and Roche, J. K.,** Modulation of T lymphocyte function by neuropeptides, *J. Immunol.*, 150,5281, 1993.
159. **Blum, A. M., Metwali, A., Cook, G., Mathew, R. C., Elliott, D., and Weinstock, J. V.,** Substance P modulates antigen-induced, IFN-gamma production in murine *Schistosomiasis mansoni, J. Immunol.*, 151,225, 1993.
160. **Cook, G. A., Blum, A. M., Ballas, Z., and Weinstock, J. V.,** Substance P does not alter interleukin-1 expression by splenic or granuloma macrophages in murine schistosomiasis, *J. Neuroimmunol.*, 33,217, 1991.
161. **Ottaway, C. A., Bernaerts, C., Chan, B., and Greenberg, G. R.,** Specific binding of VIP to human circulating mononuclear cells, *Can. J. Physiol. Pharm.*, 61,664, 1983.
162. **Rola-Pleszcynski, M., Bolduc, D., and St.Pierre, S.,** The effects of vasoactive intestinal peptide on human natural killer cell function, *J. Immunol.*, 135,2569, 1985.
163. **Moore, T. C.,** Modification of lymphocyte traffic by vasoactive neurotransmitter substances, *Immunology*, 52,511, 1984.
164. **O'Dorisio, M. S.,** Biochemical characteristics of receptors for vasoactive intestinal polypeptide in nervous, endocrine, and immune systems, *Fed. Proc.*, 46,192, 1987.
165. **Metwali, A., Blum, A., Mathew, R., Sandor, M., Lynch, R. G., and Weinstock, J. V.,** Modulation of T lymphocyte proliferation in mice infected with Schistosoma mansoni, VIP suppresses mitogen- and antigen-induced T cell proliferation possibly by inhibiting IL-2 production, *Cell. Immunol.*, 149,11, 1993.
166. **Mathew, R. C., Cook, G. A., Blum, A. M., Metwali, A., Felman, R., and Weinstock, J. V.,** Vasoactive intestinal peptide stimulates T lymphocytes to release IL-5 in murine schistosomiasis mansoni infection, *J. Immunol.*, 148,3572, 1992.
167. **Payan, D. G., Brewster, D. R., and Goetzl, E. J.,** Specific stimulation of human T lymphocytes by substance P, *J. Immunol.*, 13,1613, 1983.
168. **Muscettola, M., and Grasso, G.,** Somatostatin and vasoactive intestinal peptide reduce interferon-gamma production by human peripheral blood mononuclear cells, *Immunobiol.*, 180,419, 1990.
169. **Wagner, M., Hengst, K., Zierden, E., and Gerlach, U.,** Investigations of the antiproliferative effect of somatostatin in man and rats, *Met. Clin. Exp.*, 27,1381, 1979.
170. **Hinterberger, W., Cerny, C., Kinast, M., Pointer, H., and Trag, K. M.,** Somatostatin reduces the release of colony-stimulating activity (CSA) from PHA-activated mouse spleen lymphocytes, *Experientia*, 34,860, 1977.
171. **Weinstock, J. V., Blum, A. M., and Malloy, T.,** Macrophages within the granulomas of murine Schistosoma mansoni are a source of a somatostatin 1-14-like molecule, *Cell. Immunol.*, 131,381, 1990.
172. **Blum, A. M., Metwali, A., Mathew, R. C., Cook, G., Elliott, D., and Weinstock, J. V.,** Granuloma T lymphocytes in murine schistosomiasis mansoni have somatostatin receptors and respond to somatostatin with decreased IFN-gamma secretion, *J. Immunol.*, 149,3621, 1992.
173. **Weinstock, J. V.,** Neuropeptides and the regulation of granulomatous inflammation, *Clin. Immunol. Immunopath.*, 64,17, 1992.

174.   **Sirinek, L. P., and O'Dorisio, M. S.,** Modulation of immune function by intestinal neuropeptides, *Acta Oncologica*, 30,509, 1991.

175.   **Johnson, H. M., Farrar, W. L., and Torres, B. A.,** Vasopressin replacement of interleukin-2 requirement in gamma interferon production, Lymphokine activity of a neuroendocrine hormone, *J. Immunol.*, 129,963, 1982.

176.   **Heijnen, C. J., Kavelaars, A., and Ballieux, R. E.,** Corticotropin-releasing hormone and proopiomelanocortin-derived peptides in the modulation of the immune function, in *Psychoneuroimmunology, 2nd ed.*, Ader, R., Felten, D. L., Cohen, N., Eds., Academic Press, San Diego, 1991, pp 429.

177.   **Jain, R., Zwickler, D., Hollander, C. S., Brand, H., Saperstein, A., Hutchinson, B., Brown, C., and Audhya, T.,** Corticotropin-releasing factor modulates the immune response to stress in the rat, *Endocrinology*, 128,1329, 1991.

178.   **Lipton, J. M.,** Neuropeptide α-melanocyte-stimulating hormone in control of fever, the acute phase response, and inflammation, in *Neuroimmune Networks, Physiology and Diseases*, Alan R. Liss, Inc., 1989, pp. 243.

179.   **Smith, E. M., Hughes, T. K., Cadet, P., and Stefano, G. B.,** Corticotropin-releasing factor-induced immunosuppression in human and invertebrate immunocytes, *Cell. Mol. Neurobiol.*, 12,473, 1992.

180.   **Smith, E. M., Hughes, Jr., T. K., Hashemi, F., and Stefano, G. B.,** Immunosuppressive effects of corticotropin and melanotropin and their possible significance in human immunodeficiency virus infection, *Proc. Natl. Acad. Sci., USA*, 89,782, 1992.

181.   **McGillis, J. P., Humphreys, S., Rangnekar, V., and Ciallella, J.,** Modulation of B lymphocyte differentiation by calcitonin gene-related peptide (CGRP). I. Characterization of high-affinity CGRP receptors on murine 70Z/3 cells, *Cell. Immunol.*, 150,391, 1993.

182.   **Foster, C. A., Mandak, B., Kromer, E., and Rot, A.,** Calcitonin gene-related peptide is chemotactic for human T lymphocytes, *Ann. NY Acad. Sci.*, 657,397, 1992.

Chapter 4

THE PITUITARY GLAND, PSYCHONEUROIMMUNOLOGY AND INFECTION

Istvan Berczi
Department of Immunology,
University of Manitoba
Winnipeg, MB, Canada

Andor Szentivanyi
Departments of Internal Medicine,
Neurology and Pharmacology and Therapeutics,
University of South Florida
Tampa, FL

INTRODUCTION

Observations of glucosuria during infections were already made at the end of the 19th century, which was followed by the demonstration of hyperglycemia and insulin resistance in infectious disease. Other studies revealed that fever is due to endogenous pyrogenic substances and that a leukocyte endogenous mediator (LEM)* induces the synthesis of a number of proteins called acute phase reactants in the liver during severe infection. It has also been found that the endogenous pyrogen is capable of inducing ACTH release. By 1975 a substantial amount of information was available with regards to the metabolic response to infection and the accompanying hormonal alterations that included glucocorticoids (GC), mineralocorticoids, thyroid hormones, insulin (INS), glucagon (GLN), and growth hormone (GH).[7,51,127]

Hans Selye described first the general reaction to injury, which led to a profound involution of the thymus, spleen and lymph nodes, and the enlargement of the adrenal gland.[97] Selye also clarified that these changes were mediated by the activation of the pituitary-adrenal axis and that glucocorticoids were responsible for the lymphoid involution.[98] The physical, chemical, or emotional stimuli that could evoke this neuroendocrine response was termed by Selye as stress and the body's reaction to stress as the general adaptation syndrome. During stress the initial alarm reaction is followed by a period of adaptation when the organism shows resistance towards the stressor and the endocrine and other parameters return to normal. Eventually, with lasting stress, breakdown due to exhaustion may occur, which could lead to death.[100,102] Selye[101] (1949) described first the endocrine regulation of inflammation and

---

*A list of abbreviations used is given at the end of this chapter.

observed with his colleagues that immune reactions are also subject to stress induced alterations.[65] He also demonstrated the influence of sex hormones on lymphoid tissue.[99]

Recent advances in our understanding of the interaction of neuroendocrine and immune systems with each other, and with various organs and tissues in the body, make it possible to give a general outline of the systemic response to infection. Even though it seems clear that cytokines, hormones and neurotransmitters are the chief mediators of the pathophysiological response to infection, there are many gaps in our knowledge of this process at the present time.

THE SYSTEMIC RESPONSE TO INFECTION
AND TO ENDOTOXIN
A. The Cytokine Response to Infection

Infections are frequent in immunocompromised patients, in the elderly, in patients on immunosuppressive therapy, cancer patients and those suffering from severe trauma or shock. The specific immune response is designed to control infections locally through which the spread of pathogenic microbes is prevented while systemic immunity is established. However, when this sophisticated and highly efficient mechanism fails, viremia, bacteremia, or parasitemia will develop, which invariably triggers an acute phase reaction (APR) with the elevation of tumor necrosis factor (TNF), interleukin-1 (IL-1), IL-6, and a number of other cytokine hormones in the circulation, which in turn induce the neuroendocrine and metabolic changes characteristic of APR.[24,72,94,129]

There is much information in the literature with regards to the cytokine response to gram-negative bacteria and to lipopolysaccharide (LPS) endotoxin, which is released from these bacteria. The cytokine response to gram-positive bacteria, to viruses, and even to protozoa (malaria) has virtually identical characteristics. Pathogenic microbes stimulate the immune system through their antigenic determinants and also by nonspecific means (Table 1).

In order to survive the enormous destructive power of host immunity, pathogenic organisms must resort to various escape mechanisms. This may involve the manipulation of immune host defense by altering cytokine production and the endocrine and metabolic responses. For instance, many pathogens are capable of polyclonal lymphocyte activation,[4] which results in the nonspecific stimulation of excess cytokines, such as TNF, IL-l, IL-2, which are neurotoxic, and also elicit a powerful ACTH glucocorticoid response leading to immunosuppression. Moreover, infection with lymphocytic choriomeningitis virus leads to GH deficiency in some mouse strains which is due to the impairment of GH secreting pituitary cells by the virus.[115] Trypanosomiasis in man often exhibits the infiltration of the central nervous system, the pituitary, thyroid, adrenals and gonads by mononuclear inflammatory cells.[93] In the case of viruses it is almost obligatory to manipulate cytokines and other regulators of cell growth in the interest of viral replication. Thus TNFα, IL-1, IL-1β, -2, -6, -8, 13, and GM-CSF are all implicated in promoting HIV replication, whereas IFN-γ and TGFβ are inhibitory.[32] It is of interest that glycoprotein 120 of HIV stimulates IL-1 production in the brain, but not in the periphery, which activates the ACTH-adrenal axis and the glucocorticoids so induced cause immunosuppression.[110]

TABLE 1
Cytokines and Other Mediators Released During Infection

Source	Activator/Stimulus		Cytokines/Mediators released
	Specific	Nonspecific	
Monocyte-macrophage	Ag*-Ab complex (IgG1,-3)	C3a, C5a, CRP, IL-4,-6, IFN, LPS, oxidative insults, TGFβ, TNF	angiotensin, FGF, G-CSF, GM-CSF, M-CSF, IP-10, IFN, IL-1,-6,-8, LT, MCAF, PDGF, PGE₂, PAF, sterol hormones, TGFβ, TBXA₂, TNFα, enzymes, coagulation factors, fibronectin
Tissue mast cells and basophilic leukocytes	Ag-IgE complex	β-END, bradykinin, C3a, C5a, CGRP, Cold, dynorphin, GM-CSF, IL-3, L-ENK, LPS, neurotensin, oxidative insults, SOM, SP, VIP	ECF, GM-CSF, heparin, HIS, IL-3, LTC₄, NCF, PAF, PGD₂, -D₄, -E₄, SER, TNF, CATG, proteases
Neutrophilic granulocytes	Ag-Ab complex	C3a, C5a, CRP, formyl peptide, IL-8, kallikrein, LTB₄, PAF, SP, TNF	LTB₄, MCF, PAF, PGE₂, TGFβ, cathepsin G, DEF, HIS, PKC1, plasminogen activator, enzymes, oxygen metabolites
Platelets	Ag-IgG complex	ADP, bacteria, collagen, EP, NEP, fibrinogen, LPS, PAF, SER, thrombin, TBXA₂, tumor cells, vasopressin, viruses	β-TBG, IL-8, PAF, PDGF, PF-4, PG, SER, TGFβ, TBX(A₂), chemotactic and mitogenic factors, ATP, ADP, GTP, GDP, Ca, factor V, -VIII, fibrinogen, fibronectin, thrombin, enzymes, proteins, von Willebrand factor
Endothelial cells	-	CRP, G-CSF, GM-CSF, IFN-γ, IL-1, injury, LPS, thrombin, TNF	endothelin, GM-CSF, IL-1, -8, IP-10, PDGF, PGE₂, PG12
T lymphocytes	Antigen	EP, IFN-γ, IL-1, IL-2, lectins, adhesion molecules, microorganisms	GM-CSF, IFN-γ, IL-2, -3, -4, -5, -6, -10, TGFβ, TNFα, -β
B lymphocytes	Antigen	IL-4, -5, -6, -10, LPS, lectins, adhesion molecules, microorganisms	IgA, IgE, IgG, IgM, TNFα, IL-10

B. The Cytokine Response to Endotoxin

LPS is capable of inducing lethal shock in higher vertebrate animals, but is without ill effects in fish or frog. There is considerable difference in sensitivity to LPS even among various species of mammals.[15] When administered parenterally, LPS is capable of inducing pathological changes which are characteristic of gram-negative bacterial infections, whereas LPS administered orally is ineffective.[13,16,84] Normally bile acids present in the gastrointestinal tract detoxify endotoxin and thus oral intoxication is impossible unless bile is eliminated from the gut.[19]

LPS activates B lymphocytes polyclonally and induces proliferation and immunoglobulin secretion. Some of the antibodies secreted will combine with autoantigens indicating that clones with autoimmune potential can be turned on by LPS. In monocyte/macrophages LPS is a powerful stimulator of TNFα, IL-1, IL-6, and other biologically active molecules. LPS is capable of triggering mediator release from mast cells and basophils from endothelial cells, and it also affects neutrophilic granulocytes (Table 1). LPS activates the complement and coagulation cascades and has an effect on virtually every tissue and organ in the body.[30,82]

A lipopolysaccharide binding protein (LBP) was isolated from human and rabbit serum. LBP is a 60 kd glycoprotein and normally is present at 0.5 µg/ml concentration in the serum, but it will rise up to 50 µg/ml, during an acute phase response. LBP is capable of opsonizing of LPS bearing particles which suggests that this protein is necessary for the activation of the complement system by LPS. LBP-LPS complexes were as much as 1,000 fold more active than LPS alone in the induction of TNF or IL-1β. LBP was found to be related to another LPS binding protein of neutrophils called bacterial permeability increasing protein. A 55 kd glycoprotein called CD14 which serves as the receptor for LPS-LBP on monocytes and macrophages. CD14 does not contain a transmembrane domain but holds onto the cell surface by a phosphatidylinositol glycane anchor.[91] Other LPS binding proteins have also been identified on various cell types. B lymphocytes do not possess CD14, yet they respond to stimulation by LPS with proliferation and polyclonal immunoglobulin secretion. LPS regulates transcription of IC gene mRNA by the stimulation of the cis-acting nuclear factor NF-KB or OTF-2.[91] B lymphocytes of C3H/HeJ mice cannot be activated by LPS for proliferation and polyclonal immunoglobulin secretion. Such mice respond poorly to infection with gram-negative bacteria.[72]

A brief overview of the cytokine response to LPS is given below:

(a) Tumor necrosis factor (TNFα)

The i.v. administration of LPS to rats, mice or to man induces a sharp increase of TNFα in the blood which peaks 1-2 hr after injection and returns to normal by about 4 hr. The blood level of TNF rises over 50 times higher in response to LPS if the animals are adrenalectomized (ADX) with no alteration in the kinetics of the response. Pregnancy, growth hormone, ACTH, and insulin are known to increase the TNF response to LPS, whereas glucocorticoids, transforming growth factor beta (TGF-β), have a suppressive effect. This suppression can be antagonized by interferon gamma (IFNγ). IFNγ also stimulates the expression of TNF receptors on various cells, whereas epinephrine (EP), INS, GLN, somatostatin (SOM), ACTH, and angiotensin (AGT) had an inhibitory effect on TNF receptors.

Normal animals will not respond with similar elevation of serum TNF to a second injection of LPS, which may be given as soon as 2 hr after the first injection. This phenomenon is known as endotoxin tolerance. A similar condition can be induced by the continuous infusion to normal animals of LPS at low concentrations. Tolerant animals show an impaired production of cytokines other than TNF. ADX animals or animals intoxicated with galactosamine, which severely impairs liver function, are unable to develop endotoxin tolerance. In contrast with the rapid rise and decline of TNF after LPS injection, the blood level during experimental bacterial peritonitis is 50 to 100-fold lower and remains constant until the death of the animals. Such animals can be protected by anti-LPS, but not by anti TNF antibodies. TNF is capable of inducing its own mRNA and that of IL-1.[10,49,52,92,133]

(b) Interleukin-1

Monocyte macrophages and related cell types and even endothelial cells respond to stimulation with LPS by IL-1 production. Glucocorticoids, progesterone (PGS) and estradiol (E2) have an inhibitory effect on monocyte-derived, but not on endothelial, TNF. IL-1β mRNA was increased in the pituitary gland by treatment of rats with LPS.[10]

(c) Monocyte macrophages

They also release interleukin-6 after stimulation with LPS which is not inhibited significantly by hydrocortisone and stimulated by ACTH and INS. LPS stimulated the release of IL-6 from rat medial basal hypothalamus and stimulated IL-6 mRNA in both the hypothalamus and pituitary tissues.[10]

(d) Colony stimulating factors (CSF)

A significant elevation of CSF was detected in mice 45 min after LPS injection which rose further after 2 hr.[90]

(e) Other mediators

LPS stimulated the release of prostaglandin E2 from microvessels, of platelet activating factor (PAF), particularly in the lung, and induced the production of nitric oxide (NO) synthase in macrophages, vascular smooth muscle cells, fibroblasts, hepatocytes, in Kupffer cells, keratinocytes and megakaryocytes *in vitro*.[10,112]

C. The Neuroendocrine Response to Infection

The endocrine response to infection and to endotoxin is summarized in Table 2. The increase of glucocorticoids during infection occurs shortly before the onset of fever. The circadian rhythm is lost. If the infection becomes subacute or chronic, glucocorticoid levels generally fall to values below normal.[7]

Acute bacteremia in sheep caused a surge of plasma, β-endorphin (β-END) which was followed by an increase in plasma prolactin (PRL) and GH and a depression of plasma LH.[67] In mice infected with *Pseudomonas aeruginosa* resulting in severe infection, the fall of serum tri-iodiothyronine (T3) concentrations present in normal animals after fasting could not be induced.[29]

In 12 patients with sepsis, hormone levels were measured every 6 hours. Growth hormone levels were normal; however, the nocturnal GH surge was lost. Insulin was normal and IGF-I was decreased. Cortisol values were in the upper normal range.[121] Basal insulin levels were suppressed in 5 septic patients, whereas glucagon was elevated. These patients did not respond to treatment with human GH with elevated serum levels of IGF-I.[37]

Leprosy patients could be divided into two groups on the basis of LH response to LHRH: (i) those exhibiting an exaggerated response, (ii) showing a normal response. Basal PRL levels were normal in both groups, but in group (i) there was an increased PRL response to stimulation with either TRH or metoclopramide, whereas group (ii) patients showed a PRL response identical to controls. Both groups had increased TSH response to TRH in the presence of normal basal T4 and T3 levels. The PRL response to TRH correlated with basal and peak follicle stimulating hormone (FSH) responses to LHRH but not with LH, E2 or testosterone. The TSH response did not correlate with either gonadotropins, E2 or thyroid hormone levels.[103]

Table 2
Neuroendocrine Changes During Infection, Endotoxicosis
and the Acute Phase Response

Hormone	Infection[a]	Endotoxin[a]	APR[b]
adrenocorticotropic hormone		↑	↑
glucocorticoids	↑	↑	↑
growth hormone	↑↓	↑	↑↓
prolactin	↑	↑↓	
luteinizing hormone	↓		
estrogen			↑↓
androgen			↑↓
thyroxine		↓	↑↓
triiodothyronine	0↑		↑↓
insulin	↑	↑	↑↓
glucagon	↑	↑	↑
α-melanocyte stimulating hormone		↑	
endorphin, enkephalin	↑	↑	
epinephrine	↑	↑	↑
norepinephrine	↑	↑	↑
dopamine		↑	
arginine-vasopressin		↑	↑
aldosterone		↑	

[a]Please see text for references. [b]References.[1,14]
↑ = increased serum/plasma level; ↓ = decreased level; ↑↓ = level may be either increased or decreased; 0 = no change.
APR = acute phase response
(Table taken from **Berczi, I.,**[10] reproduced with permission of the publisher)

Sixty-six AIDS patients, of which 14 showed the wasting syndrome, were evaluated for thyroid, gonadal, and adrenal function. Total and free testosterone were significantly lower in patients with wasting than in those without wasting. Prolactin values were significantly higher and cortisol levels were also higher with borderline significance in wasting patients compared to non-wasting ones. Prolactin and cortisol levels were inversely correlated with weight loss.[35]

Patients suffering from African trypanosomiasis showed a significant impairment of thyroid function with elevated TSH concentration and low free T3 and T4 levels.[93]

D. The Neuroendocrine Response to Endotoxin

The release of ACTH in intact rats after injection of LPS was first observed by Wexler and coworkers in 1957. More recently, IL-1 has been shown to play a major role in LPS-induced ACTH release. Sub-pyrogenic doses of LPS given to rats i.p. elevated ACTH and cortisone concentrations 2 hr later, whereas plasma PRL, LH, EP, and NEP levels were not altered significantly. A small elevation of NEP was noted at 6 hr. The ACTH and cortisone response to sub-pyrogenic, but not to pyrogenic doses of LPS, could be prevented by macrophage depletion. Unknown macrophage products also participated in the immunoregulatory feedback circuit to the hypothalamic-pituitary-adrenal axis.[40,41] Corticotropin releasing factor (CRF) plays a major role in IL-1 mediated ACTH release. However, the finding that in rats with lesioned paraventricular nuclei, which are the source of CRF, an ACTH response could still be elicited by LPS indicates that other factors are also involved.[47] LPS and histamine, but not stress, induced a significant increase in the serum corticosterone level of hypophysectomized (Hypox) rats. Neither histamine nor LPS had an appreciable effect on corticosterone release by cultured adrenal cells.[111]

Circulating alpha-melanocyte stimulating hormone (α-MSH) was increased significantly in rabbits after LPS injection. A large dose of α-MSH given i.v. to LPS treated animals inhibited fever, but had no effect on the acute phase response.[73]

Mice responded with a 5-fold increase of serum PRL 1 hr after LPS injection, which returned to normal by 90 min, and was not influenced by ADX.[92] Low doses of LPS given to sows 2 days postpartum caused a decrease in PRL concentrations and 5 to 10-fold increase in cortisol levels. A similar response was observed in primaparous gilts if LPS was given on day 2, but PRL concentrations remained unchanged if LPS was given at a later time postpartum. The PRL response to injection of thyroid releasing hormone (TRH) was significantly lower in gilts treated with LPS on day 2 postpartum. Cortisol concentrations increased after endotoxin exposure on postpartum day 2 and 6.[105,106] Human volunteers injected with 4 ng/kg LPS responded with elevated growth hormone and cortisol secretion.[48]

The administration of nonlethal doses (0.01-2 mg/kg) of LPS i.v. to male Wistar rats induced large increases in estrogens (3 to 9-fold), progesterone (4-fold), and corticosterone (2 to 3-fold), and decreased serum testosterone (2-fold) with maximal response at 2 h. Low doses of LPS did not produce an increase in serum estrone and estradiol in adrenalectomized or orchidectomized rats.[34]

A single subtoxic dose of LPS given to newborn rats decreased the level of serum thyroxine and led to an impaired thyroid response to TSH in adulthood, which was associated with somatic retardation. Treatment of adult rats with a shock inducing dose of LPS induced a similar decrease in serum T4, and the inhibition of T4 response to stimulation with exogenous TSH.[20,83]

Endotoxin induced hyperglycemia is coupled with elevated insulin levels. Glucose induced insulin release was potentiated from pancreases obtained 8 hr after LPS or 30 min after IL-1 treatment. Both LPS and IL-1 increased the sensitivity of isolated pancreatic islets to glucose, but neither agent affected the maximum insulin secretory response. LPS induced hyperglycemia did not lead to elevation of plasma insulin in anesthetized rats. Exogenous glucose stimulated insulin release in both conscious and anesthetized animals, which was coupled with elevated levels of EP, NEP and dopamine (DOP). Endotoxin treatment decreased the inhibitory effect of EP on INS levels, which was more pronounced in the fed than in the fasted rat, and was potentiated by pentobarbital anesthesia.[64,130,132] The treatment of cultured hepatocytes from adult male rats with LPS decreased insulin binding and endocytosis in a time- and temperature-dependent manner.[89]

Pretreatment of rats with dexamethasone (DEX) 1 hr before LPS administration prevents hyperinsulinemia and hypoglycemia.[131] In LPS treated dogs the elevation of plasma met-enkephalin (ENK) and beta-endorphin (β-END) preceded the onset of hyperinsulinism, but the elevation of plasma leu-ENK did not. Plasma INS was elevated 100-fold by 6 hr in dogs treated with LPS and glucose. Hyperinsulinism was markedly reduced by naloxone (NAL) treatment. However, when NAL was given in conjunction with glucose, it appeared to have a stimulatory effect on insulin secretion.[77]

Plasma levels of EP and NEP increased in adult conscious rats with both the dose of LPS and with time. NEP levels increased maximally at 30 min following 1 mg/kg LPS and remained elevated for the observation period (6 hr). Plasma EP levels were also maximal at 30 min but declined thereafter while remaining significantly higher than in groups treated with lower doses of LPS (10 and 100 μg/kg) for the duration of the protocol. Peak EP levels given 10 or 100 pg/kg LPS occurred at 3 hr and then decreased by 6 hr.[63]

Dogs given a large bolus of *E. coli* LPS (5 mg/kg) followed by a continuous infusion of 2 mg/kg/h showed a marked increase of EP and NEP in the adrenal vein and in the circulation. Naloxone limited the fall of catecholamines in the adrenal vein following volume replacement which led to a sustained systemic increase in EP levels. There was a hemodynamic response to NAL. Immunoreactive met-ENK levels followed closely the changes in catecholamines in the adrenal vein, whereas arterial levels rose progressively and were unaffected by NAL.[124] In the rat, LPS activated noradrenergic, dopaminergic and epinephrine-containing neurons in the hypothalamus and dopaminergic neurons in other regions of the brain. Sympathetic nervous system activity was increased in the heart, liver, brown adipose tissue and gastrocnemius muscle 26 hr after LPS injection. Prostaglandin synthesis played an important role in the catecholamine response to LPS in all brain regions and also in other tissues examined.[2,54,74]

LPS is capable of decreasing the number of β-adrenergic binding sites on splenic lymphocytes and in lung tissue. The LPS-induced deterioration of β-adrenergic system in the lung could completely be prevented by splenectomy, suggesting that spleen-derived substances are involved in the changes of β-adrenergic reactivity of the airways.[118] Similarly, in dogs 5 hr after the injection of 1 mg/kg LPS i.v., lymphocyte β-adrenergic receptor numbers and NaF-stimulated cAMP accumulation were reduced significantly. The myocardial β-adrenergic receptor number was also reduced in LPS-treated animals.[104]

SEPSIS AND ENDOTOXIN SHOCK

Sepsis and endotoxin shock are characterized by diffuse intravascular coagulation, complement activation by both pathways and the release of immunosuppressive hormones and cytokines. Systemically released C3a and C5a is immunosuppressive and may impair chemotaxis. The activation of neutrophilic granulocytes and other phagocytic cells by complement split products in the circulation may lead to vascular damage, increased permeability with edema, impaired microvascular flow, and ultimately contribute to multiorgan failure. Basophils, mast cells and platelets also release cytokines and biologically active substances under these conditions, which may further aggravate the situation. The production of IL-1, IL-2 and IFN-γ is depressed, whereas PGE2 levels are excessively elevated throughout the course of disease. TNF is detectable with equal frequency in patients with shock from gram-negative or gram-positive bacillary sepsis and its level is highest in the initial stage and decreasing significantly over the subsequent 24 hr.

Normally, the rise in serum glucocorticoid and catecholamine levels is in direct correlation with the severity of the septic condition. Patients with subclinical hypocortisolism are more susceptible to death during sepsis than those with normal adrenal function. In severe sepsis there is a lack of skin reactivity to recall antigens, monocyte/macrophage function is depressed, and chemotaxis, phagocytosis, chemiluminescence and intracellular killing activity of neutrophils are also suppressed. Natural killer and lymphokine activated killer cell functions are depressed. Suppressor T cell function is increased, which causes long term T cell depression and reverses the T helper/ T suppressor ratio.[8,14,25,27,50,59,71,86]

The pathological process during sepsis and shock may start by various pathways and snowball into the septic state, multiple organ failure and death. A correlation could be shown between the intensity of complement activation and the blood levels of elastase with the severity of injury, the development of adult respiratory distress syndrome, multiple organ failure, and fatal outcome. Septic patients with detectable TNF also had a higher incidence and severity of adult respiratory distress syndrome and higher mortality rate.[27,71,86]

Animal experiments showed that TNF, IL-1, IFNγ, platelet activating factor, opioids, prostaglandins, and nitric oxide all contribute to the fatal outcome of endotoxin shock. During infection, the destructive effect of microorganisms in tissues and in blood, the consumption of essential nutrients, and the toxicity of their metabolic byproducts are additional factors that complicate the situation. By the time septic shock develops, usually enough damage has been done to kill the host, and therefore, it is not surprising that treatment with glucocorticoids has no beneficial effect.[14,26,71,108,112,119] However, glucocorticoid therapy could be expected to be beneficial in a minority of patients where glucocorticoid deficiency exists. In patients with normal glucocorticoid response additional high dose glucocorticoid therapy may inflict further damage by inhibition of the rapid growth of the intestinal cells, for instance[79] and by the aggravation of immunosuppression. Nevertheless, pharmacological treatment with cortisol, lidocain and calcium blockers reduced the release of bioactive complement components. Plasmapheresis has been used for the removal of anaphylatoxins from the blood.[59] Other approaches such as the administration of antibodies to LPS or to its toxic moiety, lipid A, the prevention of excess cytokine production and antagonizing the toxic effect of cytokines and other mediators are the subject of current investigations.

THE ACUTE PHASE RESPONSE

Wannemacher *et al.*[123] discovered that a leukocyte endogenous mediator (LEM) stimulated the uptake of amino acids by the liver in ADX, Hypox, thyroidectomized and diabetic rats. This stimulation could not be duplicated by pharmacological doses of a large variety of hormones. LEM induced the production of a number of acute phase plasma globulins without the mediation of other hormones, with the possible exception of corticoids.

The acute phase response (APR) is the reaction of the body to infection and to a number of other injuries. It is characterized by fever, loss of appetite, inactivity, sleepiness, and is mediated by cytokines which appear in the circulation and function as acute phase hormones.

MAJOR PATHOPHYSIOLOGICAL PATHWAYS
IN THE ACUTE PHASE RESPONSE

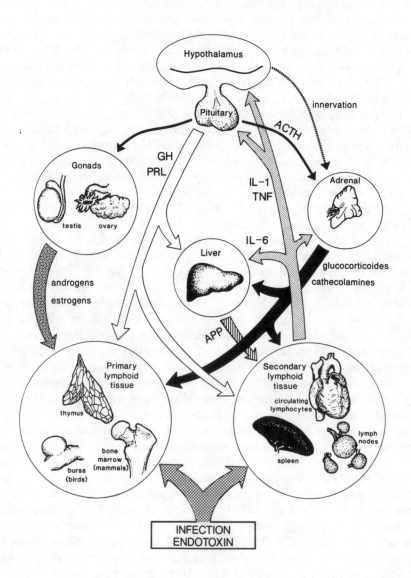

Fig 1.   Major pathophysiological pathways of the acute phase response during infection or endotoxin shock.[10] Systemic infections with bacteria, viruses and parasites induce the appearance/elevation in the circulation of IL-1, IL-6 and TNFα, which are primarily released by monocytes and macrophages. These cytokine hormones activate other leukocytes and additional cytokines and mediators are produced as detailed in the text. They also affect the brain and elicit fever, and induce pituitary hormone release and effect directly other organs such as the adrenals, liver, and muscle. During acute infection there is an early rise in serum GH and PRL, the levels returning to normal in a few hours and may become subnormal in severe sepsis and shock. GH and PRL enhance immune and inflammatory reactions and also increase nonspecific resistance to infection. In severe infections or endotoxemia these cytokine initiated interactions culminate in the acute phase response, which is characterized by catabolism, the elevation of acute phase proteins (APP) in serum, which are produced primarily in the liver at the expense of other serum proteins (e.g. albumin, transferin). Bone marrow function and the metabolic activity of leukocytes are greatly increased in acute phase reactions. The acute phase response may be regarded as an emergency reaction which is initiated in the case of failure by the immune system to control infections locally at the site of tissue penetration. Glucocorticoids and catecholamines released by the adrenals are fundamental to switching the immune system over from specific reactivity to nonspecific defense. The liver and the bone marrow are the major target organs. The production of ancient defense molecules, such as C reactive protein and endotoxin binding protein, is increased enormously. These proteins recognize surface moieties present on a wide variety of pathogenic agents and activate complement and leukocytes after combination with the specific ligand. The production of complement components and leukocytes is also enhanced, which amplifies this mechanism. The elevated fibrinogen levels are likely to serve the isolation of infectious agents by promoting blood clotting at sites of infection. Other acute phase proteins, such as enzyme inhibitors and inhibitors of inflammation, serve the function of damage control in systemic disease. This profound neuroendocrine and metabolic conversion of the host is the result of the systemic action of cytokines produced by the immune system in response to infection, intoxication and other forms of injury. The neuroendocrine system keeps cytokine production under very tight control during the acute phase response, primarily by increased glucocorticoid secretion. Excess cytokine production has fatal consequences due to neurotoxicity. Circulatory collapse may occur due to the excess release of vasoactive substances. Therefore, during the acute phase response specific immune reactivity is suppressed, whereas nonspecific resistance to infection and other noxious agents is increased.

Acute phase hormones affect the central nervous system, the neuroendocrine system, and virtually the function of every other organ and tissue in the body. IL-1, IL-6 and TNF have been identified as major mediators of endocrine and metabolic changes in APR (Figure 1).[1,14,62]

ACTH, glucocorticoids, EP, NEP, glucagon, arginine vasopressin and aldosterone are elevated during the APR, whereas GH, estrogens, androgens, insulin and thyroid hormones may be either elevated or suppressed depending on the severity of the condition (Table 2).[6,14]

The synthesis of new proteins, the so-called acute phase reactants, is most characteristic of APR, while the synthesis of some normal serum constituents, such as albumin and transferrin, is decreased. As a result, the concentration of acute phase serum proteins (APP) is increased dramatically. For example, C-reactive protein (CRP) and serum amyloid A (SAA) may increase over 1,000-fold in man during a severe reaction.

C-reactive protein combines specifically with phosphocholine and possibly also with some polysaccharides containing galactose, and with some biologic polycations, such as protamine, poly-L-lysine, and myelin basic protein. These moieties are frequent cell surface components of bacteria, fungi, parasites and damaged cells. CRP was shown to localize *in vivo* at sites of inflamed and damaged tissues. Once complexed with a specific ligand, CRP acquires the capacity to activate complement by the classical pathway and to activate neutrophils and monocytes for chemotaxis, enhanced phagocytosis, to induce tumoricidal activity in macrophages, and to stimulate the synthesis of IL-1, TNFα, to potentiate the cytotoxic activity of T lymphocytes, natural killer cells and platelets. CRP also binds and blocks the activity of PAF. IL-6, but not other cytokines, induces the synthesis of CRP by human liver cells *in vitro*. However, the production of CRP can be modulated further by IL-1, TNFα, TGF-β1 and IFN-γ. Dexamethasone potentiates the production of CRP and SAA. The clinical determination of CRP has diagnostic value for infectious and inflammatory diseases. Proteins similar to CRP have been identified in all mammals, chicken, fish and even crab.[3]

Lipopolysaccharide binding protein (LBP) also shows 100-fold increase in human and rabbit serum during APR. LBP is capable of opsonizing LPS bearing particles and thus may be required for the activation of complement by endotoxin through the alternate pathway. LBP-LPS complexes are also potent stimulators of cytokines from monocytes and macrophages after combining with CD14 on the surface of these cells.[91]

A number of APP function as proteinase inhibitors, such as α2-macroglobulin, α1-acid glycoprotein, antithrombin III, α1-acute phase globulin, and α1-proteinase inhibitor, which are abundant in the rat. Other APP play a role in blood clotting (fibrinogen) and healing. α-macrofetoprotein (α-MFP) is a strong inhibitor of inflammatory mediators, including histamine, bradykinin, serotonin, PGE2 and inhibit polymorphonuclear chemotaxis. Catecholamines and glucocorticoids induce α-MFP in normal rats. Hypox rats did not show an APR to turpentine as did normals, unless treated with dexamethasone. In ADX rats, the effect of catecholamines on α-MFP synthesis was greatly diminished, whereas the moderate effect of glucocorticoids remained. Other APP, such as haptoglobulin and α1-major acute phase protein, were affected differently by these hormones.[117] Recently cytokines other than IL-6, namely IFN-γ, leukemia inhibitory factor, TGF-β, oncostatin M, and ciliary neurotrophic factor (CNTF), were found to be inducers of acute phase proteins from the liver.[196] IL-6 activates the genes of APP through the DNA binding protein called NF-IL-6. Therefore, NF-IL-6 may be a pleiotrophic mediator of many inducible genes involved in the acute-, immune- and inflammatory responses similarly to another DNA binding protein, NF-KB. Both NF-IL-6 and NF-KB binding sites are present in the inducible genes, such as IL-6, IL-8 and several acute phase genes.[1]

Adrenalin invokes a high level of IL-6 in rats which can be antagonized by propranolol. When IL-6 release is blocked this way, the fast reacting acute phase proteins, α2-macroglobulin and cysteine protease inhibitor, are strongly depressed. Isoprenalin, a β2 receptor agonist, also induces very high levels of IL-6, indicating that β2 adrenoreceptors are involved.[117]

THE IMMUNE-NEUROENDOCRINE REGULATORY CIRCUIT

The first definite indication for the interaction of the central nervous and immune systems was the discovery of the leukocyte derived pyrogen, which later on was identified as interleukin-1. Pyrogens have also been used in endocrinology for the testing of ACTH reserve, which suggests a functional interaction between the immune and endocrine systems.[8] That immune reactions are also capable of inducing an elevation in glucocorticoid levels, which in turn act to terminate the response, was shown first by Besedovsky and coworkers.[21] These investigators and several other laboratories went on to show that IL-1 is the principal messenger to activate the ACTH-adrenal axis during immune reactions, which leads to the elevation of glucocorticoids. Current evidence indicates that a number of cytokines are capable of activating the ACTH-adrenal axis and also influence the release of most other pituitary hormones.[22,75] Inflammatory mediators, such as PAF, bradykinin, histamine, and prostaglandins, are also involved in the regulation of pituitary hormone secretion.[9,11,58] In addition, the immune system is activated nonspecifically during injury which leads to a pattern of cytokine production similar to acute infections and may culminate in endocrine and metabolic responses characteristic of the acute phase response. Moreover, there is some evidence to indicate that the endocrine response to immunization and injury may also be modulated by nerve endings located in the area of reaction/insult.[14] In turn the endocrine alterations influence cytokine production (Table 3).

PITUITARY HORMONES AND HOST DEFENSE IN INFECTION
A. Pituitary Hormones and Immune Function

Members of the growth and lactogenic hormone family, which include growth hormone, prolactin and placental lactogens, are fundamental to the development and normal function of the immune system. Bone marrow and thymus function and immunocompetence are all maintained by pituitary GH and PRL during postnatal life. These hormones are required for humoral, cell mediated, and autoimmune reactions, and promote the function of phagocytic cells and the nonspecific inflammatory process. On the basis of these facts, it has been postulated that GH and PRL are the hormones of immunocompetence.[12]

The ACTH-adrenal axis functions as the immunosuppressive pathway in hormonal immunoregulation, which is capable of antagonizing the immunostimulatory and proinflammatory effect of GH and PRL. The helper, suppressor and killer function of T lymphocytes and the production of interleukins by them are inhibited by glucocorticoids. This inhibitory effect can be abrogated by IL-2 and by insulin. IL-2 can also protect murine cytotoxic lymphocytes from glucocorticoid induced DNA fragmentation. The early stages of B lymphocyte activation are also suppressed by glucocorticoids, but the proliferative response of activated B cells does not seem to be affected. In glucocorticoid treated animals the number of B cells are decreased in the spleen and lymph nodes, but not in the bone marrow. Glucocorticoids enhance immuno-globulin production by human B cells *in vitro*, which was attributed to the selective inhibition of suppressor T cell function in the system. Glucocorticoids potentiated the synergistic

Table 3
The effect of hormones on cytokine secretion

Hormone	Cytokines				
	IL-1	TNFα	IL-6	IFNγ	IL-2
ACTH		↑	↑		
GC	↓	↓	0↓	↓	↓
GH	↑	↑			
PRL				↑	
β-END	↑			↑↓	
ENK	↑			↑	↑
ES	↑		↓	↓	
PGS	↑		↓		
INS				↑	↑
CAT		↑↓	↑		

↑ = increased serum/plasma level; ↓ = decreased level; ↑↓ = level may be either increased or decreased; 0 = no change.
ENK = enkephalin; CAT = catecholamines.
GC also inhibits platelet activating factor, IL-8, granulocyte-macrophage colony stimulating factor and chemotactic-proinflammatory factor secretion, has no effect on macrophage colony stimulating factor and stimulates the production of IL-4. Please see the details and relevant references for this table in the text. Table taken from **Berczi, I. and Nagy, E.,**[14] reproduced with permission of the publisher)

stimulatory effect of IL-1 and IL-6 on immunoglobulin production by activated B cells. Thymic hormones, IL-1, IL-2 and interferons antagonize the inhibitory effects of glucocorticoids on lymphoid tissue.[8,11]

Glucocorticoids inhibit natural killer (NK) cells, which can be abrogated by IFN-β. Killer cells mediating antibody dependent cellular cytotoxicity are also inhibited by glucocorticoids which is potentiated by PGE2 and abrogated by IFN or IL-2. Lymphokine activated killer cells destroy target cells similarly to NK cells and are suppressed by cortisone. Glucocorticoid treatment reduces the number of mast cells in both humans and animals. Immunologically triggered mediator release from mast cells and basophilic leukocytes is also inhibited. Virtually all functions of the monocyte macrophage lineage, including cell metabolism, chemotaxis, phagocytosis, cytotoxic reactions, the capacity to present antigen, to secrete cytokines and to produce enzymes and other factors, are inhibited by glucocorticoids. On the other hand, glucocorticoids increase the expression of HLA-DR, antigens and receptors for IFN-γ and FC-γ receptors. In murine mast cells dexamethasone downregulated FC-ε receptors and the

IgE dependent release of leukotrienes, whereas the release of PGD2 was increased significantly. The chemotactic response of human eosinophils was also inhibited by glucocorticoids. In neutrophils FC receptors, complement receptors, and chemotactic receptors were rendered nonfunctional, and β-adrenergic receptors were uncoupled from adenylate cyclase by glucocorticoids. The systemic administration of glucocorticoids is immunosuppressive in a variety of species, including humans, mice, rats, guinea pigs, rabbits, chickens, lizards and frogs. Immunoglobulin levels and humoral and cell mediated immunity are depressed by chronic treatment. Immunological memory and the cells responsible for the induction of graft-versus-host reaction are not affected under these conditions.[8,11]

Recent studies showed that ACTH itself has a direct effect on lymphoid cell proliferation and that a-MSH antagonizes several effects of IL-1, including pyrogenecity, thymocyte proliferation, neutrophilia, the induction of acute phase proteins, depression of TNF, and contact sensitivity and induction of PGE in fibroblasts. Although it is clear that immune function is influenced by endogenous opioids, so far the findings are controversial and the physiological significance of this regulatory mechanism is uncertain. Antibody production, NK and lymphokine activated killer (LAK) activity, cytotoxic T cell activity, the mixed lymphocyte reaction, IL-1, IL-2, prostaglandin and interferon synthesis, the regulation of mast cells, and chemotaxis of neutrophilic leukocytes are all affected by opioid peptides. Apparently, opioids are capable of immunomodulation by acting directly on lymphoid cells or indirectly through the brain.[11]

The pituitary-gonadal axis also affects immune reactions. A direct effect of gonadotrophins on immune phenomena has been demonstrated. However, the significance of these findings is uncertain at the present time. The thymus and the bursa of Fabricius (the maturation site of B lymphocytes in birds) express receptors for estrogens and androgens. Mature lymphocytes do not seem to express sex hormone receptors, with the exception of the CD8+ T cell that has receptors for estradiol. Progesterone seems to affect immune function through glucocorticoid receptors, though the existence of specific receptors for this hormone has also been proposed in lymphoid tissue. Estrogens and androgens also have the capacity to combine with glucocorticoid receptors of lymphocytes and at high concentrations may affect their function.[8,11]

Estradiol suppresses bone marrow function and causes thymic involution, and inhibits various T lymphocyte functions that include regulatory (helper and suppressor) and effector (killer and delayed hypersensitivity) T cell mechanisms. NK cells, neutrophils and mast cell degranulation are also inhibited by E2. However, phagocytosis, humoral immune reactions, and certain autoimmune diseases are enhanced in laboratory animals by this hormone.[11]

The development of the bursa of Fabricius is prevented in chicken embryos by testosterone treatment. In general, androgens have a suppressive or moderating effect on immune reactions, and antagonize the enhancing effect of estrogens in a variety of animal autoimmune disease models. The hemopoietic effect of androgens is well known. Testosterone promotes IgE secretion and the synthesis of secretory component in the lacrimal gland. E2, progesterone and PRL stimulate the mammary gland for lactation, which leads to the migration of lymphoid cells to mammary tissue and the synthesis and secretion of IgA in mice. This process can be antagonized by testosterone. Estradiol stimulates the secretion of immunoglobulin in the uterus. Progesterone inhibits this function of E2, but promotes the cervical and vaginal secretion of IgA, in the latter case synergistically with E2.[11]

The pituitary-thyroid axis also affects immune phenomena, but again the exact role of thyroid hormones in immune function remains to be clarified. The most consistent finding is that thyroid hormone deficiency leads to immunodeficiency, which can be restored by treatment with triiodothyronine. However, the findings with regards to the effect of thyroid hormones on lymphoproliferation and immune function are highly controversial.[8,11]

Arginine vasopressin (AVP) attenuated fever after central administration and stimulated the production of PGE2 by human mononuclear phagocytes, enhanced the proliferation of lymphocytes in autologous mixed cultures and the production of β-endorphin by human peripheral blood mononuclear cells. A newly isolated pituitary hormone, called suppressin, inhibited mitogen-induced and IL-2-dependent lymphocyte proliferation. Another pituitary factor isolated from bovine glands induced thymic epithelial cell proliferation *in vitro* .[11]

B. Host Defense Against Infection

In the majority of cases, infections are controlled locally by inflammatory and immune responses. Infectious agents often cause tissue injury by their metabolic effect, the production of toxins and the destruction of normal tissue structure. Successful pathogens also have ways of avoiding destruction by immune mechanisms. A detailed discussion of these aspects of infectious disease is beyond the scope of this chapter. Local irritation or tissue damage evokes inflammation, which involves the nonspecific activation of the immune system by the release of cytokines and other biologically active mediators from tissue mast cells, endothelial cells, and platelets. Sensory nerves may participate in, or even initiate, the local inflammatory response by neuropeptide (e.g. substance P) mediated mast cell discharge. A whole host of chemotactic and proinflammatory cytokines are released under these conditions (Table 1), which in turn attract phagocytic cells and other leukocytes from the circulation and thus the inflammatory process is initiated.[14]

A wide range of cells, that include macrophages, monocytes, T cells, B cells, fibroblasts, endothelial cells, keratinocytes, platelets, neutrophils, hepatocytes, chondrocytes, synovial cells, and smooth muscle cells, have been shown to produce various members of the chemotactic and proinflammatory cytokine family.[87] So far, interleukin-8, TGF-β, IL-4 and 1,25(0H)2 vitamin D3, which may be locally produced/activated, have been identified as inhibitors of chemotactic and proinflammatory cytokines. The type of insult and the nature of the responding tissue may lead to differences in the cytokine/inflammatory response. Bacterial infections often activate the complement system nonspecifically (e.g. with the aid of CRP or LBP) and the complement split products, C3a and CSa, will in turn stimulate chemotaxis and phagocytosis nonspecifically and release chemotactic and inflammatory cytokines from various cells at the site of infection. Some of the antigens released from bacteria during this initial response will be presented to lymphocytes and immune responses will be generated. This involves antigen presenting cells interacting initially with helper T lymphocytes. In turn helper T cells promote the maturation of antigen sensitive T cell subsets (which may belong to delayed type hypersensitive, killer, suppressor, or helper lineages), and B cells, which initially produce IgM and switch to other classes during maturation. Type 1 helper T cells promote cell mediated immunity, and interleukin-1, -2 and interferon-γ are the major cytokines regulating this reaction. Type 2 helper T cells induce humoral immunity and IL-1, -4, -5, -6, -10 are involved.[81]

Antibodies are capable of lysing bacteria by activating the complement cascade or opsonizing them which amplifies their engulfment and destruction by phagocytes, which are stimulated through their Fc and complement receptors. Neutrophilic leukocytes and monocyte/macrophages play a fundamental role in host defense against bacteria. Certain bacteria that live intracellularly, such as the tubercle bacillus, leprosy bacillus, and *Brucella*, preferentially stimulate cell mediated immunity, which protects the host, whereas antibody formation is weak and irregular and contributes little to host defense. Cell mediated immunity is important for host defense against viruses as the infected cells have to be destroyed in order to eliminate the virus. However, antibodies are also needed to neutralize virus particles in the circulation. Delayed type hypersensitivity reactions provide a defense against fungal infections. IgE antibodies, mast cells, basophilic leukocytes, and eosinophilic leukocytes participate in host defense against parasitic infestation.[95,114] Natural antibodies and natural killer cells also participate in host defense against infections.[44,60,95,114]

Once immunity has developed, the immune system will have the capacity to activate its various effector mechanisms, such as antibody mediated neutralization, complement fixation, chemotaxis, phagocytosis, cytotoxicity, etc. in response to the specific antigen. Even though the initiation may be specific, the pathophysiology of the specific and nonspecific inflammatory response is virtually identical. In addition to the inflammatory cells, the increased delivery of serum factors (complement components, natural antibodies, hormones, nutrients, etc.) through the dilation and increased permeability of blood vessels is also important. These changes are aimed at the restriction, destruction, and elimination of the infectious agent from the tissue of invasion. In mammals, the regional lymph nodes provide further barriers to the spread of infectious agents. Host tissues and cells also become damaged during an inflammatory response, either by the infectious agent or by enzymes and toxins released from damaged cells and activated phagocytes. Once the infectious agent is eliminated, the healing process begins in which macrophages play an important role. Macrophages produce a variety of growth factors acting on fibroblasts and other cells, one of which is TGF-β. TGF-β promotes the production of IL-1, TNF, platelet derived growth factor (PDGF), basic fibroblast growth factor, and IL-6, induces Fc-γ receptor III on monocytes. Through these and other effects, TGF-β plays a critical role in the early phase of inflammation. TGFβ also regulates angiogenesis, chemotaxis, fibroblast proliferation and controls the synthesis of the extracellular matrix necessary for repair. It also induces PDGF receptors.[122] In contrast with the local proinflammatory effect of TGF-β, it exerts an antiinflammatory effect systemically, which may be supported by other cytokines, such as IL-4, IL-5, and IL-10. TGF-β diminishes rapidly the synthesis of acute phase proteins by hepatocytes.[5,122]

Prolactin has been shown to potentiate the inflammatory response in rats to various irritants except dextran- or serotonin-induced paw edema and cotton pellet granuloma, which were not affected. Mice treated with bromocriptine (BRC), which suppresses pituitary PRL secretion, showed an increased mortality after infection with *Listeria monocytogenes*. Normal resistance could be restored by additional treatment with PRL. Treatment of mice with PRL reduced the mortality rate in a dose dependent manner after infection with *Salmonella typhimurium*. PRL treatment also increased phagocytosis and intracellular killing by peritoneal macrophages and enhanced chemotaxis by peritoneal granulocytes. The activity of blood leukocytes was not modified by PRL treatment.[18,76] Treatment of Hypox rats with either GH or PRL elevated superoxide production by macrophages and enhanced phagocytic activity towards opsonized *Listeria monocytogenes in vitro*. GH or PRL treatment also enhanced the survival of normal and Hypox rats infected with *Salmonella typhimurium*. IGF-I also primed macrophages *in vitro* in a similar manner.[45,46] The joint injection of mice with a vaccine against tick borne encephalytis with human GH produced complete protection with a single dose of vaccine.[108]

The treatment of 20 patients with septic shock with 0.1 mg/kg/day of GH improved the nitrogen balance and increased the levels of IGF-I, insulin, glycerol, free fatty acids and β-hydroxybutyrate. Despite the continuous i.v. administration, GH levels declined during the three day treatment period indicating an increased metabolic clearance.[120]

Colony forming cells from peripheral blood of children with severe HIV-I infection showed resistance to growth stimulation by IGF-I, GH and insulin, which was not related to the presence of gross malnutrition, differences in hematologic status or overwhelming illness.[55] Recombinant human GH enhanced the replication of HIV-I in acutely infected peripheral blood monocytes, which correlated with the stimulation of DNA synthesis and increase in TNF secretion.[66] Prolactin has also been shown to stimulate the production of retroviruses by lymphoid and other tissues in mice.[33]

Treatment of experimental animals with glucocorticoids increases their sensitivity to viral, bacterial and fungal infections and to parasitic infestations. In patients, gram-negative bacterial and staphylococcal infections, tuberculosis, diptheria, certain types of fungal, viral and protozoal infections are frequently encountered during treatment with glucocorticoids.[8] Clearly, an overwhelming body of experimental and clinical evidence indicates that the immunosuppressive and antiinflammatory effect of glucocorticoids leads to decreased host resistance to various forms of infection and parasitic diseases.

Although there is indication that sex hormones influence host resistance to infectious agents, much remains to be done in this respect. Treatment with estrogens seems to increase the humoral immune response and host resistance against bacterial infections, but facilitates the replication of retroviruses (murine leukemia virus), potentiates the induction of conjunctivitis in guinea pigs by the chlamydial agent, and exacerbates the disease caused in guinea pigs by *Toxoplasma gondii*. Testosterone protected mice against pneumococcus infection. In male mice, castration reduced Coxsackie B3 virus-induced necrotic lesions and T cell mediated cytotoxicity which could be reversed by testosterone treatment. Progesterone treatment of monkeys was reported to make them susceptible to the Carr-Zibler strain of Rous chicken sarcoma virus.[8]

C. The Acute Phase Response and Host Defense

If the immune system fails to control infection locally, systemic infection (e.g. viremia, bacteremia, septicemia and endotoxemia), and eventually shock may develop. This results in the systemic activation of various immune mechanisms, such as the complement system, the stimulation of cytokine production by leukocytes, the production of acute phase reactants by the liver and in some other sites, and increased leukopoesis in the bone marrow. One may suggest that the acute phase response is an adaptive switch from specific to nonspecific defense mechanisms which are more ancient and capable of providing a widespread protection, though the probability of failure is much higher than with specific immunity. C reactive protein and lipopolysaccharide binding protein are examples. The production of complement components and of fibrinogen is also elevated. Complement plays a fundamental role in nonspecific defense. Fibrinogen is necessary for blood clotting and increased levels are presumably helpful to isolate the invading organisms by widespread blood clotting in capillaries of infected tissues. In addition, a number of enzyme inhibitors are produced which are likely to play an important role in the prevention of nonspecific tissue damage, which would otherwise be inflicted on a large scale by enzymes released from phagocytic and other damaged cells. The antiinflammatory α-MFP also contributes to the control of damage by inflammatory cells. Catecholamines are known to inhibit inflammatory responses and have recently been shown to promote, even initiate, the acute phase response. The production of cytokines is also controlled during the acute phase reaction, which is illustrated by endotoxin tolerance. Some recent observations made with TNF production have now been extended to the production of other cytokines and show that cytokine production is diminished significantly after the first insult eliciting the acute response.[52] Endotoxin tolerant animals are also protected against nonspecific insults, such as oxygen intoxication, which also is related to the suppression of cytokine production in this condition.[113a] Although the exact nature of this diminished reactivity is yet to be elucidated, it appears that glucocorticoids and some acute phase proteins produced by the liver play essential roles.

Some of the characteristics of the major cytokine hormones participating in the immune, endocrine and metabolic responses to infection are summarized below.

(i) <u>Interleukin-1α and β</u> stimulate T and B lymphocytes for proliferation, stimulate IL-2, IL-6 and IL-8 production, induce fever, release ACTH and other pituitary hormones, antagonize opioid receptors in the brain, promote slow wave sleep, decrease appetite, stimulate acute phase proteins in the liver, promote proteolysis in muscle, inhibit thyroglobulin gene expression, stimulate thyroid growth, insulin release (inhibitory at high doses), inhibit steroid synthesis by the gonads and adrenals and stimulate bone resorption. IL-1a may exert bioactivity as a membrane associated molecule. An IL-1 receptor antagonist has also been identified.[17,42,68]

(ii) <u>TNFα (cachectin) and TNFβ (lymphotoxin)</u> may act as membrane bound or secreted cytokines and have many overlapping functions with IL-1, such as pyrogenicity, promotion of slow wave sleep, a strong catabolic effect, release of ACTH, GnRH, the blocking of GH release, T and B cell activation, and stimulation of bone resorption. TNF stimulates the production of IL-1, IL-6, IL-8 and endothelial adhesiveness for leukocytes. Neutrophils, eosinophils and macrophages are activated by TNF and TNF has a cytotoxic activity on tumors and on some other targets, and can cause inflammation, hemorrhage and shock, if produced in excess. TNF plays a major role in the endocrine and metabolic changes associated with trauma and sepsis. IL-1 and TNF inhibit the β-adrenergic responsiveness of cardiac myocytes.[14,22,23,68] TNF, IL-1 and other LPS induced mediators enhance nonspecific resistance to bacterial, viral and parasitic infections, autoimmune disease and radiation sickness. IL-1 given to rats either i.v. or into the cerebrospinal fluid inhibited gastric injury induced by a variety of experimental agents, such as stress, aspirin, and ethanol. Protection was mediated by the inhibition of acid secretion by prostaglandin release.[14,113]

(iii) Interleukin-6 promotes the differentiation of B lymphocytes, supports the multipotential proliferation of hemopoietic stem cells, induces acute phase proteins in the liver, plays a role in the differentiation and activation of T cells and macrophages and contributes to neural differentiation. IL-6 has been shown to release ACTH and β-endorphin from mouse pituitary adenomas, to stimulate the release of ACTH and inhibit TSH release from the pituitary by acting on the hypothalamus in the rat, and to inhibit ACTH release by acting directly on the rat pituitary gland. IL-6 has also been shown to act as a locally produced cytokine in the pituitary gland of rats.[1,14,22]

(iv) IL-2 is a product of T lymphocytes. It functions primarily as a growth factor for T cells and natural killer and lymphokine activated killer cells. IL-2 exerts a neurotoxic effect if given at high doses to animals or patients which coincided in rats with the induction of serum TNF. IL-2 was reported to inhibit GH, LH, and FSH release, and to promote ACTH, TSH, and PRL release from pituitary glands of rats and mice. In humans, IL-2 treatment increased ACTH, β-endorphin, GH and PRL in the circulation.[14,39,75]

(v) IFN-γ is produced by T lymphocytes and macrophages and has been discovered by its antiviral effect. IFN-γ promotes the differentiaton of T and B lymphocytes and is a major regulator of cytokine production by monocytes, macrophages, and T lymphocytes, and is also known to activate NK cells for cytotoxicity. IFN-γ induces class II major histocompatibility antigens, receptors for GC in murine monocytes, and acts in conjunction with GC to initiate the expression of FcRγ-I receptors on human monocytes. IFN-γ was reported to release ACTH and to inhibit GH and TSH release by acting on the hypothalamus in rats. It also has an effect on pituitary hormone release by acting directly on the pituitary gland. However, these findings are controversial at the present time.[14,68]

(vi) The colony stimulating factors, G-CSF, M-CSF, and GM-CSF, play an important role in the maturation of granulocytes and macrophages in the bone marrow and are also important activators of these cells during infection and inflammation.[78] GCSF released ACTH *in vitro* from human pituitary adenomas.[69]

(vii) Platelet activating factor is a lipid chemical mediator produced by most mammalian cells. It is a major inflammatory mediator with vasoactive properties and it plays a role in arterial thrombosis, endotoxic shock, acute allergic disease, embryo implantation, parturition and fetal lung development.[57] PAF released ACTH and β-endorphin in rats by acting on the hypothalamus, and GH and PRL through direct action on the pituitary gland.[14]

The neuroendocrine response to emotional and other forms of stress where tissue injury is not involved resembles that of the response to injury. However, it appears that the endocrine response to emotional stress is not associated with a cytokine response, which appears only if the immune system is activated either by specific or nonspecific stimulation.[14,38] Stress has many effects on the immune system and can be shown to be immunosuppressive or occasionally, immunostimulatory, depending on the conditions used. Long lasting stress is known to be associated with immunosuppression and a decreased resistance to infection. Much remains to be learned about the complex interaction of the immune and neuroendocrine systems during infectious disease.

CONCLUSIONS

Infectious agents have the capacity to activate the immune system, both in a nonspecific and specific manner, and cause inflammation and specific immune reactions. The inflammatory response may be regarded as a first line of defense against infection. Injured cells release a number of chemotactic-proinflammatory cytokines which will attract leukocytes to the site of infection that play a major role in the inflammatory process. The complement system is also activated frequently and the split products C3a and C5a function as chemotactic factors and activators for mast cells and basophils, monocytes and macrophages, and granulocytes, and thus inflammation is induced. Once specific immunity has been developed in the host, reinfection with the same microorganism may lead to immediate hypersensitivity, delayed type hypersensitivity, or Arthus type reactions, which are inflammatory reactions mediated by IgE antibodies and mast cells, by T lymphocytes and by complement fixing antibodies, respectively. Sensory nerves also have the capacity to induce inflammation by causing mast cell discharge. In turn, mast cell products act on the nervous system. This is especially important in infections of the gastrointestinal tract where diarrhea can rapidly be initiated by this mechanism which is helpful for the elimination of the infectious agent.

In inflammation a whole host of cytokines, which include IL-1, -2, -6, -8, TNF and IFN-γ and a number of growth factors, such as PDGF, EGF, and TGF-β, are produced locally. A number of other inflammatory mediators are under characterization. These factors are essential for the elimination of the infectious agent by effector cells and for the promotion of healing. TGF-β plays a role in both the initiation of inflammation and healing, whereas it exerts an antiinflammatory effect systemically and thus functions as a major coordinator of the inflammatory response. Other anti-inflammatory cytokines are IL-4, IL-5, and IL-10.

There is evidence to suggest that systemic infection or a sublethal dose of endotoxin elicits a sharp but brief elevation of growth hormone and prolactin in the serum. Because these hormones enhance the reactivity of the immune system, this initial elevation may be an important mechanism for the amplification of host defense. In severe sepsis and shock, GH and PRL are suppressed, whereas glucocorticoids and catecholamines are elevated. Under these conditions the danger is excessive cytokine production by the systemic activation of the immune system, which has lethal consequences. Glucocorticoids and catecholamines are powerful inhibitors of cytokine production and also play a key role in the generation of a refractory state in cytokine induction, which is known as endotoxin tolerance. Patients with subclinical adrenal insufficiency succumb to septic shock almost invariably if glucocorticoid therapy is not given.

Severe systemic infections elicit an acute phase response which is initiated by cytokines such as IL-1, IL-6 and TNFα (several others may be involved) that appear in the circulation and function as acute phase hormones. These cytokines elicit a neuroendocrine response and also initiate major metabolic alterations. There is protein loss from muscles and in general catabolism prevails in many other tissues and organs, whereas the synthesis of the so-called acute phase proteins in the liver, cell proliferation in the bone marrow, and protein synthesis in leukocytes, is elevated. The acute phase response may be regarded as an emergency measure to save the organism after the local immune/inflammatory response has failed to isolate and eliminate the infectious agent. In septicemia the systemic activation of complement system and of leukocyte releasing enzymes and highly toxic cytokines seriously threaten survival. Glucocorticoids suppress proinflammatory cytokine production and greatly potentiate the secretion of the acute phase proteins. Some of these proteins, such as C reactive protein or LPS binding protein, are designed to bind to a wide spectrum of microorganisms and trigger their destruction by the activation of the complement system and of phagocytes. The enhanced production of complement components also strengthens nonspecific resistance under these conditions. Increased production of fibrinogen promotes blood clotting, which can still serve to isolate the invading agent by facilitating thrombosis in damaged tissues. The enzyme inhibitors produced as acute phase proteins are likely to serve to curb the nonspecific damage caused by enzymes released from phagocytes and possibly other cells into the circulation. The antiinflammatory α-macrofetoprotein is likely to inhibit damage inflicted by inflammatory cells. Catecholamines also have an antiinflammatory effect and recently have been shown to promote the acute phase response.

Endocrine alterations are also present in patients with chronic infections, such as leprosy and AIDS. It remains to be seen whether or not hormonal changes play a role in the outcome of chronic infectious disease.

Although our understanding of the significance of neuroendocrine and metabolic responses to infection in host defense is very limited at the present time, it is clear that the immune system alone is incapable of performing this complex and difficult task. Rather, a continuous interaction between the immune and neuroendocrine systems during immune, inflammatory and acute phase responses maintains a highly adaptive and multifaceted response which assures optimal host defense. Further analysis and understanding of the nature of these complex host defense mechanisms is of utmost importance, which should provide new approaches to the prevention and treatment of infectious disease.

ABBREVIATIONS

Ab, antibody; ACTH, adrenocorticotropic hormone; ADP, adenosine diphosphate; ADX, adrenalectomy; Ag, antigen; AGT, angiotensin; APP, acute phase serum proteins; APR, acute phase reaction; ATP, adenosine triphosphate; AVP, arginine vasopressin; BRC, bromocriptine; C3a, C5a, complement components; CATG, cathepsin G; CD14, monocyte macrophage surface molecule; CGRP, calcitonin gene related peptide; CNTF, ciliary neurotrophic factor; CRF, corticotropin releasing factor; CRP, C reactive protein; CSF, colony stimulating factor; DEF, defensins; DEX, dexamethasone; DOP, dopamine; E2, estradiol; ECF, eosinophil chemotactic factor; EGF, epidermal growth factor; END, endorphin; ENK, enkaphalin; EP, epinephrine; ES, estrogen; FcR, Fc receptor; FGF, fibroblast growth factor; FOP, formyl peptide; FSH, follicle stimulating hormone; GC, glucocorticoid; G-CSF, granulocyte colony stimulating factor; GDP, guanosine diphosphate; GH, growth hormone; GLN, glucagon; GM-CSF, granulocyte macrophage colony stimulating factor; GTP, guanosine triphosphate; HIS, histamine; Hypox, hypophysectomized; IFN, interferon; IGF-I, insulin-like growth factor-I; IL, interleukin; INS, insulin; IP-10, chemotactic and proinflammatory cytokine; LAK, lymphokine activated killer; LBP, lipopolysaccharide binding protein; LEM, leukocyte endogenous mediator; L-ENK, leucine-enkephalin; LH, luteinizing hormone; LHRH, luteinizing hormone releasing hormone; LPS, lipopolysaccharide; LT, leukotriene; LTB4, LTC4, leukotriene B4, -C4; MCAF, macrophage chemotactic and activating factor; MCF, monocyte chemotactic factor; M-CSF, macrophage colony stimulating factor; a-MFP, a-macrofetoprotein; a-MSH, alpha-melanocyte stimulating hormone; NAL, naloxone; NCF, neutrophil chemotactic factor; NEP, norepinephrine; NFIL-6, nuclear factor for IL-6; NF-lκB (0TF2), nuclear factor regulating κ gene expression in B lymphocytes; NK, natural killer; NRT, neurotensin; PAF, platelet activating factor; PDGF, platelet derived growth factor; PF-4, basic platelet factor; PG, prostaglandin; PGD2, -D4, -E4, -E2, prostaglandins; PG12, prostacyclin; PGS, progesterone; PKCI, protein kinase C inhibitors; PRL, prolactin; SAA, serum amyloid A; SER, serotonin; SOM, somatostatin; SP, substance P; T3, tri-iodiothyronine; β-TBG, β-thromboglobulin; TBXA2, thromboxane A2; TGF, transforming growth factor; TNF, tumor necrosis factor; TRH, thyroid releasing hormone; TSH, thyroid stimulating hormone; VP, vasopressin; VIP, vasoactive intestinal peptide.

REFERENCES

1. **Akira, S. and Kishimoto, T.,** IL-6 and NF-IL6 in acute-phase response and viral infection, *Immunol. Rev.,* 127,26, 1992.
2. **Arnold, J., Choo, J.J., Little, R.A. and Rothwell, N.J.,** Dopamine beta-hydroxylase inhibition reveals a selective influence of endotoxin on catecholamine content of rat tissues, *Circ. Shock,* 31,387, 1990.
3. **Ballou, S.P. and Kusher, I.,** C-reactive protein and the acute phase response, in *Advances in Internal Medicine,* Mosby Year Book, St. Louis, MO, 1992, p.313.
4. **Banck, G. and Forsgren, A.,** Many bacterial species are mitogenic for human blood lymphocytes, *Scand. J. Immunol.,* 8,347, 1978.
5. **Barbul, A.,** Immune regulation of wound healing, in *Immune Consequences of Trauma, Shock and Sepsis,* Faist, E., Ninnemann, J. and Green, D., Eds., Springer Verlag, 1989, p.339.
6. **Baue, A.F.,** Neuroendocrine response to severe trauma and sepsis, in *Immune Consequences of Trauma, Shock and Sepsis,* Faist, E., Ninnemann, J. and Green, D., Eds., Springer Verlag, 1989, p.17.
7. **Beisel, W.R.,** Metabolic response to infection, *Ann. Rev. Med.,* 26,9, 1975.
8. **Berczi, I.,** *Pituitary Function and Immunity,* CRC Press, Boca Raton, FL, 1986.
9. **Berczi, I.,** The immunology of prolactin, *Sem. Reprod. Endocrinol.,* 10,196, 1992.
10. **Berczi, I.,** Neuroendocrine defense in endotoxin shock, *Acta Microbiol. Hung.,* 40,265, 1993.
11. **Berczi, I.,** Hormonal interactions between the pituitary and immune system, in *Bilateral Communication Between the Endocrine and Immune Systems,* Grossman, C.J., Ed., Springer Verlag, 1994a, 96.
12. **Berczi, I.,** The role of the growth and lactogenic hormone family in immune function (invited review), *Neuroimmunomodulation,* 1,201, 1994b.

13. **Berczi, I., Bertok, L., Baintner, K., Jr., and Veress, B.,** Failure of oral Escherichia coli endotoxin to induce either specific tolerance or toxic syndromes in rats, *J. Path. Bact.*, 96, 481, 1968.

14. **Berczi, I. and Nagy, E.,** Neurohormonal control of cytokines during injury, in *Brain Control of the Response to Injury,* Rothwell, N.J. and Berkenbosch, F., Eds., Cambridge University Press, 32, 1994.

15. **Berczi, I., Bertok, L. and Bereznay, T.,** Comparative studies on the toxicity of Escherichia coli lipopolysaccharide endotoxin in various animal species, *Can. J. Microbiol.*, 12,1070, 1966a.

16. **Berczi, I., Baintner, K. Jr., and Antal, T.,** Comparative assay of endotoxin by oral and parenteral administration, *Zbl. Vet. Med. Reihe B.*, 13,570, 1966b.

17. **Berkenbosch, F., Van Dam, A.-M., Derijk, R. and Schotanus, K.,** Role of the immune hormone interleukin-1 in brain-controlled adaptive responses to infection, in *Stress: Neuroendocrine and Molecular Approaches,* Gordon & Breach Science Pub. S.A., New York, 1992, p.623.

18. **Bernton, E.W., Meltzer, M.T. and Holaday, J.W.,** Suppression of macrophage activation and T-lymphocyte function in hypoprolactinemic mice, *Science*, 239,401, 1988.

19. **Bertok, L.,** Physico-chemical defense of vertebrate organisms. The role of bile acids in defense against bacterial endotoxin, *Perspect. Biol. Med.*, 21,70, 1977.

20. **Bertok, L. and Nagy, S.U.,** The effect of endotoxin and radio-detoxified endotoxin on the serum T4 level of rats and response of their thyroid gland to exogenous TSH, *Immunopharmacology,* 8,143, 1984.

21. **Besedovsky, H., Sorkin, E., Keller, M. and Muller, J.,** Changes in blood hormone levels during the immune response, *Proc. Soc. Exp. Biol. Med.,* 150,466, 1975.

22. **Besedovsky, H.O., del Rey, A., Klusman, I., Furukawa, H., Arditi, G.M., and Kabiersch, A.,** Cytokines as modulators of the hypothalamus-pituitary-adrenal axis, *J. Steroid Biochem. Mol. Biol,* 40,613, 1991.

23. **Beutler, B. and Cerami, A.,** Cachectin (tumor necrosis factor): A macrophage hormone governing cellular metabolism and inflammatory response, *Endocr. Rev.,* 9,57, 1988.

24. **Billiau, A. and Vandekerckhove, F.,** Cytokines and their interactions with other inflammatory mediators in the pathogenesis of sepsis and septic shock, *Eur. J. Clin. Invest.,* 21,559, 1991.

25. **Bjornson, A.B., Altemeier, W.A. and Bjornson, H.S.,** Changes in humoral components of host defense following burn trauma, *Ann. Surg.,* 186,88, 1977.

26. **Bone, R.C., Fisher Jr., C.J., Clemmer, T.P., Slotman, G.J., Metz, C.A., and Balk, R.A.,** Methylprednisolone Severe Sepsis Study Group, A controlled clinical trial of high-dose methylprednisolone in the treatment of severe sepsis and septic shock, *N. Engl. J. Med.,* 317,653, 1987.

27. **Border, J.R., Hassett, J., LaDuca, J., Seibel, R., Steinberg, S., Mills, B., Losi, P. and Border, D.,** The gut origin septic states in blunt multiple trauma (ISS=40) in the *ICU, Ann. Surg,* 206,427, 1987.

28. **Broudy, V.C., Kaushansky, K., Segal, G.M., Harlan, J.M. and Adamson, J.S.,** Tumor necrosis factor type α stimulates human endothelial cells to produce granulocyte/macrophage colony-stimulating factor, *Proc. Nat. Acad. Sci USA* 83,7467, 1986.

29. **Burgi, U., Feller, C., and Gerber, A.U.,** Effects of an acute bacterial infection on serum thyroid hormones and nuclear triiodothyronine receptors in mice, *Endocrinology,* 119,515, 1986.

30. **Burrell, R.,** Immunopharmacology of bacterial endotoxin, *EOS-J. Immunol. Immunopharmacol.,* 11,85, 1991.

31. **Bussolino, F., Wang, J.M., Defilippi, P., Turrini, F., Sanavio, F., Edgell, C.-J.S., Aglietta, M., Arese, P., and Mantovani, A.,** Granulocyte- and granulocyte-macrophage-colony stimulating factors induce human endothelial cells to migrate and proliferate, *Nature,* 337, 471, 1989.

32. **Butera, S.T.,** Cytokine involvement in viral permissiveness and the progression of HIV disease, *J. Cell. Biochem.,* 53,336, 1993.

33. **Chen, H.W., Meier, H., Heininger, H.J. and Huebner, R.J.,** Tumorigenesis in strain DW/J mice and induction by prolactin of the group specific antigen of C-type RNA tumor virus, *J. Natl. Cancer Inst.,* 49,1145, 1972.

34. **Christeff, N., Auclair, M.-C., Benassayag, C., Carli, A. and Nunez, E.A.,** Endotoxin-induced changes in sex steroid hormone levels in male rats, *J. Steroid Biochem.* 26,67, 1987.

35. **Coodley, G.O., Loveless, M.O., Nelson, H.D. and Coodley, M.K.,** Endocrine function in the HIV wasting syndrome, *J. AIDS,* 7,46, 1994.

36. **Curnutte, J.T. and Babior B.M.,** Composition of neutrophils, in *Hematology,* 4th ed., Williams, W.W., Beutler, E., Erslev, A.J. and Lichtman, M.A., Eds., McGraw-Hill, New York, 1990, p.770.

37. **Dahn, M.S., Lange, M.P. and Jacobs, L.A.,** Insulin like growth factor 1 production is inhibited in human sepsis, *Arch. Surg.,* 123,1409, 1988.

38. **Dantzer, R. and Kelley, K.W.,** Minireview: Stress and immunity: An integrated view of relationships between the brain and the immune system, *Life Sci.,* 44,1995, 1989.

39. **Denicoff, K.D., Durkin, T.M., Lotze, M.T., Quinland, P.E., Lewis, C.L., Listwak, S.J., Rosenberg, S.A. and Rubinow, D.R.,** The neuroendocrine effects of interleukin-2 treatment, *J. Clin. Endocrinol. Metab.,* 69,402, 1989.

40. **Derijk, R., Vanrooijen, N., Tilders, F.J.H., Besedovsky, H.O., Del Rey, A. and Berkenbosch, F.,** Selective depletion of macrophages prevents pituitary-adrenal activation in response to subpyrogenic, but not pyrogenic, doses of bacterial endotoxin in rats, *Endocrinology,* 129,330, 1991.

41. **Derijk, R.H., Vanrooijen, N. and Berkenbosch, F.,** The role of macrophages in the hypothalamic-pituitary-adrenal activation in response to endotoxin (LPS), *Res. Immunol.,* 143,224, 1992.

42. **Dinarello, C.A.,** Biology of interleukin-1, *FASEB J.,* 2, 108, 1988.

43. **Douglas, S.D. and Hassam, N.F.,** Morphology of monocytes and macrophages, in *Hematology,* 4th ed., Williams, W.W., Beutler, E., Erslev, A.J. and Lichtman, M.A., Eds., McGraw-Hill, New York, 1990, p.859.

44. **Drutz, D.J. and Graybill, J.R.,** Infectious diseases, in *Basic and Clinical Immunology,* 6th edition, Stiles, D.P., Stobo, J.D. and Wells, J.V., Eds., Appleton & Lange, Norwalk, Connecticut, 1987, p.534.

45.   **Edwards, C.K., Lorence, R.M., Dunham, D.M., Arkins, S., Yunger, L.M., Graeger, J.A., Walter, R.J., Dantzer, R. and Kelley, K.W.,** Hypophysectomy inhibits the synthesis of tumor necrosis factor-α by rat macrophages. Partial restoration by exogenous growth hormone or interferon-γ, *Endocrinol.,* 128,989, 1991.

46.   **Edwards, C.K., Ghiasuddin, S.M., Yunger, L.M., Lorece, R.M., Arkins, S., Dantzer, R. and Kelley, K.W.,** *In vivo* administration of recombinant growth hormone or gamma interferon activates macrophages - Enhanced resistance to experimental *Salmonella typhimurium* infection is correlated with generation of reactive oxygen intermediates, *Infect. Immun.,* 60,2514, 1992.

47.   **Elenkov, I.J., Kovacs, K., Kiss, J., Bertok, L. and Vizi, E.S.,** Lipopolysaccharide is able to bypass corticotrophin-releasing factor in affecting plasma ACTH and corticosterone levels. Evidence from rats with lesions of the paraventricular nucleus, *J. Endocrinol.,* 133, 231, 1992.

48.   **Elin, R.J. and Csako, G.,** Response of humans to gamma-irradiated reference *Escherichia coli* endotoxin, *J. Clin. Lab. Immunol.,* 29,17, 1989.

49.   **Evans, G.F. and Zuckerman, S.H.,** Glucocorticoid-dependent and glucocorticoid independent mechanisms involved in lipopolysaccharide tolerance, *Eur. J. Immunol.,* 21,1973, 1991.

50.   **Faist, E., Ertel, W., Mewes, A., Alkan, S., Walz, A. and Strasser, T.,** Trauma-induced alterations of the lymphokine cascade, in *Immune Consequences of Trauma, Shock and Sepsis,* Springer Verlag, 1989, p.79.

51.   **Fox, M.J., Kuzma, J.F. and Washam, W.T.,** Transitory diabetic syndrome associated with meningococcic meningitis, *Arch. Intem. Med.,* 79,614, 1947.

52.   **Friedman, H., Newton, C., Widen, R., Klein, T.W. and Spitzer, J.A.,** Continuous endotoxin infusion suppresses rat spleen cell production of cytokines, *Proc. Soc. Exp. Biol. Med.,* 199,360, 1992.

53.   **Galli, S.J., Dvorak, A.M. and Dvorak, H.F.,** Morphology, biochemistry and function of basophils and mast cells, in *Hematology,* 4th ed., Williams, W.W., Beutler, E., Erslev, A.J. and Lichtman, M.A., Eds., McGraw-Hill, New York, 1990, p.840.

54.   **Garcia-Barreno, P. and Suarez, A.,** Modification of catecholamine and prostaglandin tissue levels in E. coli endotoxin-treated rats, *J. Surg. Res.,* 44,178, 1988.

55.   **Geffner, M.E., Yeh, D.Y., Landaw, E.M., Scott, M.L., Stiehm, E.R., Bryson, Y.J. and Israele, V.,** *In vitro* insulin-like growth factor-I, growth hormone, and insulin resistance occurs in symptomatic human immunodeficiency virus-1-infected children, *Pediat. Res.,* 34,66, 1993.

56.   **Haak-Frendscho, M., Arai, N., Arai, K., Baeza, M.L., Finn, A. and Kaplan, A.P.,** Human recombinant granulocyte-macrophage colony-stimulating factor and interleukin-3 cause basophil histamine release, *J. Clin. Invest.,* 82,17, 1988.

57.   **Hanahan, D.J.,** Platelet-activating factor: A novel lipid agonist, *Curr. Topics Cell. Regul.,* 33,65, 1992.

58.   **Hedge, G.A.,** Roles for the prostaglandins in the regulation of anterior pituitary secretion, *Life Sci.,* 20,17, 1977.

59.   **Heideman, M.,** The role of complement injury, in *Immune Consequences of Trauma, in Shock and Sepsis,* Springer Verlag, 1989, p.215.

60. **Heyeman, D. and McKerrow, J.H.**, Parasitic diseases, in *Basic and Clinical Immunology,* 6th ed., Stiles, D.P., Stobo, J.D. and Wells, J.V., Eds., Appleton & Lange, Norwalk, Connecticut, 1987, p.634.

61. **Holmsen, H.**, Composition of platelets, in, *Hematology,* 4th ed., Williams, W.W., Beutler, E., Erslev, A.J. and Lichtman, M.A., Eds., McGraw-Hill, New York, 1990, p.1182.

62. **Jamieson, J.C., Lammers, G., Janzen, R. and Woloski, B.M.R.N.J.**, The acute phase response to inflammation: The role of monokines in changes in liver glycoproteins and enzymes of glycoprotein metabolism, *Comp. Biochem. Physiol.,* 87B,11, 1987.

63. **Jones, S.B. and Romano, F.D.**, Dose- and time-dependent changes in plasma catecholamines in response to endotoxin in conscious rats, *Circ. Shock,* 28,59, 1989.

64. **Jones, S.B. and Yelich, M.R.**, Simultaneous elevation of plasma insulin and catecholamines during endotoxicosis in the conscious and anesthetized rat, *Life Sci,* 41,1935, 1987.

65. **Karady, S., Selye, H. and Brownie, J.S.L.**, The influence of the alarm reaction on the development of anaphylactic shock, *J. Immunol.,* 35,335, 1938.

66. **Laurence, J., Grimison, B. and Gonenne, A.**, Effect of recombinant human growth hormone on acute and chronic human immunodeficiency virus infection *in vitro, Blood,* 79,467, 1992.

67. **Leshin L.S. and Malven, P.V.**, Bacteremia-induced changes in pituitary hormone release and effect of naloxone, *Am. J. Physiol.,* 247,E585, 1984.

68. **Lowry, S.F.**, Cytokine mediators of immunity and inflammation, *Arch. Surg.,* 128,1235, 1993.

69. **Malarkey, W.B. and Zvara, B.J.**, Interleukin-1-beta and other cytokines stimulate adrenocorticotropin release from cultured pituitary cells of patients with Cushing's disease, *J. Clin. Endocrinol. Metab.,* 69,196, 1989.

70. **Malone, D.G., Pierce, J.H., Falko, J.P. and Metcalfe, D.D.**, Production of granulocyte macrophage colony-stimulating factor by primary cultures of unstimulated rat microvascular endothelial cells, *Blood,* 71,684, 1988.

71. **Marks, J.D., Marks, C.B., Luce, J.M., Montgomery, A.B., Turner, J., Metz, C.A. and Murray, J.F.**, Plasma tumor necrosis factor in patients with septic shock. Mortality rate, incidence of adult respiratory distress syndrome, and effects of methylprednisolone administration, *Am. Rev. Respir. Dis.,* 141,94, 1990.

72. **Marshall, N.E. and Ziegler, H.K.**, Lipopolysaccharide responsiveness is an important factor in the generation of optimal antigen-specific T cell responses during infection with gram negative bacteria, *J. Immunol.,* 147,2333, 1991.

73. **Martin, L.W. and Lipton, J.M.**, Acute phase response to endotoxin - Rise in plasma aMSH and effects of α-MSH injection, *Am. J. Physiol.,* 259,R768, 1990.

74. **Masana, M.I., Heyes, M.P. and Mefford, I.N.**, Indomethacin prevents increased catecholamine turnover in rat brain following systemic endotoxin challenge, *Prog. Neuro Psych. Biol. Psych.,* 14,609, 1990.

75. **McCann, S.M., Karanth, S., Kamat, A., Dees, W.L., Lyson, K., Gimeno, M. and Rettori, V.**, Induction by cytokines of the pattern of pituitary hormone secretion in infection, *Neuroimmunomodulation,* 1,2, 1994.

76. **Meli, R., Gualillo, O., Raso, G.M., and DiCarlo, R.**, Further evidence for the involvement of prolactin in the inflammatory response, *Life Sci.,* 53,PL105, 1993.

77. **Merrill, G.A. and Anderson Jr., J.H.,** Involvement of endogenous opiates in glucose-stimulated hyperinsulinism of canine endotoxin shock: Inhibition by naloxone, *Diabetes,* 36,585, 1987.

78. **Metcalf, D.,** Control of granulocytes and macrophages: Molecular, cellular, and clinical aspects, *Science,* 254,529, 1991.

79. **Mochizuki, H., Trocki, O., Dominioni, L., Brackett, K.A., Joffe, S.N. and Alexander, J.W.,** Mechanism of prevention of postburn hypermetabolism and catabolism by early enteral feeding, *Ann. Surg.,* 200,297, 1984.

80. **Mohammad, S.F., Mason, R.G., Eichwald, E.J. and Shively, J.A.,** Healthy and impaired vascular endothelium, in *Blood Platelet Function and Medical Chemistry,* Lasslo, E., Ed., Elsevier, New York, Amsterdam, Oxford, 1984, p.129.

81. **Mossman, T.R., Schumacher, J.H., Street, N.F., Budd, R., O'Garra, A., Fong, T.A.T., Bond, M.W., Moore, K.W.M., Sher, A. and Fiorentino, D.F.,** Diversity of cytokine synthesis and function of mouse CD4+ T cells, *Immun. Rev.,* 123,209, 1991.

82. **Muller-Eberhard, H.J.,** Complement, *Ann. Rev. Biochem.,* 44,697, 1975.

83. **Nagy S. and Bertok, L.,** Influence of experimentally induced endotoxemias on the thyroid function of rats, *Acta Physiol. Hung.,* 76,137, 1990.

84. **Nagy, Z., Berczi, I. and Bertok, L.,** Experimental data on the pathogenesis of edema disease of swine. Clinical picture, gross and microscopic lesions related to endotoxic shock, *Zbl. Vet. Med. Reihe B,* 15,504, 1968.

85. **Ninneman, J.L.,** The immune consequences of trauma: An overview, in *Immune Consequences of Trauma, Shock and Sepsis,* Springer Verlag, 1989, p.1.

86. **Nuytinck, J.K.S., Goris, R.J.A., Redl, H., Schlag, G. and van Munster, P.J.J.,** Posttraumatic complications and inflammatory mediators, *Arch. Surg.,* 121,886, 1986.

87. **Oppenheim, J.J., Zachariae, C.O.C., Mukaida, N., and Matsushima, K.,** Properties of the novel proinflammatory supergene "intercrine" cytokine family, *Ann. Rev. Immunol.* 9,617, 1991.

88. **Packham, M.A. and Mustard, J.F.,** Normal and abnormal platelet activity, in *Blood Platelet Function and Medical Chemistry,* Lasslo, E., Ed., Elsevier, New York, Amsterdam, Oxford, 1984, p.61.

89. **Portoles, M.T., Pagani, R., Ainaga, M.J., Diaz-Laviada, I. and Municio, A.M.,** Lipopolysaccharide-induced insulin resistance in monolayers of cultured hepatocytes, *Brit. J. Exp. Pathol.,* 70,199, 1989.

90. **Quesenberry, P., Morley, A., Stohlman Jr., F., Rickard, K., Howard, D. and Smith, M.,** Effect of endotoxin on granulopoiesis and colony-stimulating factor, *N. Engl. J. Med.,* 286,227, 1972.

91. **Raetz, C.R.H., Ulevitch, R.J., Wright, S.D., Sibley, C.H., Ding, A. and Nathan, C.F.,** Gram negative endotoxin: An extraordinary lipid with profound effects on eukaryotic signal transduction, *FASEB J.,* 5,2652, 1991.

92. **Ramachandra, R.N., Sehon, A.H., and Berczi, I.,** Neuro-hormonal host defense in endotoxin shock, *Brain Behav. Immun.,* 6,157, 1992.

93. **Reincke, M., Allolio, B., Petzke, F., Heppner, C., Mbulamberi, D., Vollmer, D., Winkelmann, W. and Chrousos, G.P.,** Thyroid dysfunction in African trypanosomiasis: a possible role for inflammatory cytokines, *Clin. Endocrinol.,* 39,455, 1993.

94. **Ringwald, P., Peyron, F., Vuillez, J.P., Touze, J.E., Le Bras, J. and Deloron, P.,** Levels of cytokines in plasma during Plasmodium falciparum malaria attacks, *J. Clin. Microbiol.,* 29,2076, 1991.

95. **Rook, G.,** Immunity to viruses, bacteria and fungi, in *Immunology,* 2nd ed., Roit, I., Brostoff, J. and Male, E., Eds., Gower Med. Pub., London, 1989, 16.1.

96. **Schooltink, H., Stoyan, T., Roeb, E., Heinrich, P.C. and Rose-John, S.,** Ciliary neurotrophic factor induces acute-phase protein expression in hepatocytes, *FEBS Lett.,* 314,280, 1992.

97. **Selye, H.,** A syndrome produced by diverse nocuous agents, *Nature,* 138, 32, 1936a.

98. **Selye, H.,** Thymus and the adrenals in response of the organism to injuries and intoxications, *Br. J. Exp. Pathol.,* 17,234, 1936b.

99. **Selye, H.,** Morphological changes in the fowl following chronic overdosage with various steroids, *J. Morphol.,* 73,401, 1943.

100. **Selye, H.,** The general adaptation syndrome and the diseases of adaptation, *J. Clin. Endocrinol.,* 6,117, 1946.

101. **Selye H.,** Effect of ACTH and cortisone upon an "anaphylactoid reaction", *Can. Med. Assoc. J.,* 61,553, 1949.

102. **Selye, H.,** Stress and disease, *Science,* 122,625, 1955.

103. **Shilo, S., Livshin, Y., Zylber-Haran, E., Sheskin, J. and Spitz, I.M.,** Gonadotropin, prolactin, and thyrotropin secretion in lepromatous leprosy, *J. Andrology,* 3,320, 1982.

104. **Silverman, H.J., Lee, N.H. and El-Fakahany, E.E.,** Effects of canine endotoxin shock on lymphocytic beta-adrenergic receptors, *Circ. Shock,* 32,293, 1990.

105. **Smith, B.B. and Wagner, W.C.,** Suppression of prolactin in pigs by *Escherichia coli* endotoxin, *Science,* 224, 605, 1984.

106. **Smith, B.B. and Wagner, W.C.,** Effect of Escherichia coli endotoxin and thyrotropin-releasing hormone on prolactin in lactating sows, *Am. J. Vet. Res.,* 46,175, 1985.

107. **Smolen, J.E. and Boxer, L.A.,** Function of neutrophils, in *Hematology,* 4th ed., Williams, W.W., Beutler, E., Erslev, A.J. and Lichtman, M.A., Eds., McGraw-Hill, New York, 1990, p.780.

108. **Sprung, C.L., Caralis, P.V., Marcial, E.H., Pierce, M., Gelbard, M.A., Long, W.M., Duncan, R.C., Tendler, M.D. and Karpf, M.,** The effects of high-dose corticosteroids in patients with septic shock: A prospective, controlled study, *N. Engl. J. Med.,* 311,1137, 1984.

109. **Stephenson, J.R., Lee, J.M., Bailey, N., Shepherd, A.G. and Melling, J.,** Adjuvant effect of human growth hormone with an inactivated flavivirus vaccine, *J. Infect. Dis.,* 164,188, 1991.

110. **Sundar, S.K., Cierpial, M.A., Kamaraju, L.S., Long, S., Hsieh, S., Lorenz, C., Aaron, M., Ritchie, J.C. and Weiss, J.M.,** Human immunodeficiency virus glycoprotein (prl20) infused into rat brain induces interleukin 1 to elevate pituitary-adrenal activity and decrease peripheral cellular immune responses, *Proc. Natl. Acad. Sci. USA,* 88,11246, 1991.

111. **Suzuki, S., Oh, C. and Nakano, K.,** Pituitary-dependent and -independent secretion of CS caused by bacterial endotoxin in rats, *Am. J. Physiol.,* 250,E470, 1986.

112.   Szabo, C., Thiemermann, C., Wu, C.-C., Perretti, M. and Vane, J.R., Attenuation of the induction of nitric oxide synthase by endogenous glucocorticoids accounts for endotoxin tolerance *in vivo*, *Proc. Natl. Acad. Sci. USA*, 91,271, 1994.

113.   Tache, Y. and Saperas, E., Potent inhibition of gastric acid secretion and ulcer formation by centrally and peripherally administered interleukin-1, *Ann. NY Acad. Sci.*, 664,353, 1992.

113a.   Tang, G., Berg, J. T., White, J. E., Lumb, P.D., Lee, C. Y. and Tsan, M. F., Protection against oxygen toxicity by tracheal insufflation of endotoxin: role of Mn SOD and alveolar macrophages, *Am. J. Physiol.*, 266,L38, 1994.

114.   Taverne, J., Immunity to protozoa and worms, in *Immunology*, 2nd ed., Roitt, I., Brostoff, J., and Male, D., Eds., Gower Med. Pub., London, 1989, p.17.1.

115.   Tishon, A. and Oldstone, M.B., Perturbation of differentiated functions during viral infection *in vivo*. *In vivo* relationship of host genes and lymphocytic choriomeningitis virus to growth hormone deficiency, *Am. J. Pathol.*, 137,965, 1990.

116.   Van Gool, J., Boers, W., Sala, M. and Ladiges, N.C.J.J., Glucocorticoids and catecholamines as mediators of acute-phase proteins, especially rat α-macrofoetoprotein, *Biochem. J.*, 220,125, 1984.

117.   Van Gool, J., Van Vugt, H., Helle, M. and Aarden, L.A., The relation among stress, adrenalin, interleukin 6, and acute phase proteins in the rat, *Clin. Immunol. Immunopathol.*, 57,200, 1990.

118.   Van Oosterhout, A.J.M., Folkerts, G., Ten Have, G.A.M. and Nijkamp, F.P., Involvement of the spleen in the endotoxin-induced deterioration of the respiratory airway and lymphocyte beta-adrenergic systems of the guinea pig, *Eur. J. Pharmacol.*, 147,421, 1988.

119.   The Veterans Administration Systemic Sepsis Cooperative Study Group, Effect of high dose glucocorticoid therapy on mortality in patients with clinical signs of systemic sepsis, *N. Engl. J. Med.*, 317,659, 1987.

120.   Voerman, H.J., Strack van Schijndel, R.J.M., Groeneveld, A.B.J., de Boer, H., Nauta, J.P., van der Veen, E.A. and Thijs, L.G., Effects of recombinant human growth hormone in patients with severe sepsis, *Ann. Surgery*, 216,648, 1992.

121.   Voerman, H.J., Groeneveld, J., de Boer, H., Strack van Schijndel, R.J.M., Nauta, J.P., van der Veen, E.A. and Thijs, L.G., Time course and variability of the endocrine and metabolic response to severe sepsis, *Surgery*, 114, 951, 1993.

122.   Wahl, S.M., Transforming growth factor beta (TGF-β) in inflammation - A cause and a cure, *J. Clin. Immunol.* 12,61, 1992.

123.   Wannemacher Jr., R.W., Pekarek, R.S., Thompson, W.L., Curnow, R.T., Beall, F.A., Zenser, T.V., de Rubertis, R.F. and Beisel, W.R., A protein from polymorphonuclear leukocytes (LEM) which affects the rate of hepatic amino acid transport and synthesis of acute-phase globulins, *Endocrinol.*, 96, 651, 1975.

124.   Watson, J.D., Varley, J.G., Bouloux, P.M., Tomlin, S.J., Rees, L.H., Besser, G.M. and Hinds, C.J., Adrenal vein and arterial levels of catecholamines and immunoreactive metenkephalin in canine endotoxin shock, *Res. Exp. Med.*, 188,319, 1988.

125.   Wexler, B.C., Dolgin, A.E. and Tryczynski, E.W., Effects of bacterial polysaccharide (Piromen) on the pituitary-adrenal axis: Adrenal ascorbic acid, cholesterol and histological alterations, *Endocrinol.*, 61,300, 1957.

126. **Wieser, M., Bonifer, R., Oster, W., Lindemann, A., Mertelsmann, R. and Herrmann, F.,** Interleukin-4 induces secretion of CSF for granulocytes and CSF for macrophages by peripheral blood monocytes, *Blood*, 73,1105, 1989.

127. **Williams, J.L. and Dick, G.F.,** Decreased dextrose tolerance in acute infectious diseases, *Arch. Intern. Med.,* 50,801, 1932.

128. **Wodnar-Filipowicz, A., Heusser, C.H. and Moroni, C.,** Production of the haemopoietic growth factors GM-CSF and interleukin-3 by mast cells in response to IgE receptor mediated activation, *Nature*, 339,150, 1989.

129. **Yamamura, M., Uyemura, K., Deans, R.J., Weinberg, K., Rea, T.H., Bloom, B.R. and Modlin, R.L.,** Defining protective responses to pathogens: cytokine profiles in leprosy lesions, *Science,* 254,277, 1991.

130. **Yelich, M.R.,** *In vivo* endotoxin and IL-1 potentiate insulin secretion in pancreatic islets, *Am. J. Physiol.,* 258,R1070, 1990.

131. **Yelich, M.R., Havdala, H.S. and Filkins, J.P.,** Dexamethasone alters glucose, lactate, and insulin dyshomeostasis during endotoxicosis in the rat, *Circ. Shock,* 22,155, 1987.

132. **Yelich, M.R., Umporowicz, D.M., Qi, M. and Jones, S.B.,** Insulin-inhibiting effects of epinephrine are blunted during endotoxicosis in the rat, *Circ. Shock,* 35,129, 1991.

133. **Zanetti, G., Heumann, D., Gerain, J., Kohler, J., Abbet, P., Barras, C., Lucas, R., Glauser, M.-P. and Baumgartner, J.-D.,** Cytokine production after intravenous or peritoneal gram-negative bacterial challenge in mice. Comparative protective efficacy of antibodies to tumor necrosis factor-α and to lipopolysaccharide, *J. Immunol.,* 148,1890, 1992.

Chapter 5

NEUROENDOCRINE REGULATION OF HEMATOPOIESIS

Georges J. M. Maestroni
Center for Experimental Pathology
Istituto Cantonale di Patologia
Locarno, Switzerland

INTRODUCTION

Why should the blood forming system be controlled by neural or neuroendocrine factors? This was the question I was asked quite recently during a discussion about the clinical relevance of the findings which are discussed here below. To my answer which sounded: why not?, the clinician who asked the question explained that it was difficult for him to envisage any physiological reason which could explain the need for such control of hematopoiesis. Apparently, he ignored totally the entire field of psychoneuroimmunology and therefore he could not find any logical support to the findings we were discussing. Unfortunately, this is still a too frequent experience. The existence of a common messengers/receptors network between the central nervous system and the immune system[1] is today beyond any doubt. Cytokines which play a crucial role in the immune response may also be produced in the central nervous system and affect neuroendocrine functions.[1,2] Neurohormones may, in turn, be synthesized by immunocompetent cells and affect immune functions.[1] On the other hand, almost all cytokines have been shown to affect the blood forming system.[3,4] The link between the neuroendocrine system and hematopoiesis seems thus straightforward. However, the experimental evidence for any neuroendocrine/hematopoietic relationship is scant or virtually non-existent. This is rather surprising especially in view of the central role of hematopoiesis as a source of lymphoid and myeloid cells. Beside lymphocytes and macrophages which are involved in antigen-specific immune responses, neutrophils constitute the first defense line in case of infectious events and eosinophils have recently appeared to exert a cytotoxic action against tumor cells.[5] Granulocytes are short-lived cells with a renewing-time of about 24 hours. Alterations of myelopoietic function are thus mirrored immediately in altered blood counts as it occurs during acute infections. Relevant to the topic of this book, I am not aware of any study about a possible neuroendocrine (stress) influence on myelopoiesis either in physiological or in pathological situations (infections).

Finally, some considerations from the mere hematological point of view deserve to be mentioned. Hematopoiesis depends on a highly complex series of events in which a small population of stem cells needs to generate large populations of maturing cells. The diverse differentiative and proliferative events as well as entry of mature cells into the circulation and their selective localization in appropriate tissues require a sophisticated regulatory control. The existence of multiple hematopoietic regulators such as cytokines and growth factors is well documented.[3] The multiplicity of hematopoietic regulators seems to reflect the need for a subtle physiological control of the complex cell mixtures required in certain situations.[3] If this poses several problems in our understanding of hematopoiesis, the situation is much worse from the clinical point of view. In fact, if the reality is that regulators are more effective when used in combination, the situation for a clinician becomes almost a nightmare. Single hematopoietic regulators are already used to counteract the bone marrow toxicity of cancer chemotherapy compounds or to enhance regeneration after bone marrow transplantation.[6-8] However, such procedures remain problematic because of negative side effects.[6-8] An endogenous modulation of hematopoietic regulators is likely to present substantial advantages over exogenous administration and might also circumvent the need for a very difficult clinical testing of thousands of regulators combinations. To ascertain the existence of a neural or neuroendocrine influence on the endogenous production of hematopoietic regulators seems thus very important. Here, I present the results of two different approaches we have undertaken to investigate this fascinating field.

ADRENERGIC MODULATION OF HEMATOPOIESIS

Our studies on the immunoregulatory role of the pineal neurohormone melatonin[9] led us to investigate its role in hematopoietic reconstitution after lethal irradiation and syngeneic bone marrow transplantation (BMT) in C57BL/6 mice. To inhibit the synthesis of melatonin, we used permanent environmental lighting (L24) or surgical pinealectomy. While exposure to L24 resulted in decreased peripheral blood leukocytes (PBL) counts after syngeneic BMT, surgical pinealectomy had no effect on PBL counts. In addition, melatonin replacement in the L24 treated group did not reverse the negative effect of L24. These findings suggested the pineal gland and melatonin have no effect on hematopoietic reconstitution, at least after BMT. More recent experiments have shown that melatonin can indeed influence hematopoiesis in other situations (see below). In the BMT model, however, after having ruled out stress hormones as possible mediators of the light effect, we had no thought about an alternate mechanism for the effect of L24. Beside the main optical system, light activates the accessory optical tract which includes the suprachiasmatic nucleus and the supracervical ganglia[10] (SCG). From the latter, efferent sympathetic fibers innervate directly the pineal gland, the thymus and send projections to the spinal cord which in turn reach the bone marrow via the para- and pre-vertebral ganglia.[10-12] In the case of pineal gland, the light (day)-darkness (night) cycle modulates the release of norepinephrine by the sympathetic terminals and constitutes the main environmental factor regulating the synthesis of melatonin.[1] Although the functional significance of the sympathetic innervation of the bone marrow is unknown, it was reasonable to hypothesize that constant environmental lighting inhibited leukocytes reconstitution after BMT via a direct modulation of norepinephrine release by sympathetic terminals in the bone marrow. Thus, it may be possible to influence bone marrow reconstitution using agents known to modulate the synaptic release of norepinephrine and/or compete for binding to the adrenergic receptor. Our studies showed that adrenergic agents can indeed influence hematopoietic functions either after syngeneic BMT or in a normal condition by acting on adrenoceptors present on bone marrow cells.

A. EFFECT OF CHEMICAL SYMPATHECTOMY

Groups of mice were sympathectomized by one i.p. injection of 6-hydroxydopamine (6-OHDA) on day 2 or given saline as control. On day 0 the mice were lethally irradiated and, on day 1, transplanted with syngeneic bone marrow and exposed either to L 24 or to 12 hour light:dark cycle (L12). The results showed that at various times after BMT, blood leukocytes were significantly lower in mice exposed to L 24 than in control mice.[13] However, chemical sympathectomy by 6-OHDA abolished the effect of light and increased the leukocyte concentration even under normal conditions (L12). Such increase was mainly due to granulocytes.[13]

B. EFFECT OF ADRENERGIC ANTAGONISTS

We continued our studies using mice kept under a normal L 12 photoperiodic cycle throughout the experiments. We treated groups of mice with the α1-adrenergic antagonist prazosin and/or with the β-adrenergic blocker propranolol after syngeneic BMT. We found that the daily s.c. administration of prazosin for 14 days induced a dramatic increase of the number of peripheral blood leukocytes, platelets and spleen cells (Fig 1). The effect of prazosin was significant at doses ≥5 mg/kg b.w.[13] Differential leukocyte counts revealed that the prazosin--induced increase was mainly due to increased granulocytes and monocytes (Fig 1). The number of granulocyte-macrophage-colony-forming-units (GM-CFU) in the transplanted marrow was consistent with such effect and indicated that prazosin really enhanced granulocyte and macrophage reconstitution.[13] The number of nucleated spleen cells was also significantly increased by prazosin treatment starting on day 14. Consistent with the peripheral blood counts, histologic analysis of spleen sections from such mice revealed a clear granulocytic hyperplasia.[13]

C. EFFECT OF PRAZOSIN ON LYMPHOPOIESIS

The α1-adrenergic antagonist prazosin was administered either after syngeneic BMT or in normal mice. As expected, the daily s.c. administration of prazosin (10 mg/kg b.w.) induced a very significant increase in the number of peripheral blood leukocytes, platelets and marrow GM-CFU both in transplanted and normal mice. Differential count of blood leukocytes confirmed that prazosin enhanced myelopoiesis in contrast to lymphopoiesis.[13,14] The number of nucleated spleen cells was significantly higher in prazosin-treated mice. Again, this effect was also evident when prazosin was injected chronically into normal mice. The FACS-determined, phenotypic composition of spleen cells from prazosin-treated mice either after syngeneic BMT or in normal animals showed that the increase of spleen cellularity appeared to depend on an impressive increase of granulocytes and macrophages while T and B lymphocytes were decreased, especially after BMT (Fig 2). Such effects were in line with the blood and bone marrow data. The apparent inhibition of lymphopoiesis was also reflected by a clear inhibition of thymus reconstitution after BMT and by a reduction of thymus cellularity in normal mice.[14] A comparison between effect of prazosin on thymus cellularity and subpopulations, in both bone marrow transplanted and normal, age-matched mice with the effect of corticosterone injected in normal mice, showed that, in contrast with corticosterone, prazosin affected only the

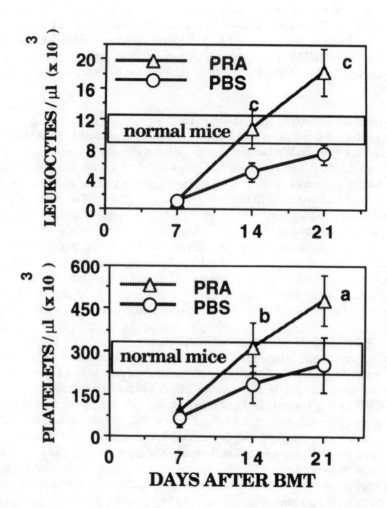

Fig 1.   Enhancing effect of prazosin on blood granulocytes and platelets after syngeneic BMT in mice. The mice were injected daily with prazosin (PRA, 10 mg/kg b.w.) or with PBS. Mean values ± standard deviation are relative to 4 experiments. The cumulative number of animals per group and point was of 35 mice at day 7, of 21 mice at day 14 and of 16 mice at day 21. Such differences were due to the timed sampling of animals to sacrifice for spleen cell count and GM-CFU assay. The horizontal bars represent the range of leukocyte and platelets concentration in normal mice.
a: $p < 0.005$; b: $p < 0.002$; c: $p < 0.001$ (ANOVA)

number of thymocytes and did not induce any change in the relative proportion of thymocyte populations.[14]

Moreover, to study whether the prazosin-induced decrease of lymphocyte regeneration after BMT was reflected by impaired cell mediated immune responses, we investigated NK and anti-viral CTL response in mice treated with prazosin or saline, 21 days after BMT. The results demonstrated that indeed NK activity and CTL response against Lancy Vaxinia virus were reduced in prazosin treated mice when compared with saline treated controls.[14]

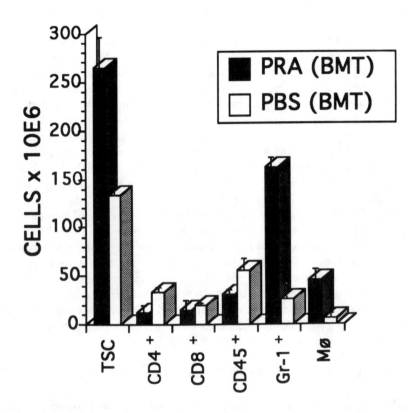

Fig 2. Effect of prazosin on number of spleen leukocyte subsets after syngeneic BMT. Prazosin (PRA) or saline (PBS) were injected once a day for 21 consecutive days after lethal irradiation and syngeneic BMT. The values represent the mean number of total spleen cells (TSC) and of the various leukocyte subsets ± the standard deviation evaluated by FACS. T-helper (CD4+) and suppressor/cytotoxic (CD8+) lymphocytes were labeled, respectively, with anti-L3T4 and anti-Ly-2, while B-lymphocytes, granulocytes and macrophages (Mo) were identified by anti-B.220 (CD45R), anti-Gr-1 and anti-macrophage monoclonal antibodies. All the PRA values were significantly different from PBS with $p < 0.01$.

D. *IN VITRO* STUDIES

Our results show that the function of the blood forming system is apparently under a sympathetic nervous regulation. Such regulation might be exerted directly on bone marrow cells or indirectly via other mechanisms. To verify these hypotheses, we investigated whether adrenergic agonists could influence hematopoietic functions *in vitro*. As a matter of fact, we found that when added directly to bone marrow cultures, adrenergic agonists proved to inhibit the number of GM-CFU. The number of colonies in control plates (0% inhibition) was 84.6 ± 18.5. The effect of the physiological adrenergic agonist noradrenaline started to be significant (p <0.01) already at a final concentration of 10^{-8} M (64.1 ± 12.4 colonies). At 10^{-4} M of noradrenaline the number of colonies was 15.6 ± 13.6. These results are shown in Fig 3.

Also the α1-selective adrenergic agonist methoxamine and to a much lesser extent the α2-agonist clonidine proved to exert an inhibitory action when added in the GM-CFU assay. The relative potency of these adrenergic agonists in inhibiting the number of GM-CFU appeared to be noradrenaline > methoxamine > clonidine.[15] In other experiments, α-adrenergic antagonists such as prazosin, phentolamine or yohimbine were added at different concentrations together with noradrenaline (10^{-6} M) in the GM-CFU assay. These antagonists were selected for their different adrenergic receptors selectivity. Prazosin is in fact an α1-antagonist, phentolamine antagonizes both α1 and α2 adrenoceptors and yohimbine is predominantly α2-selective. By comparing the concentrations at which the adrenergic antagonists were effective in counteracting noradrenaline, a relative order of potency was apparent, i.e.: prazosin $\geq$ phentolamine > yohimbine.[15]

At this point our results were consistent with the presence of α1-adrenergic receptors on bone marrow cells. Therefore we investigated whether ^{3}H-prazosin could bind specifically to bone marrow cells in saturation and competition experiments. Figure 4 shows the isotherm binding of 3H-prazosin in a saturation study. The curve shows a saturation region with concentrations of 3H-prazosin within 1-5 nM (Fig 4a) and, apparently, over 50 nM(Fig 4b). A computer-assisted non-linear regression analysis revealed that the data best fit to a model for two binding sites (Fig 4c). Competition experiments performed with a variety of α-adrenergic agonists and antagonists gave IC50 and kd values which were compatible with the presence of a high affinity α1B-adrenoceptor on bone marrow cells.[15]

CONCLUSIONS REGARDING ADRENERGIC MODULATION OF HEMATOPOIESIS

We have demonstrated that hematopoiesis is under a sympathetic neural control. Such control seems to be exerted directly on bone marrow cells via α1-adrenergic receptors. Blockade of such receptors seems to enhance myelopoiesis and to inhibit lymphopoiesis. This dual effect might imply pluripotent stem cells as targets of the adrenergic regulation. This would not necessarily mean that pluripotent stem cells bear adrenergic receptors. In fact, we do not

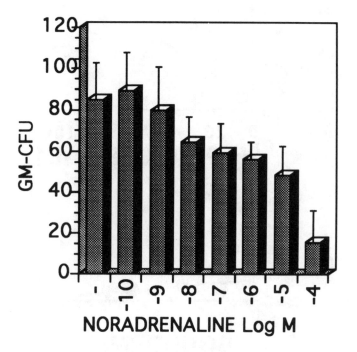

Fig 3. Inhibition of GM-CFU by noradrenaline. Noradrenaline was dissolved in α-MEM/- 20% horse serum and added at the reported concentrations in the bone marrow cell suspensions before plating. For each concentration, the mean value from five experiments ± the standard deviation is shown. a: p <0.01

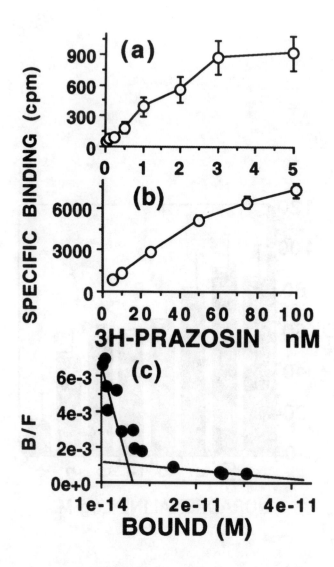

Fig 4. Saturation isotherm and Scatchard plot of ^{3}H-prazosin binding in bone marrow cells. Triplicate samples of 1 ml of viable bone marrow cell suspension (2 x 10^6) were incubated at room temperature for 60 min, in presence of the reported concentrations of ^{3}H-prazosin ± phentolamine 10µM. The curve shows two saturation regions within concentrations of ^{3}H-prazosin 1-5 nM (a) and 40 - 100 nM (b). Non linear regression analysis gave the Scatchard plot represented in (c), which describes 2 binding sites with kd 0.98 ± 0.32 nM, Bmax 5 ± 1.9 fM/1 x 10^6 cells, kd 55.9 ± 8.2 nM, Bmax 44 ± 7.7 fM/2 x 10^6 cells.

yet know the nature of the bone marrow cells which express adrenergic receptors. As bone marrow cells contain also macrophages and mature lymphoid cells, we cannot exclude that the observed noradrenergic modulation of GM-CFU occurs via adrenoceptors present on these cells which in turn might release mediators affecting GM-CFU precursors or pluripotent stem cells. For example, it has been reported that stimulation of α2-adrenoceptor on macrophages augments the production of tumor-necrosis-factor(TNF).[16] TNF has been reported to suppress proliferation of hematopoietic progenitor cells with some preference for erythroid progenitors.[17] Apart from the type of adrenoceptor, a similar mechanism might account for our findings. Other cyto- or lymphokines might also be involved or, alternatively, activation of adrenergic receptors might mediate suppression of receptors for growth factors on hematopoietic cells, consistent with what has been reported for β-adrenergic and interleukin-2 receptors on lymphocytes.[18]

HEMATOPOIETIC RESCUE VIA T-CELL-DEPENDENT, ENDOGENOUS GM-CSF INDUCED BY THE PINEAL NEUROHORMONE MELATONIN

Melatonin is an indoleamine synthesized from serotonin. Its synthesis and release which occur primarily in the pineal gland follows a circadian rhythm with the highest blood concentration occurring at night in all species.[19] In previous work we have shown that melatonin can correct immunodeficiency states which may follow acute stress, viral diseases, aging or drug treatment.[9] We also observed that melatonin was able to antagonize the effect of high dose cyclophosphamide on antibody production.[20] This finding has been then confirmed and extended by other authors.[21,22] Such interesting effects of melatonin seem to depend on activated CD4+, T-cells which upon melatonin stimulation show an enhanced synthesis and/or release of opioid peptides, IL-2 and γ-interferon.[9,22-24] On the basis of our animal studies,[9] we have investigated the clinical effect of melatonin in association with low-dose interleukin-2 in cancer patients and found that this association represents a well tolerated strategy capable of determining an apparent control of tumor growth in patients with advanced solid neoplasms.[25-28]

A. EFFECT OF MELATONIN IN MICE TREATED WITH CANCER CHEMOTHERAPY COMPOUNDS

Programmed cell death or apoptosis is a normal process by which cells are eliminated during embryonic development and in adult life. Disruption of this normal process can cause developmental abnormalities and facilitate cancer development.[29] Normal hematopoietic cells require certain viability and growth factors like CSFs and undergo apoptosis when these factors are withdrawn.[30] Programmed cell death can be induced by removal of CSFs, by cytotoxic therapeutic agents, or by the tumor suppressor gene wild-type p53.[31]

Based on the above-mentioned properties, we studied whether melatonin could rescue bone marrow cells from apoptosis induced by cancer chemotherapy compounds. Melatonin was injected in 2 months old, female C57BL/6 mice which were transplanted with Lewis lung carcinoma (LLC) and treated with the antitumor agents etoposide or cyclophosphamide. The results obtained showed that melatonin can protect bone marrow functions from the toxic effect of cancer chemotherapy compounds without interfering with their anti-cancer action *in vivo*. Such effect was apparent and highly significant on leukocytes, platelets and marrow GM-CFU. On the contrary, melatonin per se or in association with the anti-tumor compounds did not influence neither the size of primary tumor nor the number of lung metastases.[15] To investigate whether melatonin acted by preventing bone marrow toxicity or by enhancing the post-treatment recovery of hematopoiesis, experiments were devised in which melatonin was injected only during cyclophosphamide treatment or only thereafter. Melatonin was able to antagonize the hematopoietic toxicity of cyclophosphamide when injected together with the drug. Bone marrow protection appeared to be less effective if melatonin was injected after cyclophosphamide treatment. These results indicated that melatonin is able to protect the bone marrow in the course of cytotoxic anti-cancer treatments.[15]

B. *IN VITRO* EFFECT OF MELATONIN ON DRUG-INDUCED APOPTOSIS

To investigate whether melatonin acted directly on hematopoietic cells, different concentrations of melatonin (10^{-10} to 10^{-6} M) in presence of etoposide 10^{-5} M were incubated with bone marrow cells. Melatonin proved to rescue bone marrow cells from apoptosis induced by etoposide (Fig 5).

Such effect of melatonin was already evident at physiological concentration (10^{-9} M). The protective effect exerted by melatonin *in vitro* was reflected by an increased number of the lineage-committed myeloid precursors GM-CFU but not of the less differentiated, pluripotent spleen-colony-forming unit (S-CFU).[15] These results suggested that melatonin can induce the production of endogenous CSFs. In particular, the *in vivo* effect on platelets and the rescue of GM-CFU but not of S-CFU candidated GM-CSF as a possible mediator of melatonin.[30-32] To test this possibility, bone marrow cells were incubated with etoposide and melatonin in presence of monoclonal rat anti-mouse GM-CSF antibodies. Indeed, the results shown in Fig 6 demonstrate that anti-GM-CSF antibodies were capable of neutralizing the protective effect of melatonin. This indicated that melatonin exerted its effect by inducing bone marrow cells to produce a factor which is biologically and immunologically indistinguishable from GM-CSF. In bone marrow cell suspensions, the principal sources of GM-CSF are macrophages and T cells. Athymic nude mice are T-cell-deficient, a condition which is in part balanced by enhanced macrophage functions. Therefore, we chose T-cell deficient C57BL/6 nu-nu mice as donors of bone marrow cells to investigate which cell type could release GM-CSF upon melatonin stimulation. As a matter of fact, when bone marrow cells were obtained from T-cell-deficient mice, melatonin did not exert any protection against etoposide toxicity. This indicated that T lymphocytes are the target of melatonin in the bone marrow.

CONCLUSION

In this study we have shown that the pineal neurohormone melatonin can protect bone marrow functions from the toxic effect of cancer chemotherapy compounds without interfering with their anti-cancer action *in vivo*. This effect is exerted directly on bone marrow T-cells which may release or participate in the induction of a GM-CSF-like factor upon melatonin stimulation. Whether the GM-CSF-like substance which mediates the effect of melatonin is released by such T cells or by stromal cells upon a paracrine stimulation by another T-cell cytokines, remains to be established. Another unsolved problem concerns the mechanism by

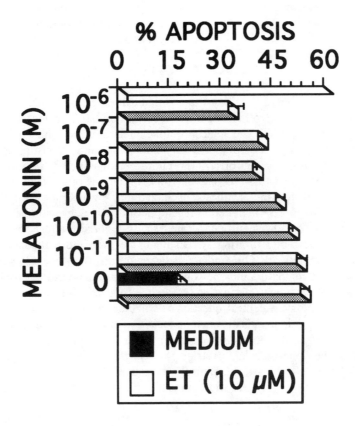

Fig 5. Prevention of etoposide-induced apoptosis by melatonin. Bone marrow cells from C57BL/6 mice were incubated with etoposide in presence of the reported concentrations of melatonin. The values are the mean ± the standard deviation of 3 experiments. The "MEDIUM" column represents the percent of apoptotic cells after 8 hr incubation in medium alone.

which melatonin can stimulate T-cells to produce cytokines. Melatonin seems, in fact, to be active only in the presence of a cytotoxic injury. In any case, this result is consistent with previous findings related to the immunoregulatory effect of melatonin.[9,21-24,33-36] In addition, it indicates that the pineal signal to the blood forming system[37] is constituted by melatonin.

The evidence of a neuroendocrine influence on endogenous GM-CSF production has important basic and practical implications. In most forms of cancer, a major barrier to achieving the best possible response to cancer chemotherapy is the hematological toxicity of available agents, which limits optimal dosing.[38] To circumvent this problem, hematopoietic rescue by CSFs and/or autologous bone marrow transplantation is increasingly used.[7,8] Handling of such procedures remains however problematic because of negative side effects or incomplete marrow regeneration, respectively.[7,8] On the contrary, melatonin is well known to have no toxic effects.[19] Our finding may thus have a straightforward clinical application.

COMMENTS

The findings presented here constitute sound evidence that hematopoiesis is under a neural and neuroendocrine control. The neural adrenergic regulation seems to affect pluripotent hematopoietic precursors, while the neurohormone melatonin acts on committed myeloid progenitor cells. We are thus in front of a very interesting and sophisticated regulation of hematopoiesis. The sympathetic neurotransmitter noradrenaline inhibits myelopoiesis by acting directly on α1-adrenergic receptors on bone marrow cells. On the other hand, it constitutes also the physiological signal for the synthesis of melatonin,[39] which, in turn, stimulates the endogenous release of GM-CSF and, hence, myelopoiesis. In analogy with the immune system, other neurotransmitters or other neuroendocrine circuits might also affect hematopoiesis. This subtle regulation of the blood forming system might be even more fundamental than that exerted by the cytokine network. Our findings indicate, in fact, that the endogenous release of multiple hematopoietic regulators may be controlled by neuroendocrine factors.

What appears as a new, fascinating research avenue calls for further studies. Relevant to psychoneuroimmunology, a central question is whether the neuroendocrine regulation of hematopoiesis plays any role in aplastic anemia, leukemia or during emergencies such as acute infections and/or stress events. Any positive answer to this question might provide a conceptual framework in which new pharmacological strategies to prevent or correct pathological situations may be devised. Last but not least, the clinical application of our melatonin findings is highly desirable in view of its feasibility and wide importance in oncology.

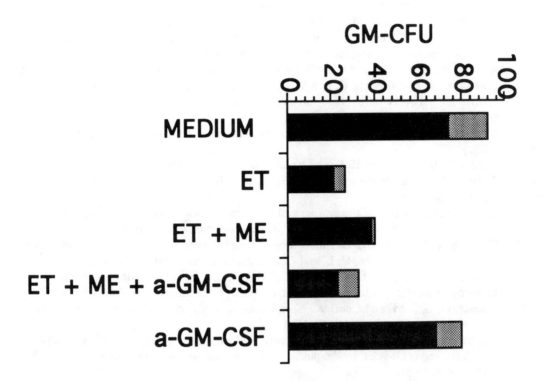

Fig 6.   Monoclonal rat anti-mouse GM-CSF neutralizes the effect of melatonin. Bone marrow cells from C57BL/6 mice were pre-incubated for 8 h in culture medium alone or in presence of etoposide (ET, 10 μM), melatonin (ME, 1 μM) and/or monoclonal rat anti-mouse GM-CSF (a-GM-CSF). The concentration of a-GM-CSF was of 150 μg/ml. The values represent the mean number plus the standard deviation (lighter area) of GM-CFU/10^5 bone marrow cells from 3 experiments.

REFERENCES

1.   **Ader, R., Felten, D. L. and Cohen, N.,** Eds., *Psychoneuroimmunology II,* Academic Press, San Diego, 1991.
2.   **Besedovsky, H. O., Del Rey, A. E., Sorkin, E. and Dinarello, A.,** Immunoregulatory feedback between interleukin-1 and glucocorticoid hormones, *Science,* 233,652, 1986.
3.   **Metcalf, D.,** Hematopoietic regulators: redundancy or subtlety? *Blood,* 82,3515, 1993.

4. **Zipori, D.,** The renewal and differentiation of hemopoietic stem cells, *FASEB J.,* 6,2691, 1992.
5. **Wenge, P.,** What is the role of the eosinophil? *Thorax,* 45,161, 1990.
6. **Nemunaitis, J., Appelbaum, F. R., Singer, K., Lilleby, K, Wolff, S., Greer, J. P., Bierman, P., Resta, D., Campion, M., Levitt, D., Zeigler, Z., Rosenfeld, C., Shadduck, R. K. and Buckner, C. D.,** Phase I trial with recombinant human interleukin-3 in patients with lymphoma undergoing autologous bone marrow transplantation, *Blood,* 82,3273,1993.
7. **Antman, K. H.,** G-CSF and GM-CSF in clinical trials, *Yale J. Biol. & Med.,* 63, 387, 1990.
8. **Sheridan, W. P., Morstyn, G. and Wolf, M.,** Granulocyte colony-stimulating factor and neutrophil recovery after high dose chemotherapy and autologous bone marrow transplantation, *Lancet,* ii,891, 1989.
9. **Maestroni, G. J. M.,** The immunoneuroendocrine role of melatonin, *J. Pineal Res.,* 14,1, 1993.
10. **Cardinali, D. P., Romeo, H. E. and Vacas, M. I.,** Neuroendocrine projections of superior cervical ganglia, in *Advances in Pineal Research 2,* Reiter, R. J., Ed., John Libbey & Co., London, 1987, p.35.
11. **Ebadi, M.,** Regulation of the synthesis of melatonin and its significance to neuroendocrinology, in *The Pineal Gland,* Reiter, R. J., Ed., Raven Press, New York, 1984, p. 1.
12. **Felten, S. Y. and Felten, D. L.,** Innervation of lymphoid tissue, in *Psychoneuroimmunology II,* 27-71, Ader, R., Felten, D. L. and Cohen, N., Eds., Academic Press, San Diego, 1991.
13. **Maestroni, G. J. M., Conti, A. and Pedrinis, E.,** Effect of adrenergic agents on hematopoiesis after syngeneic bone marrow transplantation in mice, *Blood,* 5,1178, 1992.
14. **Maestroni, G. J. M. and Conti, A.,** Noradrenergic modulation of lymphohematopoiesis, *Int. J. Immunopharmacol.,* 16,117, 1994.
15. **Maestroni, G. J. M., Covacci, V. and Conti, A.,** Hematopoietic rescue via T-cell-dependent, endogenous GMCSF by the pineal neurohormone melatonin in tumor bearing mice, *Cancer Res.,* in press., 1994.
16. **Spengler, R. N., Allen, R. M., Remick, D. G., Strieter, R. M. and Kunki, S. L.,** Stimulation of a-adrenergic receptors augments the product of macrophage-derived tumor necrosis factor, *J. Immunol.,* 145,1430, 1990.
17. **Roodman, D. G.,** TNF and hematopoiesis, in *Tumor Necrosis Factor. The Molecules and Their Emerging Role in Medicine,* Beutler, B., Ed., Raven Press, New York, 1992, p.117.
18. **Feldman, R. D., Hunninghake, G. W. and MacArdle, W. L.,** Beta-adrenergic-receptor-mediated suppression of interleukin-2 receptors in human lymphocytes, *J. Immunol.,* 139,3355, 1987.
19. **Yu, H.-S. and Reiter, R. J.,** Eds., *Melatonin. Biosynthesis, Physiological Effects, and Clinical Applications,* CRC Press, Boca Raton, 1993,
20. **Maestroni, G. J. M. and Conti, A.,** Melatonin in relation to the immune system, in *Melatonin. Biosynthesis, Physiological Effects, and Clinical Applications,* Yu, H.-S. and Reiter, R. J., Eds., CRC Press, Boca Raton, 1993, 290.
21. **Pioli, C., Caroleo, M. C., Nistico, G. and Doria, G.,** Melatonin increases antigen presentation and amplifies specific and nonspecific signals for T-cell proliferation, *Int. J. Immunopharmacol.,* 15,463,1993.

22. **Caroleo, M. C., Frasca, D., Nistico, G. and Doria, G.,** Melatonin as immunomodulator in immunodeficient mice, *Immunopharmacol.,* 23,81, 1992.

23. **Del Gobbo, V., Libri, V., Villani, N., Calio, R. and Nistico, G.,** Pinealectomy inhibits interleukin-2 production and natural killer activity in mice, *Int. J. Immunopharmacol.,* 11,567, 1989.

24. **Maestroni, G. J. M. and Conti, A.,** The pineal neurohormone melatonin stimulates activated CD4+, Thy1+ cells to release opioid agonist(s) with immunoenhancing and anti-stress properties, *J. Neuroimmunol.,* 28,167, 1990.

25. **Lissoni, P., Tisi, E., Barni, S., Ardizzoia, A., Rovelli, F., Rescaldani, R., Ballabio, D., Benenti, C., Angeli, M. A., Tancini, G., Conti, A. and Maestroni, G. J. M.,** Biological and clinical results of a neuroimmunotherapy with interleukin-2 and the pineal hormone melatonin as a first line treatment in advanced non-small cell lung cancer, *Br. J. Cancer,* 66,155, 1992.

26. **Lissoni, P., Barni, S., Rovelli, F., Brivio, F., Ardizzoia, A., Tancini, G., Conti, A. and Maestroni, G. J. M.,** Neuroimmunotherapy of advanced solid neoplasms with single evening subcutaneous injection of low-dose interleukin-2 and melatonin: preliminary results, *Eur. J. Cancer,* 29A,185, 1993.

27. **Lissoni, P., Barni, S., Tancini, G., Ardizzoia, A., Rovelli, F., Cazzaniga, M., Brivio, F., Piperno, A., Aldeghi, R., Fossati, D., Characjeius, D., Kothari, L., Conti, A. and Maestroni, G. J. M.,** Immunotherapy with subcutaneous low dose interleukin-2 and the pineal indole melatonin as a new effective therapy in advanced cancers of the digestive tract, *Br. J. Cancer,* 67,1404, 1993.

28. **Lissoni, P., Barni, S., Tancini, G., Ardizzoia, A., Ricci, G., Aldeghi, R., Brivio, F., Tisi, E., Rescaldani, R., Quadro, G. and Maestroni, G. J. M.,** A randomised study with subcutaneous low-dose interleukin 2 alone vs interleukin 2 plus the pineal neurohormone melatonin in advanced solid neoplasms other than renal cancer and melanoma, *Br. J. Cancer,* 69,196, 1994.

29. **Arends, M. J. and Willie, H. A.,** Apoptosis: Mechanisms and roles in pathology, *Inter. Rev. Exp. Pathol.,* 32,223, 1991.

30. **Williams, G. T., Smith, C. A., Spooncer, E., Dexter, T. M. and Taylor, D. R.,** Haemopoietic colony stimulating factors promote cell survival by suppressing apoptosis, *Nature,* 343,76, 1990.

31. **Sachs, L. and Lotern, J.,** Control of programmed cell death in normal and leukemic cells: New implications for therapy, *Blood,* 82,15, 1993.

32. **Garland, J. M.,** Colony stimulating factors, in *The Cytokine Handbook,* Thomson, A., Eds., Academic Press, London, 1991, p.269.

33. **Pang, S. F., Lee, P. P. N., Chan, Y. S. and Reiter, R. J.,** Melatonin secretion and its rhythms in biological fluids, in *Melatonin, Biosynthesis, Physiological Effects, and Clinical Applications,* Yu, H.-S. and Reiter, R. J., Eds., CRC Press, Boca Raton, 1993,129.

34. **Giordano, M. and Palermo, M. S.,** Melatonin-induced enhancement of antibody dependent cellular cytotoxicity, *J. Pineal Res.,* 10,117,1991.

35. **Hansson, I., Holmdahl, R. and Mattsson, R.,** The pineal hormone melatonin exaggerates development of collagen induced arthritis in mice, *J. Neuroimmunol.,* 39, 1992.

36. **Lissoni, P., Barni, S., Tancini, G., Rovelli, F., Ardizzoia, A., Conti, A. and Maestroni, G J. M.,** A study of the mechanisms involved in the immunostimulatory action of the pineal hormone in cancer patients, *Oncology*, 50,399, 1993.

37. **Haldar, C., Haussler, D. and Gupta, D.,** Response of CFUGM (colony forming units for granulocytes and macrophages) from intact and pinealectomized rat bone marrow to macrophage colony stimulating factor (rGM-CSF) and human recombinant erythropoietin (rEPO), *Pr. Brain Res.*, 91,323, 1992.

38. **DeVita, V. T. J.,** Principles of chemotherapy, in *Cancer: Principles & Practice in Oncology,* DeVita, V. T. J., Hellman, S. and Rosenberg, S. A., Eds., J.B. Lippincott Co, Philadelphia, 1989, p.276.

39. **Yu, H.-S., Tsin, A. T. C. and Reiter, R. J.,** Melatonin: History, Biosynthesis, and Assay Methodology, in *Melatonin, Biosynthesis, Physiological Effects, and Clinical Applications*, Yu, H.S. and Reiter, R. J., Eds., CRC Press, Boca Raton, 1993, p.1.

Chapter 6

PSYCHONEUROIMMUNOLOGY, STRESS AND DISEASE

Michael Schlesinger
The Hubert H. Humphrey Center for Experimental
Medicine and Cancer Research
The Hebrew University/Hadassah Medical School
Jerusalem, Israel

Yair Yodfat
Department of Family Medicine
The Hebrew University/Hadassah Medical School
Jerusalem, Israel

Stress is considered to be the intervening variable between psychosocial processes and illness. It is a very useful concept for the lay public because it helps to explain the effect of psychological distress on organic processes. However, for the study of its impact on health and illness stress usually refers to one of three components of what has been called the "stress process".[1]

1. the stressors: which are environmental events exerting an impact on the subject. Death of a close relative could be such an event;
2. the physiologic response to the stressors, such as the "flight-fight" response; and
3. the health consequences of confrontation with the stressors.

Stressful life events and other psychosocial factors may increase the susceptibility to illnesses and mortality.[2,3] In 1952 Hinkle and Plummer[4] showed that telephone company employees with high illness-absentee rates reported higher levels of dissatisfaction with life and 12 times the number of minor respiratory illnesses than a control sample of workers with low absence rate. Beautrais *et al.*[5] showed that hospital admissions and general practitioner consultations for lower respiratory tract illness, gastroenteritis, accidents, poisoning, burns/scalds, and suspected home related conditions in 1 to 4 year old children were highly correlated with the number of family life events. Children from families with more than 12 life events were 6 times more likely to have been hospitalized. According to Plaut and Friedman[6] primary and mediating variables contribute to the relation between stressors and disease resistance. The primary variables are stress, coping, and disease parameters such as weight loss, illness, behavior, etc. The mediating variables are situational factors (nutrition, circadian rhythms, temperature and extraneous stimulation), psychosocial mechanisms (e.g., differential housing and life changes), genetic and environmental history, and physiological mechanisms (e.g., hormonal mechanisms).

How people react to stress is a very complicated issue. Some stressors have an adverse effect on the individual while other stressors can even be health- or growth-promoting. Most retrospectively designed human studies in the area of psychoneuroimmunology link disease with "bad" stressors such as depression, loneliness, and hopelessness. Unresolved conceptual issues related to stress include the following problems: can stressors be chronic, unchanging conditions rather than events? Must all stressors be perceived as undesirable or distressing? Are certain physiologic processes specific to the stress process?[7] Life event instruments are complicated by considerable methodological problems and in many retrospective studies, there are serious reporting biases. There are many confounding variables like age, sex, socioeconomic status and concomitant illnesses. In most studies, a very small (less than 5% of the variance) linear correlation could be demonstrated between life event scores and subsequent illness.[8,9]

Why is it that some people become ill following exposure to stress, while others who are exposed to the same magnitude of stress remain healthy? Kobasa et al.[10] explain this different susceptibility in the personality characteristic of "hardiness". According to their findings those individuals who are "stress-hardy" and who possess a high level of control, commitment and challenge maintain or even increase their sense of well-being. There are considerable debates whether individuals with certain personality types are more prone to specific diseases than individuals with other personality types. Some studies have suggested that there may be "cancer prone", "coronary prone" or "autoimmune prone" personalities, although Solomon[11] suggested that the common denominator is that of an "immunosuppression prone" personality. Hagnell[12] found an increased risk of cancer in those with a "substable" personality. Grossarth-Maticek et al.[13,14] suggested that the "cancer prone" personality is characterized by suppression of emotion and an inability to cope with interpersonal stress, leading to feeling of hopelessness, helplessness and finally depression. He and his colleagues found that high scores of rationality and anti-emotionality predicted the mortality risk of cancer, heart disease and stroke. Williams[15] found that the hostility component of Type A behavior pattern predicted heart disease and increased risk of death from cancer and other illnesses. In contrast, Persky et al.[16] did not find any correlation between premorbid personality and the occurrence of cancer. In a cohort of medical doctors, Thomas et al.[17] correlated particular psychological characteristics and early family life experiences with a range of different diseases that these individuals experienced over 40 years. McClelland et al.[18] classified life change events as involving power (gaining power by assertiveness), affiliation (like loss of a loved one), or achievement (thoughts of performing better or of unique accomplishments). They found that individuals who scored high in the need for power, in inhibition (control of assertiveness) and in power stress (life events which challenge or threaten the ability to perform powerfully or impress others) reported more severe physical illness and effective symptoms than subjects who had low scores in their need for power, and in power stress and in inhibition.

The impact of stressful life events on an individual depends on coping skills and on conditioning against these stressors. Coping with stressful situations is determined on the one hand by the psychological profile of the individual and on the other hand by resources like social support available to "buffer" the impact of the stressors. Socially isolated persons, who as a consequence suffer from relative lack of social support, have been found to have higher rates of morbidity and mortality.[19] Moreover, social support may be one of the critical factors that can protect individuals who are exposed to stressors from getting sick.[20]

We found no correlation between the amount of stress, as determined with a stressful life events scale, and the number and activity of natural killer (NK) cells.[21] In contrast, a striking correlation was found between coping capacity and the NK system. Indeed, the lowest NK activity was observed among individuals suffering from anxiety.[21] In further studies the effect of social and family support on NK cells was determined.[22] In males the cytotoxic activity of NK cells correlated with hostility but was negatively correlated with family adhesion, adaptability and hardiness. Over a one and a half year long observation period, changes in the level of NK activity and in the number of NK cells in males were found to correlate positively with changes in demoralization and negatively with changes in family cohesion and adaptability.

The mechanism by which stress can cause illness is very complicated. Both human and animal studies indicated that stress can exert a suppressive effect on the immune function.[23, 24] At present, there is no firm conclusion concerning the clinical implication of this immunosuppression. To a large extent the effect of stress on immunity results from stress-induced alterations of a number of endocrine systems. Stress stimulates adrenocortical secretion, leading to increased glucocorticosteroid hormone levels, while activation of the sympathetic nervous system leads to an increased release of catecholamines. Within the limbic system of the central nervous system the septo-hippocampal cholinergic structure is involved in adaptation to stress.[25] Considerable evidence suggests that the hypothalamus controls the sympathetic activity and endocrine secretions activated as a consequence of stress. The hypothalamus may exert its activity through its effect on the pituitary gland, autonomic nervous system and the adrenal gland that secrete corticosteroids, β-endorphins and catecholamines, which in turn are known to inhibit the activity of both macrophages and lymphocytes.[26,27] Many other neurotransmitters and hormones possess either enhancing or suppressive immunomodulatory properties. These include the following: prolactin, enkephalins, ACTH, somatostatin, and many others.[23] Specific receptors for many of the neurohormones have been found on cells of the immune system.[28] Activated immune cells seem to be capable of producing and secreting small quantities of a variety of hormones. Moreover, immunocytes secrete cytokines which in turn may affect the neural system. Interleukin-1 (IL-1) a protein produced predominantly by activated macrophages and monocytes has an important role in the regulation of the immune defense. The administration of IL-1 to rats increased the blood levels of ACTH and corticosterone.[29] This effect was shown to be mediated by increased secretion of corticotropin releasing factor.[29] Similarly, thymosin fraction 5 produced by the thymus was shown to regulate the synthesis of ACTH and β-endorphin by cells of the pituitary gland, without any effect on the synthesis of either LH, FSH, GH, or TSH hormones.[30] Many other interactions among neural, endocrine and immune processes which bear on the stress-induced regulation of the immune system and on the state of health of the individual are still being elaborated.

It is too early to conclude which clinical impact the immunologic changes resulting from either stress or conditioning may have on health or illness. Ader[31] believes that these immunologic changes are within the homeostatic limits and only exceptional environmental circumstances can sometimes induce pathophysiologic states. According to him, in order to address the issue of clinical significance, it is necessary to consider the potential interaction between behaviorally induced physiological changes and the responses to immunogenic stimuli that will influence, if not actually determine, the adaptive response. Another complicated issue is that many confounding life habits may modulate the immune system and changes in health. Health behaviors such as sleep, smoking, physical activity, recent weight change, current health status, medications, alcohol, and caffeine intake may have immunologic and health consequences.[32]

Many attempts were made to demonstrate that the effect of stress on manifestation of illness and on premature death may be mediated through stress-mediated effects on the immune system. In this chapter we will discuss other clinical implications of psychoneuroim-munology. A number of studies demonstrated that after the death of their spouses, there is excess mortality from heart attack, infectious disease and cancer among bereaved men.[33,34] However, although in several studies immunosuppression was found after bereavement,[35,36] there is no clear evidence to support the relationship between this immunosuppression and the increased risk of dying.

Stress may affect several skin diseases such as psoriasis, urticaria, pruritus, alopecia areata, acne and eczema. According to Farber *et al.*,[37] the stress of emotional factors leads to the release of neuropeptides such as substance P from sensory nerve fibers in the skin causing local inflammatory responses, which, in turn, trigger psoriatic lesions. Koblenzer[38] predicts that similar mechanisms will be found in the entire spectrum of inflammatory dermatoses.

Acute psychological stress, such as that experienced by first-time parachute jumpers, causes a transient increase of NK activity followed by decreased activity.[39] Chronic stress usually causes immunosuppression. However, many studies, mainly in experimental animals, deal with the effect of stress on the development of autoimmune disorders. Autoimmune diseases arise from a combination of genetic, immunologic, viral, hormonal, and psychoneurologic factors.[40,41]

The effect of stress on the etiology of autoimmune disease was supported by the report of Baker *et al.*,[42] that the incidence of stressful life events was increased in the months preceding the onset of rheumatoid arthritis. The interrelationship between psychological factors, immunity and disease is highlighted by findings indicating an association between stressful life events and the pathogenesis of Graves' disease, an autoimmune disease of the thyroid. Early studies indicated that among populations of hyperthyroid patients high levels of stressful life events preceded the onset of the disease.[43,44] In two recent studies, life changes were studied in patients with Graves' disease as compared with matched control subjects. In the study of Winsa *et al.*[45] patients reported more negative events in the 12 months preceding the diagnosis of Graves' disease, and the negative life events scores were significantly higher in patients than in controls. In the study by Sonino *et al.*,[46] a semistructured research interview, covering 64 life events, was administered to 70 consecutive patients with Graves' disease and to an equal number of healthy subjects, matched for sociodemographic variables. Care was taken to administer the test during anti-thyroid drug induced remission rather than during the acute phase of the disease. Patients with Graves' disease reported significantly more life events than controls, and had more independent events including events that had an objective negative impact in comparison with controls. Stressful life events may affect the immune function in a number of ways that would precipitate an autoimmune process.[47] Autoimmunity is mediated by activation of T-lymphocytes, presumably interacting with thyroid antigens presented in the context of HLA antigens. Stressful life events could presumably lead to neuroendocrine stimulation of the production of cytokines. The cytokines could, in turn, induce the expression of HLA determinants on thyroid cells thus making them capable of presenting antigen, or lead to activation of autoimmune T lymphocytes.[47]

Many studies in this field concentrated on the therapeutic aspects of conditioning in autoimmune disease in lupus-prone mice. Ader and Cohen[48] have shown that classical Pavlovian conditioning using paired cyclophosphamide and saccharin can be therapeutic in lupus-prone mice.

Psychoneuroimmunologic factors have been reported to influence cancer risk and the progression of the disease, reviewed by Sabbioni.[49] It has been pointed out that analysis of the effect of psychological factors on the development of cancer is fraught with numerous methodological difficulties. Psychological factors may determine behavioral traits which affect the development of cancer such as cigarette smoking, alcohol consumption, and other health related habits. In addition, the possibility has been raised that specific personality traits, and in particular type C personality,[12,13,50] may be characteristic for cancer-prone individuals.

A number of studies indicated that psychological factors may determine the length of the recurrence-free period after surgery.[51,52] In contrast, in the study of Cassileth et al.,[53] no association was found between social and psychological factors and either the length of survival or the time of relapse of patients with advanced malignant disease. They concluded that the biology of the disease appears to predominate and to override the potential influence of life style and of psychological variables once the disease process is established.

Among matched pairs of women who underwent surgery for operative breast cancer an association was found between stressful life events and the relapse of cancer.[54] The relative risk of cancer relapse was significantly increased with severe life events and with severe, persistent difficulties. In a study of patients with breast cancer in stages I and II Levy et al.[55] found that the level of NK activity was correlated with perceived social support. Higher NK activity was associated with high quality emotional support either from a spouse or from another intimate individual, with perceived social support from the patient's physician, and with a major coping strategy of actively seeking social support.

꙳ Psychotherapy, counseling and the use of support groups have been reported to improve the outcome of patients with cancer.[56] Spiegel et al.[57] reported that psychosocial intervention which consisted of weekly supportive group therapy with self hypnosis for pain prolonged the survival of patients with metastatic breast cancer with a mean of 37 months as compared with 19 months in the control group.

Studies have shown that ameliorating the damage of stress may lead to immuno-enhancement but only a few studies include clinical studies outcome follow-ups. Behavioral interventions used to enhance the immune system in humans include hypnosis and relaxation,[58] conditioning,[59] motivational arousal[60] and exercise.[61] Intervention studies that have shown beneficial clinical effect in immunologically related diseases have generally lacked immunologic measures. The study of Fawzy et al.[62, 63] demonstrated that a coping-enhancement group intervention in patients with malignant melanoma showed an increase in natural-killer cell activity 6 months later and prolonged survival after 6 years. Similarly, Ironson et al.[64] demonstrated significant relations between psychological variables and both immune status and HIV-1 disease progression in a sample of gay men, enrolled in stress management intervention groups. Specifically, psychological distress, denial and low adherence during intervention signals predicted 2 year disease progress. Solomon and colleagues[65] are following a cohort of healthy independently living elderly women for 6 years in order to determine whether life stress, failure of coping, psychological distress, and immunologic changes antecede illness and death.

Prolonged psychosocial stress may cause hypertension by altering neuroadrenergic neurotransmission[66] and emotional stress may trigger acute myocardial infarction.[67] This is most probably explained by excessive secretion of catecholamines that cause overconstriction of the blood vessels, endothelial damage, increase in platelet adhesiveness and aggregation, increase of myocardial oxygen demand and decrease of insulin secretion.[68] Infusion of catecholamines in dogs caused within minutes showed microscopic evidence of contraction band lesions or overcontracted myocardial fibers. Higher doses of catecholamine caused fatal rhythm disturbances.[69] Contraction band lesions were present also in pilots who had lost control of their aircraft, could not eject and had died before their plane hit the ground.[70] It has also been shown in rabbits and rats that atherogenic uptake of low density lipoprotein is accelerated by adrenaline and noradrenaline and in human atheromatous lesions, the concentration of macrophages is greater in plaques that have fissured or ulcerated. This is compatible with the proposition that macrophages contribute to plaque fissure, the commonest immediate cause of coronary thrombosis.[71]

Much remains to be learned about the mechanisms by which psychological factors in general, and in particular those associated with stress, regulate the immune system, and in turn determine the state of health of the individual. Indeed, to some extent this may reflect our changing views on how the immune system is regulated. The lymphoid system is composed of numerous subsets of B,T, and NK-cells, subsets which either augment or suppress the immune response. Not only have we come to appreciate the opposing interactions of helper/inducer and suppressor/ cytotoxic T-lymphocytes but we have come to appreciate the opposing effects of helper inducer and helper suppressor T-cells, also known as naive and memory T-cells respectively.[72] More recently, the diverging roles of two functionally distinct populations of helper T-cells, Th1 and Th2, have become obvious.[73] This complicated network of lymphocyte subsets is, in turn, regulated by various cytokines which have different effects on different lymphocyte subsets. Moreover, the immune system is regulated by processes that, while allowing response to a particular antigenic stimulus, at the same time prevent overreaction and commitment of the entire immune system to this particular antigen. It has recently become apparent that one of the regulatory pathways is controlled cell death. Stimulation of lymphocytes and the generation of memory cells seems to set into motion a process that induces controlled apoptosis of these cells.[74] Psychoneural and endocrine factors exert an additional regulatory effect on the immune system. However, many of the mechanisms by which psychosocial factors affect the intricate regulation of immunity in health and disease remain to be elucidated.

REFERENCES

1. Pearlin L. I., Menaghan E.G., Lieberman, M.A. and Mullan, J.T., The stress process, *J. Health Soc. Behav.*, 22,337, 1981.
2. Dohrenwend, B.S. and Dohrenwend, B.P., *Stressful Life Events: Their Nature and Effects*, John Wiley & Sons, New York, 1974.

3. **Rahe, R.H.,** Life changes and near-future reports, in *Emotions: Their Parameters and Measurement*, L. Levi, Ed., Raven Press, New York, 1975.

4. **Hinkle, L.E., Jr. and Plummer, N.,** Life stress and industrial absenteeism, *Ind. Med. Surg.*, 21,363, 1952.

5. **Beautrais, A.L., Fergusson, D.M., and Shannon, F.T.,** Life events and childhood morbidity: A prospective study, *Pediatrics*, 70,935, 1982.

6. **Plaut, S.M. and Friedman, B.F.,** Psychosocial factors in infectious disease, in *Psychoneuroimmunology*, Ader, R., Ed., Academic Press, New York, 1981, pp.3-30.

7. **Campbell, T.L.,** Family's impact on health: A critical review, *Fam. Syst. Med.*, 4,135,1986.

8. **Cohen, F.,** Personality, stress, and the development of physical illness, in *Health Psychology: A Handbook*. Stone, G.C., Cohen, F., and Adler, N.E., Eds., Jossy-Bass, San Francisco, 1982, pp. 77-111.

9. **Rahe, R.H.,** The pathway between subject's recent life changes and their near future illness reports: Representative results and methodological issues, in *Stressful Life Events: Their Nature and Effects*, Dohrenwend, B.S., and Dohrenwend, B.P., Eds., New York: Wiley, 1974.

10. **Kobasa, S., Maddi, S. and Kahn, S.,** Hardiness and health: A prospective study, *J. Person. Soc. Psychol.*, 42,168, 1982.

11. **Solomon, G.F.,** The emerging field of psychoneuroimmunology: Hypotheses, supporting evidence and new directions, *Advances*, 2,6, 1985.

12. **Hagnel, O.,** The premorbid personality of persons who develop cancer in a total population investigated in 1947 and 1957, *Ann. N.Y. Acad. Sci.*, 125,846, 1966.

13. **Grossarth-Maticek, R., Eysenck, H.J. and Vetter, H.,** Personality type, smoking habit and their interaction as predictors of cancer and coronary heart disease, *Personality and Individual Differences*, 9,479, 1988.

14. **Grossarth-Maticek, R., Bastiaans, J. and Kanazir, D.T.,** Psychosocial factors as strong predictors of mortality from cancer, ischemic heart disease and stroke: The Yugoslav prospective study, *J. Psychosom. Res.*, 29,167, 1985.

15. **Williams, R.,** *The Trusting Heart: Great News About Type A Behavior*, Random House, New York, 1989.

16. **Persky, V.W., Kempthorne-Rawson, J. and Shekelle, R.B.,** Personality and risk of cancer: 20-year follow-up of the Western Electric Study, *Psychosom. Med.*, 49,435, 1987.

17. **Thomas, C.,** *et al. The Precursor Study: A Prospective Study of A Cohort of Medical Students*, Vol. 5, Johns Hopkins University Press, Baltimore, 1983.

18. **McClelland, D.C. and Jemmott, J.B., III,** Power motivation, stress and physical illness, *J. Hum. Stress*, 6,6, 1980.

19. **House, J.S., Landis, K.R. and Umberson, D.,** Social relationships and health, *Science*, 241,540, 1988.

20. **Pilisuk, M. and Parks, S.H.,** *The Healing Web: Social Networks and Human Survival*, University Press of New England, Hanover, NH, 1986.

21. **Schlesinger, M. and Yodfat, Y.,** The impact of stressful life events on natural killer cells, *Stress Medicine*, 7,53, 1991.

22. **Schlesinger, M., Yodfat, Y., Rabinowitz, R., Bronner, S. and Kark, J.D.,** Psychological stress and NK cells among members of a communal settlement, in *Drugs of Abuse,*

Immunity, and AIDS, Friedman, H., Klein, T.W. and Specter, S., Eds., Plenum Publishing Corp., New York,1993, pp. 247-254.

23.   **Khansari, D.N., Murgo, A.J. and Faith, R.E.,** Effects of stress on the immune system, *Immunology Today*, 11,170, 1990.

24.   **Ader, R., Felten, D.L. and Cohen, N.,** Eds., *Psychoneuroimmunology*, 2nd ed., Academic Press, New York, 1991.

25.   **Gilad, G.M.,** The stress-induced response of the septohippocampal cholinergic system. A vectrial outcome of psychoneuroendocrinological interactions, *Psychneuroendocrinology*, 12,167, 1987.

26.   **Borysenko, J.,** Psychoneuroimmunology: Behavioral factors and immune response, *Revision*, 7,56, 1984.

27.   **Jemmott, J.B. III and Locke, S.E.,** Psychosocial factors, immunological mediation, and human susceptibility to infectious diseases: How much do we know? *Psychol. Bull.*, 95,78, 1984.

28.   **Plaut, M.,** Lymphocyte hormone receptors, *Ann. Rev. Immunol.*, 5,621, 1987.

29.   **Berkenbosch, F., Van Oers, J., Rey, A.D., Tilders, F. and Besedovsky, H.,** Corticotropin-releasing factor producing neurons in the rat activated by interleukin-1, *Science*, 238,524, 1987.

30.   **Malaise, M.G., Hazee-Hagelstein, M.T., Reuter, A.M., VrindsGevaert, Y., Goldstein, G. and Franchimont, P.,** Thymopoietin and thymopentin enhance the levels of ACTH, β-endorphin and β-lipotropin from rat pituitary cells *in vitro*, *Acta Endocrinol.*, 115,455, 1987.

31.   **Ader, R.,** On the clinical relevance of psychoneuroimmunology, *Clin. Immunol. Immunopathol.*, 64,6, 1992.

32.   **Kiecolt-Glaser J.K. and Glaser, R.,** Psychoneuroimmunology: Can psychological interventions modulate immunity? *J. Consulting Clin. Psychiat.*, 60,569, 1992.

33.   **Kraus, A.S. and Lillenfield, A.M.,** Some epidemiological aspects of the high mortality in the young widowed group, *J. Chron. Dis.*, 10,207, 1959.

34.   **Stein, M., Miller, A.H. and Trestman, R.L.,** Depression, the immune system, and health and illness, *Arch. Gen. Psychiatry*, 48,171, 1991.

35.   **Bartrop, R.W., Lockhurst, E., Lazarus, L., Kiloh, L.G. and Penny, R.,** Depressed lymphocyte function after bereavement, *Lancet*, 1,834, 1977.

36.   **Schleifer, S.J., Keller, S.E., Camerino, M., Thornton, J.C. and Stein, M.,** Suppression of lymphocyte stimulation following bereavement, *J.A.M.A.*, 250,374, 1983.

37.   **Farber, E.M., Rein, G. and Lanigan, S.W.,** Stress and psoriasis: psychoneuroimmunologic mechanisms, *Int. J. Dermatol.*, 30,8, 1991.

38.   **Koblenzer, C.S.,** Stress and the skin, *Advances*, 5,391, 1988.

39.   **Schedlowski, M., Jacobs, R., Stratmann, G., Richter, S., Hadicke, A., Tewes, U., Wagner, T.O.F. and Schmidt, R.E.,** Changes of natural killer cells during acute psychological stress, *J. Clin. Immunol.*, 13,119, 1993.

40.   **Talal N.,** The etiology of systemic lupus erythematosus, in *Monograph on Lupus Erythematosus*, Dubois, E., and Wallace, D., Eds., 3rd Edition, Lea and Febiger, Philadelphia, 1987, Chapter 3, pp. 39-43.

41. Grade, M. and Zegans, L.S., Exploring systemic lupus erythematosus: Autoimmunity, self-destruction, and psychoneuroimmunology, *Advances*, 3,16, 1986.
42. Baker, G.H.B. and Brewerton, D.A., Rheumatoid arthritis: a psychiatric assessment, *Br. Med. J.*, 282,2014, 1981.
43. Alexander, W.D., Harden, R. and Shimming, J., Emotion and non-specific infection as possible etiological factors in Graves' disease, *Lancet*, II,196, 1962.
44. Forteza, M.E., Precipitating factors in hyperthyroidism, *Geriatrics*, 28,123, 1973.
45. Winsa, B., Adami, H.O., Bergstrom, R., Gamstedt, A., Dahlberg, P.A., Adamson, U., *et al.*, Stressful life events and Graves' disease, *Lancet*, 338,1475, 1991
46. Sonino, N., Girelli, M.E., Boscaro, M., Fallo, F., Busnardo, B. and Fava, G.A., Life events in the pathogenesis of Graves' disease. A controlled study, *Acta Endocrinol.*, 128,293, 1993.
47. Utiger, R.D., The pathogenesis of autoimmune thyroid disease, *N. Engl. J. Med.*, 325,278, 1991.
48. Ader, R. and Cohen, N., The influence of conditioning on immune responses, in *Psychoneuroimmunology*, Ader, R., Felten, D.L., and Cohen, N., Eds., 2nd ed., Academic Press, New York, 1991, pp. 611-646.
49. Sabbioni, M.E.E., Psychoneuroimmunological issues in psycho-oncology, *Cancer Investigation*, 11,440, 1993.
50. Eysenck, H.J., Personality, stress and cancer: Predictions and prophylaxis, *Br. J. Med. Psychol.*, 61,57, 1988.
51. Greer, S., Morris, T., Pettingale, K.W. and Haybittle, J.L., Psychological response to breast cancer and 15 year outcome, *Lancet*, 335,49, 1990.
52. Hislop, T.G., Waxler, N.E., Coldman, A.J., Elwood, J.M. and Kan, L., The prognostic significance of psychological factors in women with breast cancer, *J. Chronic Dis.*, 40,729, 1987.
53. Cassileth, B.R., Lusk, E.J., Miller, D.S., Brown, L.L. and Miller, C., Psychological correlates of survival in advanced malignant disease? *New Engl. J. Med.*, 312,1551, 1985.
54. Ramirez, A.J., Craig, T.K.J., Watson, J.P., Fentiman, I.S., North, W.R.S. and Rubens, R.D., Stress and relapse of breast cancer, *Br. Med. J.*, 298,291, 1989.
55. Levy, S.A., Herberman, R.B., Whiteside, T., Sanzo, K., Lee, J. and Kirkwood, J., Perceived social support and tumor estrogen/progesterone receptor status as predictors of natural killer cell activity in breast cancer patients, *Psychosomatic Med.*, 52,73, 1990.
56. Reichlin, S., Neuroendocrine-immune interactions, *New Engl. J. Med.*, 329,1246, 1993.
57. Spiegel, D, Bloom, B. Jr, Kraemer, H.C. and Gottheil, E., Effect of psychosocial treatment on survival of patients with metastatic breast cancer, *Lancet*, 2,888, 1989.
58. Kiecolt-Glaser, J.K., Glaser, R., Strain, E., Stout, J., Tarr, K., Holliday, J. and Speicher, C.E., Modulation of cellular immunity in medical students, *J. Behav. Med.*, 9,5, 1986.
59. Smith, G. R., Jr. and McDaniel, S.M., Psychologically mediated effect on the delayed hypersensitivity reaction to tuberculin in humans, *Psychosom. Med.*, 45,65, 1983.
60. McClelland, D.C. and Kirshnit, C., The effect of motivational arousal through films on salivary immune function, 1984, Unpublished manuscript, Harvard University.

61. **Antoni, M.H., Schneiderman, N., Fletcher, M.A., Goldstein, D., Ironson, G. and Laperrier, A.,** Psychoneuroimmunology and HIV-1, *J. Consultina. Clin. Psychol.*, 58,38, 1990.

62. **Fawzy, F.I., Kemeny, M.E., Fawzy, N.W., Elashoff, R., Morton, D., Cousins, N. and Fahey J.L.,** A structured psychiatric intervention for cancer patients, II. Changes over time in immunological measures, *Arch. Gen. Psychiatry*, 47,729, 1990.

63. **Fawzy, F.I., Fawzy, N.W., Hyun, C.S., Elashoff, R., Guthrie, D., Fahey, J.L. and Morton, D.L.,** Malignant melanoma: effects of an early structured psychiatric intervention, coping, and affective state on recurrence and survival 6 years later, *Arch. Gen. Psychiatry,* 50,681, 1993.

64. **Ironson, G., Friedman, A., Klimas, N., Antoni, M. H., Fletcher, M.A., LaPerriere, A., Simoneau, J. and Schneiderman, N.,** Distress, denial, and low adherence to behavioral interventions predict faster disease progression in HIV-1 infected gay men, *Internatl. J. Behavioral Medicine*, 1,90, 1994.

65. **Solomon, G.F.,** Whither psychoneuroimmunology? A new era of immunology, of psychosomatic medicine, and of neuroscience, *Brain Behav. Immun.*, 7,352, 1993.

66. **Lake, C.R., Gullner, H.G., Polinsky, R.J., Eber, M.H., Ziegler, M.G. and Bartter, F.C.,** Essential hypertension: central and peripheral norepinephrine, *Science*, 211,955, 1981.

67. **Gelernt, M.D. and Hochman, J.S.,** Acute myocardial infarction triggered by emotional stress, *Am. J. Cardiol.*, 69,1512, 1992.

68. **Eliot, R.S.,** Stress and the heart, *Postgrad. Med.*, 92,237, 1992.

69. **Eliot, R.S., Todd G.L., Clayton F.C. and Pieper, G.M.,** Experimental catecholamine-induced acute myocardial necrosis, *Adv. Cardiol.*, 25,107, 1978.

70. **Baroldi, G., Faizi, G., and Mariani, F.,** Sudden coronary death: a postmortem study in 208 selected cases compared to 97 "control" subjects, *Am. Heart J.*, 98,20, 1979.

71. **Born, G.V.,** Recent evidence for the involvement of catecholamines and of macrophages in atherosclerotic processes, *Ann. Med.*, 23:569, 1991.

72. **Beverly, P.C.L., Merkenschlager, M. and Wallace, D.L.,** Identification of human naive and memory T cells, in *Progress in Immunology,* Melchers, F., Albert, E.D., von Boehmer, H., *et al.*, Eds., Vol 7. Springer-Verlag, Berlin, 1989, pp. 432-438.

73. **Mosmann, T.R., and Moore, K.W.,** The role of IL-10 in crossregulation of TH1 and TH2 responses, *Immunol. Today*, 12,A49, 1991.

74. **Salmon, M., Pilling, D., Borthwick, N.J., Viner, N., Janossy, G., Bacon, P.A., and Akbar, A.N.,** The progressive differentiation of primed T cells is associated with an increasing susceptibility to apoptosis, *Eur. J. Immunol.*, 24,892, 1994.

Chapter 7

CONDITIONED IMMUNITY TO *L. MAJOR* IN YOUNG AND AGED MICE

R. M. Gorczynski
Liver Program Project Group
The Toronto Hospital
University of Toronto
Toronto, Ont., Canada

INTRODUCTION

(i) Infection with, and resistance to, *Leishmania* parasites:

The leishmanias represent a group of obligate intracellular parasites, causing significant morbidity and mortality in Third-World countries. Transmission occurs after humans are bitten by infected vectors (e.g., the phlebotomine sandfly) containing promastigotes, with subsequent growth of the amastigote form of the organism, and resistant to attack by host lysosomal enzymes, in parasitophorous vacuoles.[45] Resistance to infection essentially can be described as being of two types, innate (to macrophages) and acquired, dependent upon activation of host immunity (T-cell-mediated) with secondary enhanced killing of amastigotes within infected cells.[9] Over the past several years it has become increasingly apparent that this T cell immunity results in the main from interferon gamma (IFNγ) activation, as well perhaps IL-12 activation, of macrophage killing. Ineffective immunity can result not only in death of the host, but often in the expression of antigenic variation in the parasite, as has been observed in other parasite systems.[23]

IFNγ is produced by a unique population of T cells, so-called Th1 cells, which are also producers of IL-2 and other cytokines.[42] It has also been surmised that in those individuals, both human and experimental animals, which are unable to clear infection by these parasites, the problem is at least in part due to a preferential activation of a distinct set of T cells, so-called Th2 cells, which do not produce the afore-mentioned cytokines but instead liberate IL4, IL-10 and IL-13.[37] The latter cytokines are of particular interest in this regard since they seem to cause further inhibition of IFNγ production.[52] In mice at least it has proven possible to describe two distinct strains with a resistant (CBA) or susceptible (BALB/c) pattern of parasite growth which can be understood in terms of the cytokines produced by those animals after infection.[41]

(ii) Stress, conditioning, and the immune response:

Separate studies have focussed on the notion that the immune system does not function autonomously, but is subject to regulation by the (neuro)hormonal milieu of the intact organism. Fundamental to such analyses have been the data coming from experiments initiated by Ader and Cohen, which documented that under defined circumstances the immune function of experimental animals can be "conditioned" to "respond" to nonimmunological cues.[1] Developing on these themes other groups have explored in more detail the phenomena of stress-induced immune alteration, where stress can be defined as any stimulus perceived as a threat to the (psycho)biological integrity of the responding organism.[11] There is now a large literature on this phenomenon as it pertains to the susceptibility to infectious disease, and reference is found elsewhere in this volume to many aspects of these studies.

These studies confirm that stress in human and experimental subjects increases susceptibility to Epstein-Barr virus infection and respiratory infections,[11] as well as herpes virus infection.[18,35] In at least one of these studies[18] a role for stress-induced changes in cytokine production (IFNγ) was suggested to be contributing to the effects observed. The work described below summarizes our own studies on conditioning, stress and susceptibility to infection with *L. Major* in mice, with particular reference to the role of altered cytokine production to the observed phenomena.

(iii) Aging and behavioural manipulation of immunity:

An additional feature of the study of innate and acquired resistance mechanisms with which we and many others have been concerned is the manner in which the age of the responding organism affects resistance.[14,22,47] We commented elsewhere on the decreased ability to observe conditioned immunosuppression in aged mice.[24,25] It thus became of interest to us to extend our investigations to explore the role of conditioned immunosuppression on susceptibility to infection in aged animals.

We and others have described results from a number of investigations which suggest that there are profound changes in the ability of CD4+ T lymphocytes of different mice to produce a variety of cytokines during the aging process.[28] In particular, we observed that following non-specific stimulation of splenic CD4+ T cells of young (8 weeks of age) mice, the predominant lymphokines produced were of the Th1-type (IL-2, IFNγ). In contrast, when older (≥75 weeks) mice were used, the major cytokines produced were of the Th2-type (IL4, IL-10). Based on such results we hypothesized that infection of older "resistant" CBA mice with *L. Major* might lead to a non-healing lesion, with demonstrable activation of Th2 cells (rather than the Th1 cells seen after the self-healing infection of young mice). We were, in turn, also interested in the possibility that further conditioning of immunity in young/aged animals pre-infection would alter susceptibility to disease. Experiments are described below which investigate these phenomena.

CONDITIONED IMMUNITY TO L. *MAJOR*

(i) Leishmania growth, and *in vitro* IL-2/IL-4 production from susceptible (BALB/c) and resistant 9CBA) mice of different ages:

The data in Figures 1,2 show results from one study (of three), in which groups of six 8-week or 75-week old CBA mice were infected at the base of the tail with 2 x 10[6] promastigotes of *L. Major*. Also included in this study were BALB/c mice (susceptible). Data in Figure 1 show the arithmetic mean lesion size for the different groups at various times post infection, while in Figure 2 are shown results from analysis of cytokines (IL-2/IL-4) produced on restimulation of peripheral lymph node lymphocytes from infected animals at 12 or 24 days post infection in 72 hr of culture with irradiated (150Krad) promastigotes.

It is clear from inspection of Figure 1 that older CBA animals are indeed more susceptible to a challenge which causes minimal self-healing lesions in younger CBA animals, and in fact have a susceptibility intermediate between young CBA (resistant) and young BALB/c mice (a known susceptible strain). When in separate studies mice from each group were sacrificed at 12 and 24 days post infection and lymph node cells restimulated with parasite antigen, quite different patterns of cytokine production were observed. (Note that at the earlier time point for this study there was no significant difference in lesion diameter between the groups: see Figure 1.) Young CBA mice showed predominantly IL-2 production at both times investigated, while young BALB/c mice, and the aged CBA, showed a bias towards IL4 production, even at the earliest time point (12 days post infection), which became more pronounced at the later time (24 days). These data (Figure 2) are consistent with numerous other studies which suggest the inability to clear infection with this parasite is related to diminished stimulation of infected macrophages (by e.g. IFNγ) by activated Th1 type cells.[37]

(ii) Conditioned stress suppresses immunity to *L.Major* in young, but not aged CBA mice, and spares parasite-stimulated IL-4 production:

In a conditioning model in which mice were exposed to physical stressors (rotation on a turntable at 45 rpm), along with a number of environmental cues associated with that stressor, it was found subsequently that mere exposure to the cues at the same time as the animals received an immunological challenge (injection with sheep erythrocytes) would produce documented inhibition of the immune response when compared with nonstressed animals, or animals stressed but not re-exposed to the cues.[19,20,26] In order to examine whether the same conditioned immunosuppression would affect susceptibility to infection with *L. Major*, the following study was performed. Groups of 10 young or aged CBA mice were subject to rotational stress (30 minutes/45 rpm) on 3 occasions at 12 day intervals. Fifteen days after the last trial mice were infected with 2×10^6 promastigotes of *L.Major*, and 5 mice/group were re-exposed to the environment in which rotational stress had previously been experienced. Repeated exposure to "stress-related cues" took place at 4 day intervals for the duration of the study. Data from one of four such studies are shown in Figure 3. In Figure 4 are shown data from a separate experiment in which identically treated mice were sacrificed at 20 days post infection, and their lymph nodes used for analysis of cytokine production after antigen-specific restimulation *in vitro*.

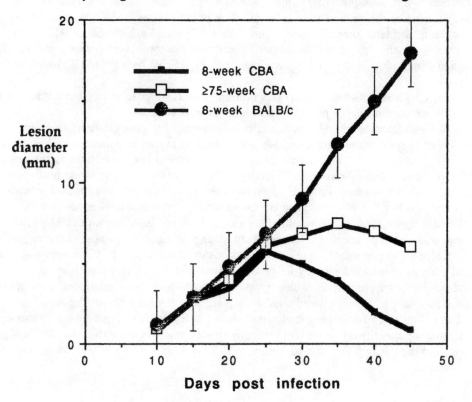

Fig 1.   Growth of *L. Major* in mice of different strains and ages. Groups of 6 mice of the type shown were infected at the base of the tail with 2 x 10⁶ cloned promastigotes of *L. Major*. Lesion size was measured at 3-5 day intervals with calipers. Data show arithmetic mean lesion size for the group on the days shown ±SD, shown for BALB/c only to retain clarity (SD≤20% in all cases here and subsequently).

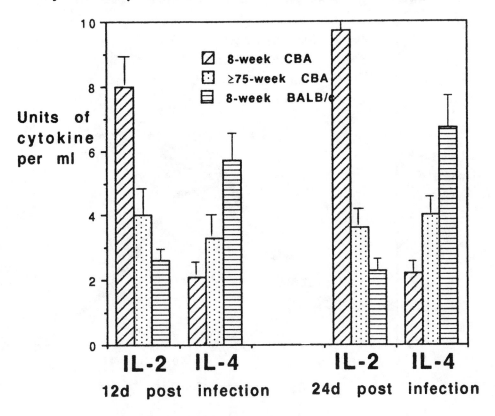

Fig 2.   a) Cytokine production at 72 hr culture using draining lymph node cells pooled from 3 mice/group of the type shown at 12/24 days post infection with *L.Major* as in Fig 1. 5 x 10^6 cells were incubated in triplicate in microtitre plates with 2 x 10^6 150Krad promastigotes. Supernatants were pooled from replicate cultures and assayed in triplicate in cytokine bioassays using CTLL-2 (IL-2) or CT4.S (IL4) as described elsewhere;[20a] b) Recombinant cytokines were used to standardize the assays. Data are shown as arithmetic means ± SD for each group.

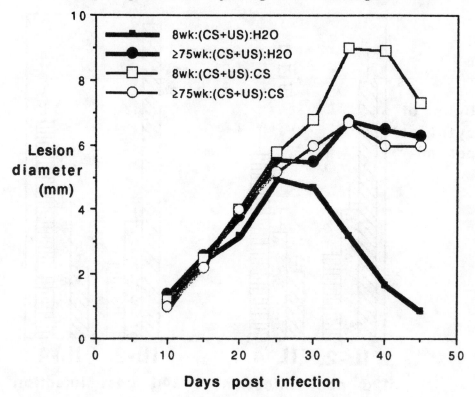

Fig 3.   Growth of *L.Major* in conditioned mice exposed/not re-exposed to stressor cues (see text for more details). CS represents cues associated with rotational stress (including saccharin, a gustatory cue, in the drinking water), US the rotational stress itself (30 min, 45 rpm). After three trials of the type shown mice were infected with 2×10^6 parasites and re-exposed to environmental cues with saccharin, or left in the home cage with water. All groups contained 6 animals; SD ≤20%.

There are a number of points of interest in these studies. Firstly, conditioned mice not re-exposed to cues associated with conditioned immunosuppression show equivalent susceptibility to parasite growth as untreated animals (see Figure 1), with aged CBA mice again more susceptible than young mice. As expected, these aged (non-cued) mice also produce preferentially IL-4 on restimulation *in vitro* (Figure 4). Interestingly, conditioned young CBA mice re-exposed to stress-related cues show a pronounced increase in susceptibility to parasite growth, which exceeds that seen even in conditioned re-exposed aged mice (Figure 3). In keeping again with these *in vivo* data, maximum IL-4/IL-2 production was seen in conditioned young mice. The relatively minor changes in parasite growth, and cytokine production, brought about by conditioning in the aged animals is reminiscent of our earlier observations which were taken to imply a decreased ability to produce conditioned immunosuppression (but not

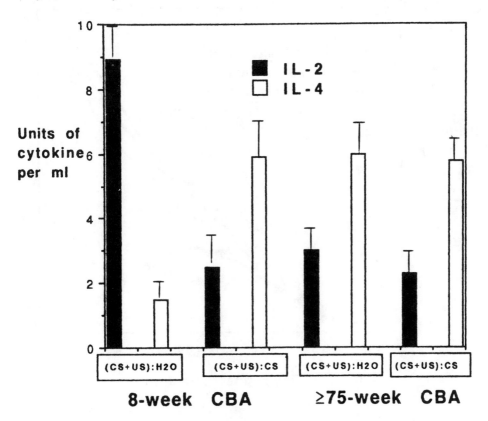

Cytokine production from conditioned infected mice

Fig 4. Cytokine production using draining lymph nodes pooled from 3 mice/group at 20 days post infection with *L. Major*, and re-challenged with antigen *in vitro*. See text, and legends to Figures 2,3 for more details.

immunosuppression *per se*) in aged mice.[24,25] Interestingly, these data imply that the conditioned immunosuppression protocol produced inhibition of the Th1-type response (and hence increased susceptibility in young mice), but relatively little inhibition of Th2-type responses, thus in turn, there was little effect on the resistance pattern of aged mice.

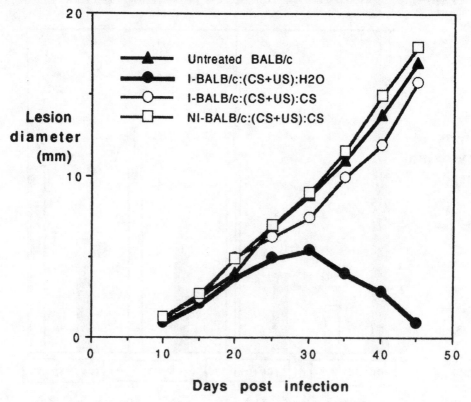

Fig 5.   Growth of *L. Major* in conditioned immunized (3 weekly injections with 1 x 10⁷ 150Krad promastigotes) BALB/c mice (I-BALB/c). See text and legend to Figure 3.

(iii) Conditioned stress response inhibits acquired immunity to
L.Major in immunized susceptible BALB/c mice:

We have examined one final model of conditioning and immunity to *L. Major* infection, in this case using manipulation of the susceptible BALB/c strain. It has been known for some time that resistance to infection can be produced in this strain by a number of means, including sublethal whole body irradiation and/or drug-curing,[31] and preimmunization with irradiated parasites.[21,23] Using this latter model, we have examined whether the conditioning model outlined above can further manipulate immunity to the parasite.

Ten BALB/c mice received three weekly immunizations with irradiated (150 Krads) *L. Major* in the lateral tail vein. A control group received saline injections only. Fourteen days after the last immunization, all mice were exposed to rotational stress as described above, and these conditioning trials were repeated at 12 day intervals. Fifteen days after the final trial all mice, along with 5 age-matched, previously untreated, BALB/c mice, were injected with 2×10^6 promastigotes. Five of each of the conditioned groups (immunized and nonimmunized) were re-exposed to conditioning cues at 4 day intervals throughout the subsequent time of parasite growth. Data for one of two such studies examining parasite growth in the different groups are presented in Figure 5. In Figure 6 are data from a separate experiment (one of three studies) in which lymph node cells from mice of equivalent groups were tested at 20 days post infection for cytokine production (IL-2/IL4) in response to restimulation with parasite antigens.

It is clear from inspection of these data that these results reinforce the observations made in Figures 3 and 4. Immune BALB/c mice are resistant to parasite growth, by comparison with naive mice (or conditioned nonimmune mice), and produce preferentially IL-2 on restimulation *in vitro*. When immune mice are conditioned by a stressor known to produce immunosuppressive effects, re-exposure of immune mice to the stress-related cues after infection leads to loss of resistance, and again now a preferential bias towards IL-4 production. Once more, it seems that conditioned immunosuppression has less impact on the immune response of Th2-type cells.

DISCUSSION

(i) Conditioning and changes in immune cells:

We discussed above some of the other data supporting a role for behaviour in resistance to infectious disease. Physical restraint has been shown to increase susceptibility to herpes simplex virus[8] and Moloney Sarcoma virus,[48] but apparently decreases susceptibility to allergic encephalomyelitis.[36] In a similar vein it has been reported that electric shock treatment can increase susceptibility to Coxsackie B virus[17] but decrease susceptibility to encephalomyocarditis virus.

Many of these apparent anomalies become less confusing when the mechanisms of immunological change are studied. Feng *et al.*[16] recently reported decreased levels of specific antibody in influenza infected mice exposed to restraint-induced stress. It appears that resistance to herpes virus is most likely a function of NK cell activity, or of cytotoxic T cells. Bonneau *et al.*[8] found that restraint induced stress could indeed suppress the development of NK cells or CTL to herpes virus, leading to higher recovery of virus from infected animals.[49] reported that inescapable, but not escapable, shock decreased NK activity in rats; in contrast

Irwin and Custeau[32] suggested that signaled shock was more immunosuppressive than unsignaled shock. We ourselves have published studies concerning conditioned suppression of NK cell activity,[20] while other groups have reported controversial studies suggesting an enhancement of NK activity.[1,30] Macrophage function has also been shown to be diminished in stressed animals;[33,55] the decreased MHC class II antigen expression on macrophages in these models in turn leads to impaired antigen-presenting-function, decreased lymphocyte activation, and thus decreased immunity to infection.

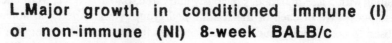

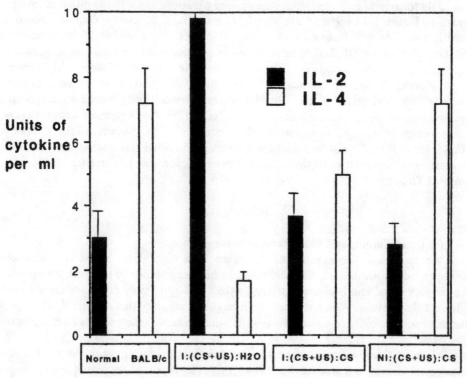

Fig 6.   Cytokine production at 20 days from mice shown in Figure 5. See text and legends to previous Figures for more details.

There have been few studies on the (cell) biological mechanisms behind these phenomena. In the studies referred to above concerning stress induced inhibition of macrophage function Sonnenfeld *et al.*[51] suggested that there was a stress-induced decrease in IFNγ production which in turn led to altered MHC class II expression and decreased T and B cell immunity. In our earlier analyses of conditioned immunosuppression of antibody responses in young and old mice we found that conditioning changes were most easily explained as due to alterations in T lymphocytes in the conditioned host, and the environment of the host itself (secondary to exposure to conditioning cues). These conclusions were reached from studies of reciprocal transplantation of T/B cells from conditioned animals into conditioned or nonconditioned hosts.[24,25] In a similar vein it has proven easier to produce conditioned suppression of T rather than B cell responses in experimental animals.[34,40,43] In more detailed studies of stress-induced changes, decreased IL-2 production was reported by Weiss *et al.*,[53] (1989), Batuman *et al.*,[3] and Hardy *et al.*[29] These data are supported by the observations described above in the *L. Major* model.

(ii) Neurohormonal changes in conditioned- and/or stress-induced immunosuppression:

Investigations of possible biochemical mechanisms behind the phenomenon of conditioning of immune responses, or stress-induced alteration of immunity, are rare. Given that when these changes are investigated *in vitro* there often seems to be marked differences between cell populations isolated from different organs, there is some basis for the belief that factors influencing the regulation of cell trafficking *in vivo* may underlie some of the key observations.[12,39,46] Note that there is evidence for direct innervation of immune organs;[15] for neurohormonal receptors on lymphoid cells themselves;[6] for production of cytokines by cells within the central nervous system,[10] and of neuroendocrine mediators by cells of the immune system;[5] and for neurohormonal regulation of immune reactions.[4] It comes as no surprise therefore that there also exists a body of data indicating that neurohormonal manipulation of the whole animal can often modify behaviourally mediated changes in immunity.

There is controversial data concerning the role of corticosteroids in regulation of stress-induced change/conditioning. Stress-induced suppression of NK activity and IL-2 production are not blocked by adrenalectomy;[53] however, Okimura *et al.*[44] reported that adrenalectomy abolished restraint-induced suppression of antibody responses in rats, and Blecha *et al.*[7] found an attenuated decrease in DTH responses in restrained adrenalectomized mice. Similar differences were found in conditioning studies by Gorczynski *et al.*[19] and Ader and Cohen.[1] To date no satisfactory explanation for these differences has been offered.

Chlorpromazine and amitriptyline have been reported to block conditioned immunosuppression of antibody responses in mice.[22a] Propanolol and nadolol have been shown to block stress induced suppression of spleen mitogen responses;[13] stress-induced suppression of IFNγ production;[51] and conditioned immunosuppression of IFNγ production.[38] Opioids acting within the CNS are also implicated in these phenomena. Naltrexone and naloxone, opioid receptor antagonists, block stress-related and conditioned immune responses of NK activity,[20,40,50] while N-methylnaltrexone, which does not cross the blood-brain barrier, does not. A recent report claims that serum from rats exposed to restraint or shock contains factors which suppress *in vitro* mitogen responses.[54] Adrenalectomy and naltrexone were ineffective in modifying this inhibitory activity.

(iii) Summary:

The data above show that young CBA mice, normally resistant to infection with *L. Major*, become susceptible to that parasite if re-exposed to cues previously signaling stress-induced immunosuppression. This same regimen of stress-induced immunosuppression can even abolish the acquired immunity in susceptible BALB/c mice which is achieved by immunization with irradiated parasites. In each case susceptibility is associated with preferential production of the Th2-type cytokine IL-4, over that of the Th1 type cytokine IL-2, on re-exposure to parasite antigens *in vitro*, as predicted from the known immunobiology of infection with this organism. These data in turn suggest that Th2-type cells may be less responsive to those (neurochemical) signals which trigger conditioned immunosuppression than are Th1-type cells, a feature which may have heuristic value in any future application of these models to clinically relevant scenarios.

REFERENCES

1. **Ader, R. and Cohen, N.,** CNS-immune system interactions: conditioning phenomena, *Behav. Brain Sci.,* 8,379, 1985.
2. **Ader, R. and Cohen, N.,** The influence of conditioning on immune responses, in *"Psychoneuroimmunology"*, 2nd. ed., R. Ader, D. L. Pelten and N. Cohen, Eds., Academic Press, New York, 1991, pp. 611-646.
3. **Batuman, O.A., Sajewski, D., Ottenweller, J.E., Pitman, D.L. and Natelson, B.H.,** Effects of repeated stress on T cell numbers and function in rats, *Brain Behav. Immun.,* 4,105, 1990.
4. **Besodovsky, H.O. and del Ray, A.,** Physiological implications of the immune-neuroendo-crine network, in *"Psychoneuroimmunology"*, 2nd. ed., R.Ader, D.L.Felten and N. Cohen, Eds., Academic Press, New York 1991, pp. 589-608.
5. **Blalock, J.E.,** Production of neuroendocrine peptide hormones by the immune system, *Prog. Allergy,* 43,1, 1988.
6. **Blalock, J.E., Ed.,** Production of peptide hormonal neurotransmitters by the immune system, *Neuroimmunoendocrinol. Chem. Immunol.,* 52,1, 1992.
7. **Blecha, F., Kelley, K.W. and Satterlee, D.G.** Adrenal involvement in the expression of delayed-type hypersensitivity to SRBC and contact sensitivity to DNFB in stressed mice, *Proc. Soc. Exp. Biol. Med.,* 169,247, 1982
8. **Bonneau,R.H., Sheridan, J.F., Feng, N. and Glaser, R.,** Stress induced suppression of herpes simplex virus (HSV)-specific cytotoxic T lymphocyte and natural killer cell activity and enhancement of acute pathogenesis following local HSV infection, *Brain Behav. Immun.,* 5,170, 1991
9. **Bryceson, A.D.M.,** Immunological aspects of cutaneous Leishmaniasis, in *"Essays on Tropical Dermatology"*, J. Marshall, Ed., Excerpta Medica, Amsterdam, 1972, pp. 230-245.

10. **Carr, D., J. J.,** Neuroendocrine peptide receptors on cells of the immune system. *Neuroimmunoendocrinol. Chem. Immunol.*, 52,84, 1992.

11. **Cohen, S. and Williamson, G.M.,** Stress and infectious disease in humans, *Psychol. Bull.*, 109,5, 1991.

12. **Cunnick, J.E., Lysle, D.T., Armfield, A. and Rabin, B.S.,** Shock induced modulation of lymphocyte responsiveness and natural killer activity: differential mechanisms of induction, *Brain Behav. Immun.*, 2,102, 1988.

13. **Cunnick, J.E., Lysle, D.T., Kucinski, B.J. and Rabin, B.,** Evidence that shock-induced suppression is mediated by adrenal hormones and peripheral, Sadrenergic receptors, *Pharmacol. Biochem. Behav.*, 36,645, 1990.

14. **Fabris, N.,** Body homeostatic mechanisms and ageing of the immune system, in *"CRC handbook on Immunology of Ageing"*, M.B. Kay and T. Makinodan, Eds., CRC Press, Boca Raton, FL, 1981, pp.61-68.

15. **Felten, S.Y. and Felten, D.L.,** Innervation of lymphoid tissue, in *"Psychoneuroimmunology"*, 2nd. edition, R.Ader, D.L.Felten and N. Cohen, Eds., Academic Press, New York, 1991, pp.27-69.

16. **Feng, N., Pagniano, R., Tovar, A., Bonneau, R. H., Glaser, R. and Sheridan, J.F.,** The effect of restraint stress on the kinetics, magnitude, and isotype of the humoral immune response to influenza virus infection, *Brain Behav. Immun.*, 5,370, 1991.

17. **Freidman, S.B., Ader, R. and Glasgow, L.A.,** Effects of psychological stress in adult mice inoculated with Coxsackie B viruses, *Psychosom. Med.*, 27,361, 1965.

18. **Glaser, R., Rice, J., Speicher, C. E., Stout, J. C., and Kielcolt-Glaser, J. K.,** Stress depresses interferon production by leukocytes concomitant with a decrease in natural killer cell activity, *Behav. Neurosci.*, 100,675, 1986.

19. **Gorczynski, R. M., MacRae, S. and Kennedy, M.,** Factors involved in the classical conditioning of antibody responses in mice, in *"Breakdown in Human Adaptation to Stress: Towards a Multidisciplinary Approach"*, Vol. II, R.E. Ballieux, J.F. Fielding and A. L'Abbute, Eds., Martinus Nijhoff Publisher, The Hague, The Netherlands, 1984, pp.704-712.

20. **Gorczynski, R.M. and Kennedy, M.,** Associative learning and regulation of immune responses, *Prog. Neuro-Psychopharmacol. Biol. Psychiat.*, 8,593, 1984

20a. **Gorczynski, R.M.,** Adoptive transfer of unresponsiveness to allogeneic skin grafts with hepatic $\gamma 8+$ T cells, *Immunology*, 81,27, 1984.

21. **Gorczynski, R.M.,** Do sugar residues contribute to the antigenic determinants responsible for protection and/or abolition of protection in leishmania infected BALB/c mice? *J. Immunol.*, 137,1010, 1986

22. **Gorczynski, R.M.,** Diversity in the lymphocyte recognition repertoire is altered during ageing, *Biomed. Pharmacotherapy*, 41,124, 1987.

22a. **Gorczynski, R.M. and Holmes W.,** Neuroleptic and anti-depressant drug treatment abolishes conditioned immunosuppression in mice, *Brain, Behav., and Immun.*, 3,312, 1989.

23. **Gorczynski, R.M.,** Analysis of antigenic variation in leishmania parasites occurring as a result of preimmunization prior to infection of susceptible murine hosts, *Infection and Immunity (Life Sci. Adv.)*, 10,1, 1991a.

24.   **Gorczynski, R.M.,** Conditioned immunosuppression: analysis of lymphocytes and host environment of young and aged mice, in *"Psychoneuroimmunology"*, 2nd. ed., Ader,R., Felten, D.L. and Cohen,N., Eds., Academic Press, New York, 1991b, pp.647-662.

25.   **Gorczynski, R.M.,** Towards an understanding of the mechanism(s) of classical conditioning of antibody responses, *J. Gerontol. Psychchol. Sci.*, 46,152, 1991c.

26.   **Gorczynski, R.M.,** Conditioned stress responses by pregnant and/or lactating mice reduce immune responses of their offspring after weaning, *Brain Behav. Immun.*, 6,87, 1992a.

27.   **Gorczynski, R.M.,** Immunosuppression induced by hepatic portal venous immunization spares reactivity in IL-4 producing T lymphocytes, *Immunol. Lett.*, 33,67, 1992b.

28.   **Gorczynski,R.M., Dubiski, S., Munder, P.G., Cinader, B. and Westphal, O.,** Age-related changes in interleukin production in BALB/cNNia and SJL/J mice and their modification after administration of foreign macromolecules, *Immunol. Lett.*, 38,243, 1993.

29.   **Hardy, C., Quay, J., Livnat, S. and Ader, R.,** Altered T lymphocyte response following aggressive encounters in mice, *Physiol. Behav.*, 47,1245, 1990.

30.   **Hiramoto, R.N., Hiramoto, N.S., Solvason,H.B. and Ghanta, V.K.,** Regulation of natural killer immunity (NK activity) by conditioning, *Ann. NY Acad. Sci.*, 496,545, 1987.

31.   **Howard, J.G., Micklem, S., Hale, C. and Liew, F.Y.,** Prophylactic immunization against experimental leishmaniasis. 1. Protection induced in mice genetically vulnerable to fatal Leishmania tropica infection, *J. Immunol.*, 129,220, 1982

32.   **Irwin, J. and Custeau, N.,** Stressor predictability and immune function, *Soc. Neurosci. Abstr.*, 15,298,1989.

33.   **J'ang, C.G., Morrow-Tesch, J.L., Beller, D.S., Levy, E.M. and Black, P.H.,** Immunosuppression in mice induced by cold water stress, *Brain Behav. Immun.*, 4,278, 1990.

34.   **Kusnecov, A.V., Husband, A.J. and King, M.G.,** Behaviourally conditioned suppression of mitogen-induced proliferation and immunoglobulin production: effect of time span between conditioning and reexposure to the conditioned stimulus, *Brain Behav. Immun.*, 2,198, 1988.

35.   **Kusnecov, A.V., Grota, L.J., Schmidt, S.G., Bonneau, R.H., Sheridan, J.F.,** *et al.*, Decreased herpes simplex viral immunity and enhanced pathogenesis following stressor administration in mice, *J. Neuroimmunol.*, 38,129, 1992.

36.   **Levine, S., Strebel, R., Wenk, E. J. and Harman, P.J.,** Suppression of experimental allergic encephalomyelitis by stress, *Proc. Soc. Exp. Biol. Med.*, 109,294, 1962.

37.   **Liew, F.Y.,** Functional heterogeneity of CD4 T cells in leishmaniasis, *Immunol. Today*, 10,40, 1989.

38.   **Luecken, L. and Lysle, D.T.,** Evidence for the involvement of betaadrenergic receptors in conditioned immunomodulation, *J. Neuroimmunology*, 38,209, 1992

39.   **Lysle, D.T. and Maslonek, K.A.,** Immune alterations induced by a conditioned aversion stimulus: evidence for a time-dependent effect, *Psychobiology*, 19,339, 1991.

40.   **Lysle, D. T., Luecken, L. J. and Maslonek, K. A.,** Modulation of immune function by a conditioned aversive stimulus: evidence for the involvement of endogenous opioids, *Brain Behav. Immun.*, 6,179, 1992.

41.   **Mitchell, G.F., Curtis, J.M., Handman, E., and McKenzie, I.F.C.,** Cutaneous leishmaniasis in mice: disease patterns in reconstituted nude mice of several genotypes infected with *Leishmania tropica*, *Aust. Exp. Biol. Med. Sci.*, 58,521, 1980.

42. **Mosmann, T.R., Cherwinski, H., Bond, M.W., Giedlin, M.A. and Coffman, R.L.,** Two types of murine helper T cell clone: definition according to profiles of lymphokine activities and secreted proteins, *J. Immunol.*, 136,2348, 1986.

43. **Neveu, P.J., Dantzer,R. and Le Moal, M.,** Behaviourally conditioned suppression of mitogen-induced lymphoproliferation and antibody production in mice, *Neurosci. Lett.*, 65,293, 1986.

44. **Okimura, T., Ogawa, M., Yamauchi, T. and Satomi-Sasaki, Y.,** Stress and immune responses. IV. Adrenal involvement in the alteration of antibody responses in restraint-stressed mice, *Jpn. J. Pharmacol.*, 41,237, 1986.

45. **Preston, P.M. and Dumonde, D.C.,** in *"Immunology of Parasitic Infection"*, S.Cohen and E.Sadun, Eds., Blackwell Sci. Public., London, 1976, pp.167-188.

46. **Rinner,L., Schauenstein, K., Mangge, H. and Porta, S.,** Opposite effects of mild and severe stress on *in vitro* activation of rat peripheral blood lymphocytes, *Brain Behav. Immun.*, 6,130, 1992.

47. **Roberts-Thomson, I., Whittingham, S., Youngchaiyud, U. and MacKay, I.R.,** Ageing, immune response and mortality, *Lancet*, 2,368, 1974.

48. **Seifter,E., Rettura,G., Zisblatt,M., Levinson,S., Levine,N.,** *et al.*, Enhancement of tumor development in physically stressed mice incubated with an oncogenic virus, *Experientia*, 29,1379, 1973.

49. **Shavit,Y., Ryan,S.M., Lewis, J.W., Laudenslager, M.L., Terman, G.W.,** *et al.*, Inescapable but not escapable stress alters immune function. *Physiologist*, 26,A64, 1983.

50. **Solvason, H.B., Hiramoto, R.N. and Ghanta,Y.K.,** Naltrexone blocks the expression of the conditioned elevation of natural killer cell activity in BALB/c mice, *Brain Behav. Immun.*, 3,247, 1989.

51. **Sonnenfeld, G., Cunnick, J.E., Armfield, A.V., Wood, P.G. and Rabin, B.S.,** Stress-induced alterations in interferon production and class II histocompatibility antigen expression, *Brain Behav. Immun.*, 6,170, 1992.

52. **de Waal Malefyt, R., Figdor, C.G., HuilMens, R., Mohan-Peterson, S., Bennett, B., Culpepper, J., Dang, W., Zuraawski, G. and de Vries, J.E.,** Effects of IL-13 on phenotype, cytokine production, and cytotoxic function of human monocytes: comparison with IL-4 and modulation by IFNγ or IL-10, *J. Immunol.*, 151,6370, 1993.

53. **Weiss, J.M., Sundar, S.K., Becker, K.J. and Cierpial, M.A.,** Behavioural and neural influences on cellular immune responses: effects of stress and interleukin-1, *J.Clin. Psychiat.*, 50,43, 1989.

54. **Zha, H., Ding, G. and Fan, S.,** Serum factors induced by restraint stress in mice and rats suppresses lymphocyte proliferation, *Brain Behav. Immun.*, 6,18, 1992.

55. **Zwilling,B.S., Brown,D., Christner, R., Faris,M., Hilberger,M.,** *et al.*, Differential effect of restraint stress on MHC class II expression by murine peritoneal macrophages, *Brain Behav. Immun.*, 4,330, 1990.

Chapter 8

NEUROIMMUNOLOGY OF HOST-MICROBIAL INTERACTIONS

David H. Brown
Department of Microbiology
The Ohio State University
Columbus, OH

Bruce S. Zwilling
Department of Microbiology
The Ohio State University
Columbus, OH

I. INTRODUCTION

The study of the effects of neuroendocrine-immune interactions on microbial pathogenesis and immunity during the course of an infectious disease has emerged as a new interdisciplinary research area, termed psychoneuroimmunology. Over the last two decades, studies have found that psychological stress and psychiatric illness can compromise immune function. The historical basis for studying the influence of stress on the immune response stems from early clinical observations that individuals became sick following stressful situations. Benjamin Richardson writes, in Diseases of Modern Life (c. 1882), about diseases arising from excessive mental strain or from mental shock that are found mainly in four classes of the community: (1) persons engaged in art, science, or literature; (2) those engaged in political life; (3) those who are occupied in commerce, exchange, and speculation; and (4) in the too laborious scholars or students.[1] Richardson goes on to say that "the diseases induced are limited in number, and, physiologically, hang closely together - links, as it were, of one chain. They all depend primarily upon a deficiency of power or paralysis of the organic nervous system, of that part of the nervous organism which sustains the motion of the heart, the stomach, and digestive system, which governs the secretions, and which, in a word, ministers to the involuntary and instinctive, as distinguished from the voluntary and intellectual life."[1] Although there exists difficulties associated with the quantitation of stress and its ultimate association with the onset of illness, it is widely accepted that stress can have an impact on susceptibility to several infectious diseases, most notably, tuberculosis.[2,3]

STRESS INDUCED IMMUNOMODULATION

The concept of stress is not sharply defined. A stressor can be defined as any sort of external or internal challenge, visual, tactile or emotional, that disrupts the physiological equilibrium or homeostasis of an individual.[4] Regardless of the nature of the stress, the common mammalian response to these stressors results in: (1) the stimulation of the hypothalamic-pituitary-adrenal (HPA) axis producing adrenocortical secretions followed by consequential increases in serum glucocorticoids and (2) the activation of the sympathetic nervous system (SNS) followed by the release of both tissue and plasma catecholamines.[5-8] SNS activation results in the local release of norepinephrine and epinephrine from chromaffin cells of the adrenal medulla.[7] Norepinephrine and epinephrine, through their α and β-adrenergic receptors, mediate known cardiovascular and metabolic effects under conditions of stress.[9,10] There is increasing evidence that the SNS plays a role in the modulation of the immune response. Lymphoid tissues are known to receive an extensive intraparenchymal innervation. A wide range of neurotransmitters, via receptors on these lymphoid tissues, influence lymphocyte and monocyte function *in vitro*.[11] Treatment of animals with 6-hydroxydopamine, a drug used to ablate the SNS by destruction of noradrenergic neurons causing a temporary, yet functional axotomy, results in a reduction of primary antibody response to T-dependent antigens and suppression of alloantigen-induced cytotoxic T-lymphocyte (CTL) activity.[12,13]

Neuroendocrine control of a wide range of immune responses occurs via a host of neuropeptides and neurohormones. A direct effect of these neuroendocrine-derived factors on immune function has been suggested based upon the presence of β-adrenegic receptors on immunocompetent cells. Receptors for these neurohormones have been identified and include: ACTH, vasoactive intestinal peptide (VIP), substance P, somatostatin, prolactin, growth hormone, steroid hormones, a number of hormone releasing factors, and catecholamines (norepinephrine and epinephrine).[11,13,14,15] These receptors are expressed on both T and B lymphocytes, macrophages, neutrophils and natural killer (NK) cells.[15-18] Any cells possessing these receptors could respond to norepinephrine and epinephrine released during the SNS response to a stressor. The interaction of the neuroendocrine factors and their receptors on immunocompetent cells could alter the cellular activity through the activation of second messengers including cAMP and cGMP.[11] Some of the direct effects of these catecholamines on cellular immune responses include suppression of lymphocyte migration and proliferation (in response to mitogen), suppression or enhancement (dependent on concentration, target cell and immune function) of cytokine production, NK activity, antibody synthesis, and macrophage activation including the inhibition of the cytokine induced major histocompatibility complex class II antigen expression on antigen presenting cells.[11,13,15,19]

Activation of the HPA axis represents the second physiologic response to stress. The hypothalamus is considered to be the efferent arm of the visceral brain. It receives information from the periphery, integrates it with that of the internal milieu and adjusts important functions, such as sympathetic activity and endocrine secretions. Ultimately, the activation of the HPA axis leads to increases in plasma glucocorticoids. Glucocorticoids exert many different effects, including effects on cardiovascular function, metabolism, muscle function, behavior and the immune system.[20-23] These effects are grouped into two categories defined as permissive and regulatory. "Permissive" effects of glucocorticoids function to "permit" other hormones or immunological factors to accomplish their function at a normal level. These permissive effects are often observed primarily in the resting state of an individual. The permissive role of the glucocorticoids holds and maintains the homeostasis of the individual at a basal state. "Regulatory" effects of glucocorticoids are exerted normally by stress-induced levels of these hormones. These elevated levels of hormone are thought to be necessary to prevent overreaction of the components of the immune system, which, if unchecked, can lead to tissue injury.[24-26]

The production of glucocorticoids from cells of the zona fasciculata of the adrenal cortex is stimulated by pituitary adrenocorticotropin hormone (ACTH).[5] Hypothalmic control of ACTH secretion is via the paraventricular nuclei which have projections into the posterior pituitary that are ultimately responsible for controlling the secretion of a number of peptide hormones, including vasopressin, oxytocin, and corticotropin-releasing hormone (CRH).[27,28] It has been shown that following stress ACTH release can be inhibited by anti-CRH antiserum. CRH secretion, conversely, is controlled by a sequential process involving catecholamines and α_1-adrenoreceptors.[29,30] It has been demonstrated that the hypothalmic injection of acetylcholine increases the secretion of CRH in portal blood which ultimately stimulates increased production of ACTH.[31] In addition to catecholamines, other neuropeptide hormones, such as vasopressin and oxytocin, mediate ACTH release.[31] The relationship between the HPA axis and the SNS is bridged here by the fact the glucocorticoids have been shown to be necessary *in vivo* for normal β-adrenoreceptor function.[32] The threshold for catecholamine receptor stimulation increases dramatically in the absence of glucocorticoids. Steroids, or more precisely, lipocortin, has been shown, in several studies, to potentiate the relaxing effects of β_2-stimulants.[33,34] Although the exact role of opioid peptides derived from proopiomelanocortin gene expression has yet to be elucidated in the stress response, these CRH induced factors including β-endorphin and methionine-enkephalin have been implicated in pain control responses and in learned behavior.[35]

Stress studies on immune function have led many to believe that adrenal glucocorticoids are the only stress induced biological modifiers of immune response. This misinterpretation is in part due to the well known effects of administration of both natural and synthetic analogs of glucocorticoid hormones. Although the majority of stress studies focus on the immunomodulatory effects of the glucocorticoids, other hormones in addition to corticosterone are altered in response to stress. Growth hormone, gonadotropin, and prolactin are just a few examples of hormones that may mediate immunomodulation by altering antibody synthesis, macrophage activation and IL-2 production.[27]

The ability of glucocorticoid hormones to effectively modulate an immune response has been and continues to be widely studied. As mentioned earlier, glucocorticoid production and release from the adrenal cortex is stimulated primarily by ACTH, which in turn is controlled by CRH derived from the hypothalamus in conjunction with catecholamines. The glucocorticoids exert negative feedback on both CRH and ACTH production which in turn results in the ultimate regulation of glucocorticoids themselves.[27] Circadian rhythms or "episodic" increases and decreases occur during each day. Persistently high levels of glucocorticoids can result in diseases such as Cushing's syndrome, suggesting that relief from glucocorticoids is needed for "normal" bodily function.[9] Stress overrides these feedback controls and results in elevated levels of glucocorticoids resulting in either an enhancement or suppression of an organism's defense mechanisms.

Glucocorticoid effects are produced when the hormone, which is assumed to freely penetrate the cell, binds to its receptor to form a non-activated complex.[36] Once activated (or "transformed"), this complex, characterized by an enhanced affinity for DNA, forms the nuclear-bound complex, which, by binding to regulatory elements associated with certain genes, can activate (and sometimes inhibit) transcription of those genes. Glucocorticoids more than likely have primary and secondary targets.   Primary targets (cells) are affected directly by glucocorticoids through the binding of the hormone to its receptor, whereas other cells, the secondary targets, are affected by mediators (cytokines) produced by primary target cells and regulated by glucocorticoids.[37] Mononuclear cells represent one of the best studied primary target cells of glucocorticoids.[37-39] Mononuclear cells possess high affinity receptors (type II) for glucocorticoids.[37,40] Many elements of the cellular immune response are altered by glucocorticoids.[41] Antigen processing and presentation by macrophages in association with MHC class II expression is inhibited by glucocorticoids.[42] The proinflammatory cytokines IL-1, IL-6 and TNF-α produced by the activated macrophages are blocked at the transcriptional and post-transcriptional levels.[43-45] The down regulation of these potent mediators of inflammation as a result of stress underlies the immunosuppressive and anti-inflammatory actions of the glucocorticoids.   Glucocorticoids also inhibit interleukin-2 (IL-2) gene expression by lymphocytes at the transcriptional level.[44] Inhibition of interleukin-2 (IL-2) receptor expression also occurs.[45] Glucocorticoids exert anti-inflammatory effects by increasing synthesis of proteins which inhibit phospholipase A_2 activity which in turn leads to a further inhibition of the arachidonic acid cascade and platelet activating factor (PAF) synthesis.[47,48] Glucocorticoids have also been shown to alter lymphocyte trafficking and inhibit T-cell binding to endothelial cells by modulating expression of adhesion molecules, e.g. intracellular adhesion molecule-1 (ICAM-1), on these cells.[49]

In addition to the immunomodulatory effects produced by glucocorticoids, a direct effect of ACTH on several types of immunocompetent cells including mononuclear cells has been observed.[50-52] ACTH, interacting with a receptor found on mononuclear cells, suppresses antibody production and enhances proliferation by B lymphocytes. ACTH has also been shown to inhibit T-lymphocyte IFN-γ production and IFN-γ induced macrophage activation and MHC class II expression.[42,53,54]

Detection of opioid binding sites present on lymphocytes, polymorphonuclear leukocytes, and platelets raises the possibility that opioids are also immunomodulatory.[55,56] However, the effects of the opioids are not purely suppressive. Enhancement of natural killer (NK) cell function has been noted depending on the particular stressor.[57,58]   In addition, increased synthesis of IL-2 and IFN-γ by lymphocytes as well as NK cells has been observed as a consequence of *in vivo* stress-induced release of opioid peptides.[59-61]

STRESS AND MICROBIAL IMMUNITY

One of the first observations that stressful life events affected pathogenesis of disease was reported by Ishigami, who studied the opsonization of tubercle bacilli among chronically ill tuberculous Japanese school children and their teachers during both active and inactive phases of the disease.[62] Ishigami found decreased phagocytic cell activity during periods of emotional distress and therefore postulated that the stressful school environment of these children and their teachers led to their immunodepressed state and consequently to an increased susceptibility to tuberculosis. This observation has been suggested to be the instigator of the now commonly held belief that certain stressful situations can serve as cofactors in the development of active tuberculosis infections. Until recently, very little solid evidence was available that suggested that stress affected the pathogenesis of certain diseases.

Early *in vivo* studies on the effect of stress on microbial pathogenesis focused primarily on the modulatory effects of the products of HPA axis activation, or glucocorticoids. Unfortunately, a good number of the studies were conducted with little or no controls and without the experimental certainty that the animals used in these studies were actually being stressed to the point that significant activation of the HPA axis occurred. Without that knowledge, easily determined in present day studies, any effect of stress reported in some of these studies must be noted with some reservation.

Many of the early stress studies were limited in their ability to indicate if HPA axis activation was a factor in modulation of certain infectious diseases. Exogenously injected adrenal cortical extracts served as the immunomodulator (stressor). These studies, conducted primarily on rodents, led to the conclusion that the injection of exogenous glucocorticoids resulted in the animal's increased susceptibility to the particular pathogen being studied (Table 1). The availability of synthetic analogs of glucocorticoids, as well as glucocorticoid agonists and antagonists, have led to the identification of the components of the immune system that are affected by products of HPA axis activation (Table 1).

While exogenous glucocorticoids served as the "stressor" in many of the early defining stress experiments, various stress paradigms were used in order to modulate the immune response to certain pathogens. Friedman *et al.*[87] showed that novel environment stress (crowding) altered the resistance of mice to malarial infections. Crowding stress, in another study, resulted in an increase in susceptibility of mice to infection with *Salmonella typhimurium*.[88] Predator stress (a large cat) reduced the resistance of immune mice to subsequent reinfection by the cestode *Hymenolepis nana*.[89] Green reported that cold stress and hypoxia (oxygen deprivation) resulted in the inhibition of clearance of *Staphylococcus albus* and *Proteus mirabilis* from the lungs of mice.[90]

Stress effects on resistance in farm animals have been reported by several laboratories. Gross *et al.*, showed that social stress (rotating cage mates) increased the susceptibility of chickens to aerosol challenge with *Escherichia coli* and to challenge with *Mycobacterium avium*.[91-93] Controlled food deprivation (fasting) resulted in the increased susceptibility of chickens to *Salmonella enteritidis* infection and exacerbated the infection.[94,95] Zamri-Saad *et al.*[96] found that transportation stress in combination with treatment of dexamethasone was associated with an increased susceptibility of goats to infection with *Pasteurella haemolytica*. Animals treated with dexamethasone alone, but not subjected to transportation stress, displayed no increase in susceptibility to the *Pasteurellae*. The specific *in vivo* stress effects on microbial pathogenesis of these and other studies are summarized in Table 2.

Table 1
Effect of glucocorticoids on microbial immunity

Organism[ref.]	Effect of Glucocorticoids
Mycobacterium tuberculosis[64,81]	increased cellular infiltration, increased bacterial growth
Mycobacterium paratuberculosis[64,81]	Increased susceptibility, increased shedding
Mycobacterium leprae[65]	alters IFN-γ efficacy of monocytes
Mycobacterium terrae[66]	increases resistance by stabilization of drug interaction
Mycobacterium avium[67,68]	impairs macrophage function
Yersinia pestis[69] *(pasteurella pestis)*	increases invasive infection, bacteremia
Escherichia coli[70,71,72]	decreases inflammation-does not affect bacterial clearance
Candida albicans[73]	exacerbates infection
Pseudomonas aeruginosa[74,75]	increases susceptibility
Staphylococcus aureus[76]	ameliorates infection (when infected via I.P. route)
Bordetella pertussis[77]	promotes initial protective effect, then promotes microbial survival
Trichinella spiralis[78]	decreases nematode clearance, cellular infiltration
Listeria monocytogenes[79,80,85]	increases susceptibility, impairs monocyte function
Pneumocystis carinii[82]	increases susceptibility

Table 1 continued

Propionibacterium acnes[83]	exacerbates infection
Neisseria gonorrheae[84]	increases rate of infection
Streptococcus faecalis[86]	increases resistance
Plasmodium berghei[87]	decreases parasitemia

Table 2
Effects of Stress on Microbial Immunity

Pathogen	Animal	Stressor	Effect
M. avium[67]	Mouse	Restraint	increases susceptibility to infection, impairs macrophage function
	Chicken	Social	increases granuloma formation
M. tuberculosis[97]	Rabbit	Tumbling	increases susceptibility
P. berghei[87]	Mouse	Electric	increases survival
	Mouse	Environmental	"
S. typhimurium[88]	Mouse	Crowding	increases susceptibility
S. enteritidis[94,95]	Chicken	Fasting	decreases Ab response
H. nana[89]	Mouse	Predator	increases reinfection rate
E. coli[91,92]	Chicken	Cold	increases resistance
	Chicken	Social	"

Table 2 continued

P. mirabilis[90]	Mouse	Cold	inhibits clearance and exacerbates infection
S. albus[90]	Mouse	Cold	inhibits clearance
P. aeruginosa[97]	Rabbit	Tumbling	increases susceptibility
P. haemolytica[96]	Goat	Transport	increases susceptibility

STRESS AND VIRAL INFECTIONS

Although there has been substantial documentation regarding the association of emotional stress and increased rates and duration of various viral infections in humans, the scope of this chapter must be limited to the effects of stress on microbial (bacterial, parasitic) infections in animal models. However, before dismissing stress-induced effects on viral pathogenesis altogether, it must be noted that experimental animal models are used frequently to study the effects of stress on viral infections. There are similarities between studies of stress on viral infection and studies on the effects of stress in microbial infections. For example, farm animals exposed to social (crowding or novel environment) stress demonstrated increased susceptibility to Newcastle disease virus infection.[98] Transportation stress in cattle resulted in increased susceptibility to an initial challenge with bovine herpesvirus-1 and increased incidence of reactivation of latent bovine herpesvirus-1 infection.[99,100] Most studies conducted on the effects of stress on viral immunity have utilized rodent models. A wide variety of stress paradigms including restraint, cold, foot shock, immobilization and isolation have been shown to differentially affect the pathogenesis of a number of viral infections in these animal models.[67,101-106] Interestingly, the ultimate outcome effects of physical restraint stress on viral pathogenesis (influenza virus and HSV) and microbial pathogenesis (*Mycobacterium avium*) in mice are similar. Recent studies have shown that mice must be subjected to restraint stress, resulting in activation of the HPA axis, **prior** to or **concurrently** with infection in order to detect a stress induced alteration in pathogenesis.[67,101] Infection prior to restraint results in no apparent alteration in the course of the mycobacterial infection or the viral infections.

STRESS AND MYCOBACTERIAL DISEASE

The incidence of mycobacterial disease in the United States has significantly increased in the last decade, due, in part, to infection of people also infected with human immunodeficiency virus as well as a rise in tuberculosis infections not associated with HIV infections.[2,107,108] Tuberculosis continues to be a significant worldwide public health problem.

Tuberculosis is an extremely tenacious infectious disease that engages the defense mechanisms of the host for a prolonged period of time. Although successful antibiotic treatment of tuberculosis infection results in the inhibition of growth of the tubercle bacilli,

complete recovery of an individual and immunity to reinfection depends on a precise functioning of the immune system.

When *Mycobacterium tuberculosis* first invades the respiratory system (or other organs of the body), certain immunologic changes occur in order to defend the host from the development of the current infection and from possible future infections by *M. tuberculosis*. The possible development of disease is related to both the virulence of the mycobacteria and to the genetic and acquired factors of host resistance to the infection. Generally, most healthy individuals, who become infected with *M. tuberculosis*, develop a delayed hypersensitivity and a protective immune response that ultimately controls the growth of the organism.[109] During initial stages of the disease, the tubercle bacilli grow within alveolar macrophages that are eventually killed by the bacteria. Delayed hypersensitivity responses develop within a period of 2 to 3 weeks which result in destruction of the infected macrophages and ultimately in the formation of caseous necrosis which no longer supports mycobacterial growth. Cell-mediated immunity prevents the further multiplication of the tubercle bacilli. In most infected individuals, growth of the organism is controlled unless the cell-mediated immune response is compromised. Reactivation of disease has been attributed to several cofactors that can compromise this acquired cellular immunity. Disease reactivation has been attributed to infection with HIV, cancer chemotherapy, immunosenescence due to aging, protein malnutrition associated with alcoholism, homelessness, and stress.[2,3,110-115] The role of genetic factors in the immune response to *M. tuberculosis* has been extensively investigated in animals.[116-118] These factors that play a role in controlling resistance to tuberculosis include sex, race, the major histocompatibility complex, and innate resistance.[119] One of the most important genetic factors connected to tuberculosis resistance can be identified as the congenital ability of macrophages to control the growth of the tubercle bacilli.[119,120] There is considerable individual variation in the response of the macrophage to *M. tuberculosis*. In some cases, the mycobacteria are destroyed by macrophages; in others, the macrophages ingest but do not kill the mycobacterium, which survives without growing within the phagocytic cells. Macrophages which are unable either to phagocytize **or** destroy *M. tuberculosis* allow the uncontrolled growth of the bacilli and the disease results.

Injection of mice with mycobacteria results in two phenotypic patterns of growth. Intravenous infection of mice with mycobacteria results in the exponential growth of the bacilli in some strains of mice and controlled growth in others. Thus, strains of mice can be classified as either resistant or susceptible based on the differential growth patterns of the mycobacteria.[120,121] The gene that controls resistance to mycobacterial growth in mice, termed *Bcg*, maps to a gene on chromosome 1.[121] A group of genes that are linked to *Bcg* in mice form a systemic group of genes on human chromosome 2q. Studies by Denis *et al.*,[122] have demonstrated that the differential growth patterns of the mycobacteria are due to differences in macrophage activation. Thus, mature macrophages from BCG-resistant (*Bcg*ʳ) mice can control the growth of the mycobacteria, whereas macrophages from BCG-susceptible (*Bcg*ˢ) mice cannot. Recently, Vidal and coworkers isolated a candidate Bcg gene, designated *Nramp* (natural resistance-associated macrophage protein), which encodes a polytopic integral membrane protein that has structural features common to prokaryotic and eukaryotic transporters.[123] Until further studies can be carried out to determine the relationship between resistance and the different structural/functional domains of *Nramp*, the genetic basis for disease susceptibility regarding mycobacterial infections remains unknown.

Based upon previous observations in which restraint stress suppressed MHC class II expression by macrophages from *Bcg*[s] mice but not macrophages from *Bcg*[r] mice[124], we extended our studies to include an assessment of the role of HPA axis activation by restraint stress, on the growth of *M. avium* in both BCG resistant and BCG susceptible mice.[67] The timing of the stress was found to be a critical factor in altering macrophage function. The effect of stress was ameliorated by the intensity of the immunological stimulus.[67]

Initial studies showed that stress resulted in the increased susceptibility of BCG-susceptible mice to mycobacterial growth in both the lungs and spleens (Figure 1).[67] In contrast, HPA axis activation did not affect the growth of the mycobacteria in BCG resistant mice. Adrenalectomy and treatment with the glucocorticoid receptor antagonist RU486 abrogated the effect of stress. HPA axis activation or the addition of exogenous corticosterone also resulted in the increased permissiveness of macrophages from BCG-susceptible mice to mycobacterial growth (Table 3). Macrophages from BCG-resistant mice were not affected by HPA axis activation or by the addition of corticosterone to the culture medium and were able to control the intracellular growth of *M. avium*.

In contrast to the differential effect of HPA axis activation on mycobacterial growth, activation of the HPA axis suppressed the anti-microbial mechanisms of macrophages from both the BCG susceptible and BCG resistant mice.[67,68] The production of TNF-α and of reactive nitrogen intermediates (nitric oxide production) was suppressed. Our findings suggest that the innate resistance mechanism(s) of BCG resistant mice is not affected by HPA axis activation and shows that mycobacterial growth in susceptible individuals may be exacerbated by stress. The observations that HPA axis activation or addition of glucocorticoids resulted in the suppression of TNF-α and NO production by macrophages from HPA axis were activated by 10 18-hr restraint periods. Splenic macrophages were isolated from congenic *Bcg*[s] and *Bcg*[r] mice and were infected *in vitro* with *M. avium*. The data are the counts per minute of both BCG resistant and BCG susceptible mice, suggesting that the mechanism of action of *Bcg* (candidate *Nramp*) probably does not involve the production of either mediator.[67,68] The macrophages from the BCG resistant mice continue to control the growth of the mycobacteria despite the inhibition of this important anti-mycobacterial pathway. Therefore, other antibacterial mechanisms may account for resistance to mycobacterial growth.

CONCLUSION

The differential effects of stress on mycobacterial growth *in vivo* may have important implications in human disease. Racial differences in susceptibility to *M. tuberculosis* infections have been reported in several studies. Results of these findings indicate that blacks are more susceptible to infection with *M. tuberculosis* than the white population.[125,126] In addition, macrophages from blacks are more permissive to the growth of mycobacteria as compared to those from the white population.[125,126] Clearly, there is increasing evidence of the association between stress and the pathogenesis of infectious disease. Extrapolation of the effect of restraint stress on mycobacterial resistance in the mouse model to humans suggests that disease reactivation in susceptible populations may be affected by life stressors. Stress effects mediated by neuroendocrine-immune interactions can result in increased susceptibility to disease. The results of several studies indicate that the timing of the stressor and the exposure to the infectious agent will ultimately determine the outcome of the disease process.[67,101] The results of our studies and those of others suggest that the effect of a stressor on reactivation of a latent disease may have important health consequences.

Table 3
Differential Effect of HPA Activation on Mycobacterial Growth in Macrophages from BCG-Resistant and BCG Susceptible Mice

Macrophage Source	Treatment	[³H]Uracil Incorporation
BCG-Resistant	Restraint	7,100 +/- 363
	Control	7,450 +/- 287
BCG-Susceptible	Restraint	13,225 +/- 425
	Control	8,850 +/- 305

HPA axis was activated by 10 18-hr restraint periods. Splenic macrophages were isolated from congenic *Bcg*^{s} and *Bcg*^{r} mice and were infected *in vitro* with *M. avium*. The data are the counts per minute of [³H]Uracil taken up by the bacteria released from the macrophages after 5 days of *in vitro* culture. The data are from a representative experiment. The effect of HPA axis activation is significant ($P < 0.004$).

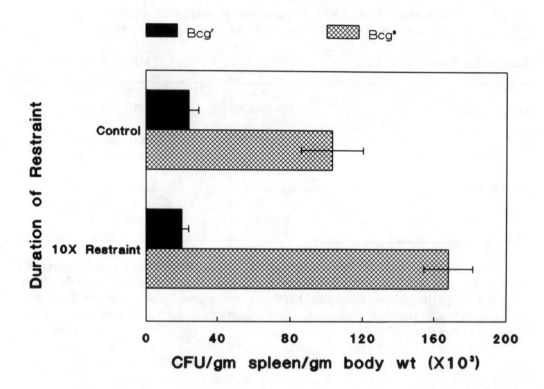

Fig 1. Differential effect of HPA axis activation on mycobacterial resistance of BALB/c.*Bcg*[r] and BALB/c *Bcg*[s] mice. Mice were infected with *M. avium* and restrained for 10-18 hr cycles. The numbers of CFU of *M. avium* in the spleens were determined 12 days after the infection. The difference between the growth of the mycobacteria in the spleens of the BCG-resistant and BCG-susceptible mice was significant (p<0.001). Similar observations were made concerning differences in the effect of HPA activation on growth of the mycobacteria in the lungs (not shown).

REFERENCES

1. **Richardson, B. W.,** Diseases from worry and mental strain, in *Diseases of Modern Life,* Bermingham and Co. Press, Union Square, New York, 73, 1882.

2. **Collins, F.M.,** Mycobacterial disease: immunosupression and acquired immunodeficiency syndrome, *Clin. Microbiol. Rev.* 2,360, 1989.

3. **Wiegeshaus, E., Balasubramanian, V., and Smith, D.W.,** Immunity to tuberculosis from the perspective of pathogenesis, *Infect. Immun.,* 57,3671, 1989.

4. **Ramsey, J.M.,** *Basic pathophysiology: modern stress and the disease process,* Addison-Wesley Publishing Co., Menlo Park, Ca., 1982.

5. **Axelrod, J., and Reisine, T.D.,** Stress hormones: their interaction and regulation, *Science,* 224,452, 1984.

6. **Bateman, A., Singh, A., Thomas, K. and Solomon, S.,** The immune-hypothalmic-pituitary-adrenal axis, *Endocrine rev.,* 10(1),92, 1989.

7. **Sheridan, J.F., Dobbs, C., Brown, D. and Zwilling, B.,** Psychoneuroimmunology: stress effects on pathogenesis and immunity during infection, *Clin. Microbiol. Rev.,* 7(2),200, 1994.

8. **Berkenbosch, F., Wolvers, D.A.W. and Derijk, R.,** Neuroendocrine and immunological mechanisms in stress induced immunomodulation, *J. Steroid Biochem. Mol. Biol.,* 40(6), 639, 1991.

9. **Gustafsson, J., Carlstedt-Duke, J. and Poellinger, L.,** Biochemistry, molecular biology, and physiology of the glucocorticoid receptor, *Endocr. Rev.,* 8, 1987.

10. **Gehring, U.,** Cell genetics of glucocorticoid responsiveness, in *Biochemical Actions of Hormones,* Academic Press, London, 311, 1985.

11. **Madden, K.S. and Livnat, S.,** Catecholamine action and immunologic reactivity, in Ader, R., Felten, D.L., and Cohen, N., Ed., *Psychoneuroimmunology,* 2nd ed. Academic Press, Inc., San Diego, Ca., 283, 1991.

12. **Hall, N.R., McClure, J.E., Hu, S.K., Tare, N.S., Selas, C.M. and Goldstein, A.L.,** Effects of 6-hydroxydopamine upon primary and secondary thymus dependent immune response, *Immunopharmacology,* 5,39, 1982.

13. **Livnat, S., Felten, S.Y., Carlson, S.L., Bellinger, D.L. and Felten, D.L.,** Involvement of peripheral and central catecholamine systems in neural-immune interactions, *J. Neuroimmunol.,* 10,5, 1985.

14. **Abrass, C.K., O'Connor, S.W., Scarpace, P.J. and Abrass, I.B.,** Characterization of the β-adrenergic receptor of the rat peritoneal macrophage, *J. Immunol.,* 135,1338, 1985.

15. **Fuchs, B.A., Albright, J.W. and Albright, J.F.,** β-adrenergic receptors on murine lymphocytes: density varies with cell maturity and lymphocyte subtype and is decreased after antigen administration, *Cell. Immunol.,* 114,231, 1988.

16. **Galant, S.P., Durisetti, L., Underwood, S. and Insel, P.A.,** β-adrenergic receptors on polymorphonuclear leukocytes: adrenergic therapy decreases receptor number, *N. Engl. J. Med.,* 299,933, 1978.

17. **Loveland, B.E., Jarrot, E.B. and McKenzie, I.F.C.,** β-adrenergic receptors on murine lymphocytes, *Int. J. Immunopharmacol.,* 3,45, 1981.

18.   **Motulsky, H.J. and Insel, P.A.,** Adrenergic receptors in man, *N. Engl. J. Med.,* 307,18, 1982.

19.   **Frohman, E.M., Vayuvegula, B., Gupta, S. and van den Noort, S.,** Norepinephrine inhibits gamma-interferon-induced major histocompatibility class II (Ia) antigen expression on cultured astrocytes via β-2-adrenergic signal transduction mechanisms, *Proc. Natl. Acad. Sci. USA,* 85,1292, 1988.

20.   **Fauci, A.S., Dale, D.C. and Balow, J.E.,** Glucocorticosteroid therapy: mechanisms of action and clinical considerations, *Ann. Intern. Med.,* 84,304, 1976.

21.   **Baxter, J.D. and Forsham, P.H.,** Tissue effects of glucocorticoids, *Am. J. Med.,* 53,573, 1972.

22.   **Baxter, J.D. and Harris, A.W.,** Mechanism of glucocorticoid action: general features, with reference to steroid-mediated immunosuppression, *Transplant. Proc.,* 7,55, 1975.

23.   **Chrousos, G.P., Laue, L., Nieman, L.K., Kawai, S., Udelson, R.U., Brandon, D.D. and Loriaux, D.L.,** Glucocorticoids and glucocorticoid antagonists: lessons from RU 486, *Kidney Int.* (suppl. 26),18, 1988.

24.   **Ingle, D.J.,** Permissibility of hormone action: a review, *Acta. Endocrinol.,* 17,172, 1954.

25.   **Munck, A. and Guyre, P.M.,** Glucocorticoid physiology, pharmacology and stress, in, Chrousos, G.P., Loriaux, D.L., Lipsett, M.B., Eds., *Steroid Hormone Resistance,* Plenum Press, New York, 1986, p.81.

26.   **Rivier, C. and Vale, W.,** Modulation of stress-induced ACTH release by CRF, catecholamines, and vasopressin, *Nature,* 305,325, 1983.

27.   **Johnson, E.O., Kamilaris, T.C., Chrousos, G.P. and Gold, P.W.,** Mechanisms of stress: a dynamic overview of hormonal and behavioral homeostasis, *Neurosci. Biobehav. Rev.,* 16,115, 1992.

28.   **Sawchenko, P.E., Swanson, L.W. and Joseph, S.A.,** The distribution and cells of origin of ACTH (1-39) stained varicosities in the paraventricular and supraoptic nuclei, *Brain Research,* 232,365, 1982.

29.   **Rivier, C., Brownstein, M., Speiss, J., Rivier, J. and Vale, W.,** In vivo corticotropin-releasing factor-induced secretion of adrenocorticotropin, beta-endorphin, and corticosterone, *Endocrinology,* 110,272, 1982.

30.   **Vale, W., Speiss, J., Rivier, C. and Rivier, J.,** Characterization of a 41-residue ovine hypothalmic peptide that stimulates secretion of corticotropin and β-endorphin, *Science,* 213,1394, 1981.

31.   **Eipper, B.A. and Mains, R.E.,** Structure and biosynthesis of proadrenocorticotropin/endorphin and related peptides, *Endocrinol. Rev.,* 1,1, 1980.

32.   **Brodie, B.B., Davies, J.I., Hynie, S., Krishna, G., and Weiss, B.,** Interrelationships of catecholamines with other endocrine systems, *Pharmacol. Rev.,* 18,273, 1966.

33.   **Townley, R.G., Daley, D. and Selenke, W.,** The effects of corticosteroids on the β-adrenergic receptors in bronchial smooth muscle, *J. Allergy,* 45,118, 1970.

34.   **Dvorsky-Gebauer, R.J.,** Potentiation of bronchodilators by glucocorticoids, *Lancet,* 2,306, 1976.

35.   **Krieger, D.T.,** Brain peptides: what, where, and why? *Science,* 222,975, 1983.

36. **Payvar, F., Wrange, O., Carlstedt-Duke, J., Okret, S., Gustafsson, J.A. and Yamamoto, K.R.,** Purified glucocorticoid receptors bind selectively *in vitro* to a cloned DNA fragment whose transcription is regulated by glucocorticoids *in vivo*, *Proc. Natl. Acad. Sci. USA,* 78,6628, 1981.

37. **Munck, A., Mendel, D.B., Smith, L.I. and Orti, E.,** Glucocorticoid receptors and actions, *Am. Rev. Respir. Dis.,* 141,S2, 1990.

38. **Crabtree, G.R., Munck, A. and Smith, K.A.,** Glucocorticoids and lymphocytes. I. Increased glucocorticoid receptor levels in antigen stimulated lymphocytes, *J. Immunol.,* 124,2430, 1980.

39. **Werb, Z., Foley, R. and Munck, A.,** Interaction of glucocorticoids with macrophages-identification of glucocorticoid receptors in monocytes and macrophages, *J. Exp. Med.,* 147,1684, 1978.

40. **Miesfeld, R.L.,** Molecular genetics of corticosteroid action, *Am. Rev. Respir. Dis.,* 141, S11, 1990.

41. **Munck, A. and Guyre, P.M.,** Glucocorticoids and immune function, in Ader, R., Felten, D.L., and Cohen, N., Eds., *Psychoneuroimmunology,* 2nd edition, Academic Press, Inc, San Diego, Ca., 1991, 447.

42. **Zwilling, B.S., Brown, D. and Pearl, D.,** Induction of major histocompatibility complex class II glycoproteins by interferon-γ: attenuation of the effects of restraint stress, *J. Neuroimmunol.,* 37,115, 1992.

43. **Kern, J.A., Lamb, R.J., Reed, J.C., Daniele, R.P., and Nowell, P.C.,** Dexamethasone inhibition of interleukin-1β production by human monocytes, *J. Clin. Invest.,* 81,237, 1988.

44. **Lee, S.W., Tso, A.P., Chan, H., Thomas, J., Petrie, K., Eugui, E.M. and Allison, A.C.,** Glucocorticoids selectively inhibit the transcription of the interleukin 1-β gene and decrease the stability of interleukin 1-β mRNA, *Proc. Natl. Acad. Sci. USA,* 85,1204, 1988.

45. **Northop, J.P., Crabtree, G.R. and Mattila, P.S.,** Negative regulation of interleukin 2 transcription by the glucocorticoid receptor, *J. Exp. Med.,* 175,1235, 1992.

46. **Gurye, P.M., Girard, M.T., Morganelli, P.M. and Manganiello, P.D.,** Glucocorticoid effects on the production and actions of immune cytokines, *J. Steroid Biochem.,* 30(1-6), 89, 1988.

47. **Svedmyr, N.,** Action of corticosteroids on Beta-adrenergic Receptors, *Am. Rev. Respir. Dis.,* 141,S31, 1990.

48. **Fraser, C.M., Venter, J.C.,** Beta-adrenergic receptors: relationship of primary structure, receptor function and regulation, *Am. Rev. Respir. Dis.,* 141,S22, 1990.

49. **Eguchi, K.A., Kawakami, A., Nakashima, M., Ida, H., Sakito, S., Matsuoka, N., Terada, K., Sakai, M., Kawabe, Y., Fukuda, T., Ishimaru, T., Kurouji, K., Fujita, N., Aoyagi, T., Maeda, K. and Nagatki, S.,** Interferon-α and dexamethasone inhibit adhesion of T cells to endothelial cells and synovial cells, *Clin. Exp. Immunol.,* 88,448, 1992.

50. **Bost, K.L., Smith, E.M., Wear, L.B. and Blalock, J.E.,** Presence of ACTH and its receptors on a B lymphocytic cell line: a possible autocrine function for a neuroendocrine hormone, *J. Biol. Regul. Homeostat. Agents,* 1,23, 1987.

51. **Johnson, E.W., Blalock, J.E. and Smith, E.M.,** ACTH receptor-mediated induction of leukocyte cyclic AMP, *Biochem. Biophys. Res. Commun.,* 157,1205, 1988.
52. **Smith, E.M., Bronsan, P., Meyer, W.J. and Blalock, J.E.,** An ACTH receptor on human mononuclear leukocytes, *N. Engl. J. Med.,* 317,1266, 1987.
53. **Johnson, H.M., Torres, B.A., Smith, E.M., Dion, L.D. and Blalock, J.E.,** Regulation of lymphokine (interferon-γ) production by corticotropin, *J. Immunol.,* 132,246, 1984.
54. **Koff, W.C. and Dunegan, M.A.,** Modulation of macrophage-mediated tumoricidal activity by neuropeptides and neurohormones, *J. Immunol.,* 135,350, 1985.
55. **Ausiello, C.M. and Roda, L.G.,** Leu-enkephalin binding to cultured human T-lymphocytes, *Cell. Biol. Int. Rep.,* 8,353, 1984.
56. **Mehrishi, J.N. and Mills, I.H.,** Opiate receptors on lymphocytes and platelets in man, *Clin. Immunol. Immunopathol.,* 27,240, 1983.
57. **Irwin, M.R. and Hauger, R.L.,** Adaptation to chronic stress: temporal pattern of immune and neuroendocrine correlates, *Neuropsychopharmacology,* 1,239, 1988.
58. **Shavit, Y., Lewis, J.W., Terman, G.W., Gale, R.P. and Liebeskind, J.C.,** Opiod peptides mediate the suppressive effect of stress on natural killer cell cytotoxicity, *Science,* 223, 188, 1984.
59. **Gilmore, W. and Weiner, L.P.,** β-endorphin enhances interleukin-2 (IL-2) production in murine lymphocytes, *J. Neuroimmunol.,* 18,125, 1988.
60. **Brown, S.L. and Van Epps, D.E.,** Opiod peptides modulate production of interferon-γ by human mononuclear cells, *Cell. Immunol.,* 103,19, 1986.
61. **Mandler, R.N., Biddison, W.E., Mandler, R., and Serrate, S.A.,** β-endorphin augments the cytolytic activity and interferon production of natural killer cells, *J. Immunol.,* 136, 934, 1986.
62. **Ishigami, T.,** The influence of psychic acts on the progress of pulmonary tuberculosis, *Am. Rev. Tuberculosis,* 2,470, 1919.
63. **Robson, J.M. and Didcock, K.A.,** The action of cortisone on corneal tuberculosis studied with the phase contrast microscope, *Am. Rev. Tuberc. Pulm. Dis.,* 74,1, 1956.
64. **Follett, D.M. and Czuprynski, C.J.,** Cyclophosphamide and prednisolone exacerbate the severity of intestinal paratuberculosis in *Mycobacterium paratuberculosis* monoassociated mice, *Microb. Path.,* 9(6),407, 1990.
65. **Vachula, M., Holzer, T.J., Nelson, K.E. and Anderson, B.R.,** Effect of glucocorticoids and interferon-gamma on the oxidative responses of monocytes from leprosy patients and normal donors, *Int. J. Leprosy & other Mycobacterial Dis.,* 59(1),41, 1991.
66. **Petrini, B., Svartengren, G., Hoffner, S.E., Unge, G. and Widstrom, O.,** Tenosynovitis of the hand caused by *Mycobacterium terrae, Eur. J. Clin. Micro. and Infect. Dis.,* 8(8),722, 1989.
67. **Brown, D.H., Sheridan, J.F., Pearl, D. and Zwilling, B.S.,** Regulation of mycobacterial growth by the hypothalamus-pituitary-adrenal axis: differential responses of *Mycobacterium bovis* BCG-resistant and -susceptible mice, *Infect. Immun.,* 61,4793, 1993.
68. **Brown, D.H. and Zwilling, B.S.,** Activation of the hypothalmic-pituitary-adrenal axis differentially affects the anti-mycobacterial activity of macrophages from BCG-resistant and susceptible mice, *J. Neuroimmuno.,* 53,181, 1994.

69. **Payne, F.E., Larson, A., Walker, D.L., Foster, L. and Meyer, K.F.,** Studies on immunization against plague. IX. The effects of cortisone on mouse resistance to attenuated strains of *Pasteurella pestis*, *J. Infect. Dis.*, 96,168, 1955.

70. **Anderson, K.I., Hunt, E. and Davis, B.J.,** The influence of anti-inflammatory therapy on bacterial clearance following intramammary *Escherichia coli* challenge in goats, *Vet. Res. Commun.*, 15(2),147, 1991.

71. **Gillissen, G.,** "Inverse" effects of cortisone in experimental infection of mice, *Adv. Exp. Med. Biol.*, 319,137, 1992.

72. **Neuwelt, E.A., Lawrence, M.S. and Blank, N.K.,** Effect of gentamicin and dexamethasone on the natural history of the rat *Esherichia coli* brain abscess model with histopathological correlation, *Neurosurg.*, 15(4),475, 1984.

73. **Wong, B., Brauer, K.L., Clemens, J.R. and Beggs, S.,** Effects of gastrointestinal candidiasis, antibiotics, dietary arabinitol, and cortisone acetate on levels of the candida metabolite D-arabinitol in rat serum and urine, *Infect. Immun.*, 58(2),283, 1990.

74. **Jones, W.G., Barber, A.E., Kapur, S., Hawes, A.J., Fahey, T.J., Minei, J.P., Shires, G.T., 3rd., Calvano, S.E. and Shires,G.T.,** Pathophysiologic glucocorticoid levels and survival of translocating bacteria, *Arch. of Surg.*, 126(1),50, 1991.

75. **Baltch, A.L., Hammer, M.C., Smith, R.P., Bishop, M.B., Sutphen, N.T., Egy, M.A. and Michelsen, P.B.,** Comparison of the effect of three adrenal corticosteroids on human granulocyte function against *Pseudomonas aeruginosa*, *J. Trauma*, 26(6),525, 1986.

76. **Badenoch, P.R., Hay, G.J., McDonald, P.J. and Coster, D.J.,** A rat model of bacterial keratitis. Effect of antibiotics and corticosteroids, *Arch. Opthamol.*, 103(5),718, 1985.

77. **Parton, R.,** Effect of prednisolone on the toxicity of *Bordetella pertussis* for mice, *J. Med. Micro.*, 19(3),391, 1985.

78. **Coker, C.M.,** Cellular factors in acquired immunity to *Trichnella spiralis* as indicated by cortisone treatment of mice, *J. Infect. Dis.*, 98,187, 1956.

79. **Schaffner, A.,** Therapeutic concentrations of glucocorticoids suppress the antimicrobial activity of human macrophages without impairing their responsiveness to gamma interferon, *J. Clin. Invest.*, 76(5),1755, 1985.

80. **Pung, O.J., Luster, M.I., Hayes, H.T. and Rader, J.,** Influence of steroidal and nonsteroidal sex hormones on host resistance in mice: increased susceptibility to *Listeria monocytogenes* after exposure to estrogenic hormones, *Infect. Immun.*, 46(2),301, 1984.

81. **Wentink, G.H., Rutte, V.P., Jaarsveld, F.H., Zeeuwen, A.A. and Van Kooten, P.J.,** Effect of glucocorticoids on cows suspected of subclinical infection with *M. paratuberculosis*, *Vet. Quarterly*, 10(1),57, 1988.

82. **Walzer, P.D., Powell, R.D., Jr. and Yoneda, K.,** Experimental *Pneumocystis carinii* pneumonia in different strains of cortisonized mice, *Infect. Immun.*, 24(3),939, 1979.

83. **Gloor, M., Funder, H. and Franke, M.,** Effect of topical application of dexamethasone on propionibacteria in the pilosebaceous duct, *Eur. J. Clin. Pharmacol.*, 14(1),53, 1978.

84. **Arko, R.J.,** *Neisseria gonorrhoeae*: experimental infection of laboratory animals, *Science*, 177(55),1200, 1972.

85. **Wesley, I.V., Bryner, J.H., Van Der Maaten, M.J. and Kehrli, M.,** Effects of dexamethasone on shedding of *Listeria monocytogenes* in dairy cattle, *Am. J. Vet. Res.*, 50(12),2009, 1989.

86. **Gross, W.B.,** Use of corticosterone and ampicillin for treatment of *Streptococcus faecalis* infection in chickens, *Am. J. Vet. Res.,* 52(8),1288, 1991.
87. **Friedman, S.B., Ader, R. and Grota, L.J.,** Protective effect of noxious stimulation in mice infected with rodent malaria, *Psychosom. Med.,* 35,535, 1973.
88. **Edwards, E.A. and Dean, L.M.,** Effects of crowding of mice on humoral antibody formation and protection to lethal antigenic challenge, *Psychosom. Med.,* 39,19, 1977.
89. **Hamilton, D.R.,** Immunosuppressive effects of predator induced stress in mice with acquired immunity to *Hymenolepis nana, J. Psychosom. Med.,* 18,143, 1974.
90. **Green, G.M. and Kass, E.H.,** The influence of bacterial species on pulmonary resistance to infection in mice subjected to hypoxia, cold stress and ethanol intoxication, *Br. J. Exp. Pathol.,* 46,360, 1965.
91. **Gross, W.B.,** Effect of a range of social stress severity on *Escherichia coli* challenge infection, *Am. J. Vet. Res.,* 45,2074, 1984.
92. **Gross, W.B.,** Effect of environmental stress on the response of ascorbic-acid treated chickens to *Escherichia coli* challenge infections, *Avian Dis.,* 32,432, 1988.
93. **Gross, W.B., Falkinham, J.D. and Payeur, J.B.,** Effect of environmental-genetic interactions on *Mycobacterium avium* challenge infection, *Avian Dis.,* 33,411, 1989.
94. **Holt, P.S.,** Effect of induced molting on the susceptibility of white leghorn hens to a *Salmonella enteritidis* infection, *Avian Dis.,* 37(2),412, 1993.
95. **Holt, P.S. and Porter, R.E.,** Microbiological and histopathological effects of an induced-molt fasting procedure on a *Salmonella enteritidis* infection in chickens, *Avian Dis.,* 36, 610, 1992.
96. **Zamri-Saad, M., Jasni, S., Naridi, A.B. and Sheikh-Omar, A.R.,** Experimental infection of dexamethasone-treated goats with *Pasteurella haemolytica* A2, *Br. Vet. J.,* 147,565, 1991.
97. **Lockard, V.G., Grogan, J.B. and Brunson, J.G.,** Alterations in the bactericidal ability of rabbit alveolar macrophages as a result of tumbling stress, *Am. J. Pathol.,* 70(1),57, 1970.
98. **Mohamed, M.A. and Hansen, R.P.,** Effects of social stress on Newcastle disease virus (LaSota) infection, *Avian Dis.,* 24,908, 1980.
99. **Filion, L.G., Willson, P.J., Bielefeldt-Ohmann,H., Babiuk, L.A. and Thomson, R.G.,** The possible roles of stress in the induction of pneumonic pasteurellosis, *Can. J. Comp. Med.,* 48,268, 1984.
100. **Thiry, E., Saliki, J., Bublot, M. and Pastoret, P.P.,** Reactivation of infectious bovine rhinotracheitis virus by transport, *Comp. Immunol. Microbiol. Infect. Dis.,* 10,59, 1987.
101. **Bonneau, R.H., Sheridan, J.F., Feng, N. and Glaser, R.,** Stress-induced suppression of herpes simplex virus (HSV)-specific cytotoxic T lymphocyte and natural killer cell activity and enhancement of acute pathogenesis following local HSV infection, *Brain Behav. Immun.,* 5,170, 1991.
102. **Kusnecov, A.V., Grota, L.J., Schmidt, S.G., Bonneau, R.H., Sheridan, J.F., Glaser, R. and Moynihan, J.A.,** Decreased herpes simplex viral immunity and enhanced pathogenesis following stressor administration in mice, *J. Neuroimmunol.,* 38,129, 1992.
103. **Ozherelkov, S.V., Khonzinsky, V.V. and Semenov, B.F.,** Replication of Langat virus in immunocompetent cells of mice subjected to immobilization stress, *Acta Virol.,* 34,291, 1990.

104. **Chetverikova, L.K., Frolov, B.A., Kramskaya, T.A. and Polyak, R.Y.A.,** Experimental influenza infection: influence of stress, *Virol.,* 31,424, 1987.

105. **Friedman, S.B., Glasgow, L.A. and Ader, R.,** Differential susceptibility to a viral agent in mice housed alone or in groups, *Psychosom. Med.,* 32,285, 1970.

106. **Ben-Nathan, D. and Feuerstein, G.,** The influence of cold or isolation stress on resistance of mice to West Nile virus encephalitis, *Experentia,* 46,285, 1990.

107. **Kochi, A.,** The global tuberculosis situation and the new control strategy of the World Health Organization, *Tubercle,* 72,1, 1991.

108. **Pitchenik, A.E. and Fertel, D.,** Tuberculosis and non-tuberculosis mycobacterial disease, *Med. Clin. N. Am.,* 76,121, 1992.

109. **Danneneberg, A.M.,** Immunopathogenesis of pulmonary tuberculosis, *Hosp. Pract.,* Jan.15, 33, 1993.

110. **Powell, K.E. and Farer, L.S.,** The rising age of the tuberculosis patient, *J. Infect. Dis.,* 142,946, 1980.

111. **Feingold, A.O.,** Association of tuberculosis with alcoholism, *South Med. J.,* 69,1336, 1976.

112. **Hodolin, V.,** Tuberculosis and alcoholism, *Ann. N.Y. Acad. Sci.,* 252,353, 1975.

113. **Nagami, P.H. and Yoshikawa, T.T.,** Tuberculosis in the geriatric patient, *J. Am. Geriatr. Soc.,* 31,356, 1983.

114. **Orme, I.,** A mouse model of the recrudescence of virulent tuberculosis in the elderly, *Am. Rev. Respir. Dis.,* 137,716, 1988.

115. **Pincock, T.A.,** Alcoholism in tuberculosis patients, *Can. Med. Assoc. J.,* 91,851, 1964.

116. **Lurie, M.B.,** *Resistance to tuberculosis: experimental studies in native and acquired defense mechanisms,* Harvard University Press, Cambridge, MA, 1964.

117. **Lurie, M.B. and Dannenberg, A.M., Jr.,** Macrophage function in infectious disease with inbred rabbits, *Bacteriol. Rev.,* 29,466, 1965.

118. **Dannenberg, A.M., Jr.,** Pathogenesis of pulmonary tuberculosis, *Am. Rev. Respir. Dis. (suppl),* 125,25, 1982.

119. **Scordamaglia, A., Bagnasco, M. and Canonica, G.W.,** Immune response to mycobacteria, in Bendinelli, M., and Friedman, H., Eds., *Mycobacterium Tuberculosis: Interactions with the Immune System,* Plenum Press, New York, 1988, p.81.

120. **Forget, A., Skamene, E., Gros, P., Miailke, A.C. and Turcotte, R.,** Differences in response among inbred mouse strains to infection with small doses of *Mycobacterium bovis, Infect. Immun.,* 32,42, 1981.

121. **Gros, P., Skamene, E. and Forget, A.,** Genetic control of natural resistance to *Mycobacterium bovis* (BCG) in mice, *J. Immunol.,* 127,2417, 1981.

122. **Denis, M., Forget, A., Pelletier,M. and Skamene, E.,** Pleitropic effects of the Bcg gene. I. Antigen presentation in genetically susceptible and resistant congenic mouse strains, *J. Immunol.,* 140,2395, 1988.

123. **Vidal, S.M., Malo, D., Vogan, K., Skamene, E. and Gros, P.,** Natural resistance to infection with intracellular parasites: isolation of a candidate for *Bcg, Cell,* 73, 469, 1993.

124. **Zwilling, B.S., Dinkins, M., Christner, R., Faris, M., Griffinn, A., Hilburger, M., McPeek, M. and Pearl, D.,** Restraint stress induced suppression of major histocompatibility complex class II expression by murine peritoneal macrophages, *J. Neuroimmunol.,* 29,125, 1990.

125. **Crowle, A.J. and Elkins, N.,** Relative permissiveness of macrophages from black and white people for virulent tubercle bacilli, *Infect. Immun.,* 58,632, 1990.
126. **Stead, W.W., Senner, J.W., Reddick, W.T. and Lofgren, J.P.,** Racial differences in susceptibility to infection by *Mycobacterium tuberculosis, N. Engl. J. Med.,* 322,422, 1990.

Chapter 9

STEROIDS AND INFECTION

Yoshimasa Yamamoto
Department of Medical Microbiology and Immunology
University of South Florida College of Medicine
Tampa, FL 33612

Herman Friedman
Department of Medical Microbiology and Immunology
University of South Florida College of Medicine
Tampa, FL 33612

INTRODUCTION

The steroids secreted by the adrenal cortex are crucial for survival in that they are necessary for the response to a threatening environment. The main adrenal steroids are those such as mineralocorticoid, glucocorticoid and some sex steroids, mainly androgens. Since glucocorticoids affect metabolism, water and various electrolyte balance, and organ systems, and have anti-inflammatory as well as immunosuppressive activities, they have been widely used as a therapeutic agent. The main endogenous glucocorticoids are hydrocortisone and corticosterone. The importance of glucocorticoids in homeostasis has been recognized in diseases of abnormal glucocorticoid production. Addison's disease is caused by a deficiency in production characterized by muscular weakness, low blood pressure, depression, anorexia, loss of weight and hypoglycemia.   In contrast, an excess of glucocorticoid activity results in Cushing's syndrome. This is usually due to hyperplasia of the adrenal glands but a somewhat similar picture can be produced by prolonged or continued administration of steroids. High level of glucocorticoid in patients is frequently associated with severe infections which may be due to the immunosuppressive effect of glucocorticoids. In this chapter, susceptibility of individuals to infections caused by steroids, such as glucocorticoids, will be discussed.

CLINICAL STUDIES

There is a large body of clinical data suggesting that excessive steroid levels usually alter the susceptibility of patients to infections.[1] Patients undergoing steroid therapy[2,3] or suffering from Cushing's syndrome[4] may show altered resistance to a broad range of microorganisms. For instance, steroid therapy has recently been suggested for a variety of complications related to human immunodeficiency virus (HIV) infection, including pneumocystis infections[5-7] and esophageal and oral ulcerations,[8] because steroid treatment decreases both morbidity and mortality due to *Pneumocystis carinii* pneumonia. However, both animal[9] and clinical studies[10,11] have shown that steroid therapy results in severe impairment of host defenses against cryptococci and this is considered a major risk factor for development of cryptococcal infection in HIV infected patients who received steroid therapy. In the early stages of cryptococcal infection, administration of steroids for treatment of pneumocystis infection may cause further deterioration of host defenses, resulting in a rapidly progressive disease process due to this fungus.[11] Treatment with steroids has also been shown to increase the risk of herpes infections of patients.[12] The study of risk factors for aspergillosis in organ transplant recipients also has demonstrated a significant relation between the total dose of steroid administered and subsequent development of invasive aspergillosis.[13] In contrast, the use of other immunosuppressive drugs, the administration of antibiotics, and leukopenia were not found to be important risk factors for aspergillosis. Similar effects of steroids on the host defense system have also been observed in animal studies, which show that steroid treatment predisposes animals to development of invasive aspergillosis by interfering with the function of alveolar macrophages.[14,15]

Steroid therapy is often used for treatment of immune diseases such as rheumatoid arthritis because steroids have marked anti-inflammatory as well as immunosuppressive activities. In a reasonably well controlled study of patients with rheumatoid arthritis it was found that steroid therapy was associated with a higher incidence of bacteriuria.[16] Furthermore, use of steroid therapy in organ transplant patients also showed that treated patients developed more infections than non-treated ones.[17] Thus, the use of steroids for treatment is often associated with a higher risk for infection. Several studies have shown that infections are more common in patients receiving greater than 20 to 40 mg steroid daily, such as prednisone.[18]

The endogenous Cushing's syndrome, initially described by H. Cushing as a "malady which appears to leave patients with a definite susceptibility to infections",[19] is a hypercortisolism and associated with frequent infections, such as mucocutaneous fungal infections, postoperative wound infections, bacterial infections, and reactivation tuberculosis.[20-22] That is, patients with endogenous Cushing's syndrome can have the same spectrum of infections that occur in patients treated with pharmacologic steroids.[22] It is clear that enhanced levels of steroid in an individual's blood caused by either endogenous or therapeutically administered steroids may result in a greater susceptibility to a wide variety of infections.

The levels of steroid in body fluids are dependent on the amount used for therapy. In general, these substances are carried in the plasma bound to corticosteroid binding globulin (CBG) and to albumin. Both CBG-bound and albumin-bound steroids are biologically inactive. Therefore, the physiological plasma cortisol levels from 8 to 220 ng/ml during the diurnal cycle in healthy subjects may provide a biologically available concentration of free cortisol of 0.8 - 28 ng/ml.[23] In patients with Cushing's syndrome, severe infections occur at cortisol levels ranging from 0.4 to 1.8 μg/ml.[4] When steroid is administered in a dose of 100-500 mg, drug levels can reach 1-2 μg/ml for sustained periods.[24,25]

ANIMAL STUDIES

A number of studies concerning steroid effects on resistance to infections have been reported using a variety of animal models. For example, Gerald & Easmon[26] studied the effects of anti-inflammatory agents, including hydrocortisone, on chronic *Salmonella typhimurium* infection using a mouse model. The mice were infected subcutaneously with a relatively low dose of bacteria (5×10^3 organisms per mouse), which did not cause any infectious death within

30 days. Treatment with 25 mg of hydrocortisone per kg orally was begun on day 30 after infection and continued for 4 weeks. Such steroid treatment induced 100% mortality within 25 days, with a corresponding increase in salmonellae to lethal levels in the drug-treated mice. This experimental model clearly demonstrated the reactivation of a latent infection after hydrocortisone treatment. That is, treatment with hydrocortisone appears to cause a serious impairment of host defense system to the bacterial infection. North[27] also demonstrated that cortisone treatment dramatically lowered resistance of mice to *Listeria monocytogenes* and provoked a rapid proliferation of the bacteria in the liver of steroid-treated animals.

The effects of cortisone treatment on experimental fungal infections in mice have been examined by several groups.[28-31] Mice given 5 mg of cortisone per animal administered subcutaneously two days before infection by inhaled *Aspergillus fumigatus* conidia showed 100% mortality and the lungs of all mice had histological evidence of invasive pulmonary aspergillosis. Normal mice did not die and when killed 60 days after exposure showed no evidence of fungi in their lungs, either by histology or culture.[31] These results indicate that cortisone treatment renders mice susceptible to fatal invasive infection by aspergilli. Thus, the deleterious effect of steroid on host resistance to microbial infections has been demonstrated *in vivo* by these experimental animal models.

EFFECT OF STEROID ON THE IMMUNE DEFENSE SYSTEM

The mechanisms of increased susceptibility to infections induced by steroids are complicated, since steroids affect the function of a wide variety of host cells, including immune cells. However, the general consensus in understanding the mechanisms for steroid induced susceptibility to infections is that steroids suppress the inflammatory and immunological response systems, which is a major reason why steroids are often used as a therapeutic agent. When cortisone treatment was first introduced by Hench and colleagues in 1949 for treatment of rheumatoid arthritis, this revolutionized the treatment of immunologically mediated diseases.[32] Thus, steroid treatment was shown to have both beneficial as well as non-beneficial effects for disease treatment. In this chapter we will discuss only the effects of steroids on phagocytic cells (macrophages and neutrophils) in relation to susceptibility to infections, since phagocytic cells are considered a major defense cell against invading microbes.

A. MACROPHAGES

One of the most important host defense mechanisms against invading microorganisms is phagocytosis by macrophages, which is then followed by intracellular killing and digestion of the microbe. It is also well recognized that macrophages are an important immune cell type during infections, since phagocytes serve as an antigen presentation cell, a source of cytokines and as an effector cell. Therefore, the effect of steroids on the function and number of macrophages is critical in the final outcome of an infection.

1. Antimicrobial activity

Studies of antimicrobial activity of macrophages from steroid treated animals or after direct treatment of the cells with steroids indicate there are major deleterious effects. For example, peritoneal macrophages from hydrocortisone-treated (2.5 mg/mouse x 3) C3H mice are 1,000 times more susceptible to challenge with mouse hepatitis virus *in vitro* as compared with macrophages from untreated mice.[33] Furthermore, resistant macrophages from normal mice were found to succumb to viral destruction when cultured with spleen cells from cortisone-treated mice but not with spleen cells from normal mice. Such data suggested that some product from lymphoid cells may be involved in increased susceptibility of macrophages by steroids.

The direct effect of steroids on macrophages regarding antimicrobial activity has also been examined using human and animal macrophages. Study of the effect of steroids on human peripheral blood monocyte functions showed a marked reduction of killing of *Staphylococcus aureus* when monocytes were treated *in vitro* with 16 μg/ml (3×10^{-5}M) of hydrocortisone succinate (HCS) for only a brief period (60-90 min).[34] This concentration of steroids can be reached *in vivo* after administration of 100-500 mg of HCS[35,36] and does not have any toxicity against monocytes, since steroid treatment did not affect glass adherence, resting or stimulated hexosemonophosphate (HMP) shunt activity, or trypan blue exclusion, all of which are parameters of cell viability.[34]

The specific mechanisms involved in steroid impairment of anti-*S. aureus* activity of monocytes are unclear. In general, intracellular killing of bacteria requires lysosomal migration to and fusion with bacteria containing phagosomes. A sequence of metabolic events is associated with this process, including stimulation of the HMP shunt.[37] Involvement of such metabolic events in HCS induced impairment of anti-*S. aureus* activity seems unlikely, because HCS did not prevent stimulation of the HMP shunt activity of monocytes after ingestion of latex particles. On the other hand, the microfilament-microtubule systems in phagocytic cells are thought to play a role in lysosomal migration to phagosomes.[38] Therefore, a possible mechanism of steroid induced impairment of anti-*S. aureus* activity of monocytes may be due to stabilization of lysosomes by steroids,[39] which could prevent their fusion with the phagosome.

A different mechanism for steroid induced impaired antimicrobial activity of macrophages to *Listeria monocytogenes* has been proposed by Schaffner and Schaffner.[40] That is, human blood-derived macrophage cultures exposed to pharmacologic concentrations of steroids (dexamethasone, 2.5×10^{-7}M) *in vitro* lost their listericidal activity without undergoing any apparent nonspecific damage, as evident by parallel studies of viability, phagocytic rate, and secretion of lysozyme.[41] However, such suppressive effect required a long incubation period with steroid, such as 24-36 hr. Their hypothesis was that steroids act by suppressing the synthesis of one or several specific proteins of the non-oxidative killing systems, since the observations reported indicate that macrophages exposed to steroids ingest microorganims at a normal rate and that after phagocytosis they secrete the usual reactive oxygen intermediates and exhibit normal fusion of lysosomes and phagosomes. That is, oxidative killing systems of steroid-treated macrophages react to interferon-γ (IFNγ) and are normally primed to secreted reactive oxygen intermediates upon challenge. This hypothesis is also supported by their findings that steroids do not affect the activation of antilisterial activity by IFNγ, even after macrophages are pretreated for prolonged periods with pharmacologic concentrations of dexamethasone.[41]

In contrast to the above reports, there are several negative reports regarding a direct deleterious effect of steroids on the antimicrobial activity of macrophages. The *in vivo* treatment of mice with 15 mg of hydrocortisone acetate three days prior to harvesting peritoneal macrophages did not cause impairment of macrophage phagocytic and killing activities against *Staphylococcus albus, S. aureus, Escherichia coli,* and *Pseudomonas aeruginosa.*[42] The treatment of human alveolar macrophages with pharmacologic but not cytotoxic doses of hydrocortisone (≤ 10 µg/ml, 2.8×10^{-5}M) for up to 48 hr *in vitro* also did not influence the rate or extent of intracellular *Legionella pneumophila* multiplication.[43] Moreover, hydrocortisone treatment of macrophages did not reduce the cytokine-induced anti-*L. pneumophila* activity of macrophages. In this regard, it has also recently been reported that IFNγ activated fungicidal activity of mouse macrophages to *Candida albicans, Blastomyces dermatitidis* and *Paracoccidioides brasiliensis* was not affected by therapeutic concentrations of hydrocortisone, such as ≤ 5 µg/ml (1.4×10^{-5}M).[44] Thus, there are divergent findings concerning the influence of steroids on the antimicrobial activity of macrophages. This may be due to the different target microorganisms and/or experimental conditions used.

The mechanisms of antimicrobial activity of macrophages to different microorganisms, such as fungi, bacteria and virus, may be different. For example, the susceptibility of microorganisms to the antimicrobial activity of macrophages is different for different organisms. Therefore, theoretically, if steroids have some effect on a specific antimicrobial mechanism, such as inhibition of nitric oxide (NO) production, all antimicrobial activities may not be affected, since susceptibility to NO dependent antimicrobial mechanism is different for different microorganisms.[45] Thus, divergent findings concerning the influence of steroids on different microorganisms seem likely. Nevertheless, most of the experiments regarding effects of steroids on antimicrobial activity of macrophages *in vitro* and *in vivo* show some inhibitory effects, even though the mechanisms are unclear. Thus, the direct effect of steroids on antimicrobial activity of macrophages may be one of the reasons for increased susceptibility of patients to infections after receiving steroid treatment.

2. Reactive oxygens

The importance of oxygen-dependent components including the key oxygen metabolites superoxide anion (O_2^-), hydrogen peroxide (H_2O_2), hydroxyl radical, and singlet oxygen in the antimicrobial activity of macrophages is widely recognized. Consequent to their activation following phagocytosis or chemical stimuli, macrophages release larger amounts of O_2^- and H_2O_2 than do normal cells.[46] Once reactive oxygens are produced, these contribute to the killing and inhibition of growth of microorganisms in macrophages as well as polymorphonuclear neutrophils due to the strong oxidative activity of the reactive oxygens. In this regard, it seems likely that steroids inhibit reactive oxygen metabolism and eventually cause impaired antimicrobial activity by the macrophages. However, there are diverse reports about reactive oxygen production by steroid treated macrophages. Dexamethasone treatment of human blood macrophages for 36 hr did not or only minimally affected generation of hydrogen peroxide or chemiluminescence in a luminol- or lucigenine-amplified system upon stimulation with phorbol myristate acetate (PMA) or opsonized zymosan.[40] Similar observations were also reported by two other groups. That is, treatment of human monocytes with 10^{-3}M prednisolone caused only minimum suppressive effects on the production of reactive oxygen intermediates.[47] Other reports noted that 24 hr treatment with 2×10^{-4}M hydrocortisone did not cause any decreased release of reactive oxygens from activated mouse peritoneal macrophages.[48] In contrast, treatment of human blood monocytes with hydrocortisone for a sustained exposure lasting 3-4 days suppressed H_2O_2 release by 93% in response to PMA (50%-inhibitory concentration, $1.9 \pm 0.3 \times 10^{-7}M$).[49] Probably the need for such prolonged exposure may explain why little or no suppressive effects were noted in other studies. Thus, under certain conditions, such as a relatively long or sustained exposure of macrophages to steroids, production of reactive oxygens, which are important for antimicrobial activity as a direct effector molecule, might be reduced.

3. Nitric oxide

Nitric oxide (NO), a simple and relatively unstable radical under aerobic conditions, has been identified in recent years as a potent and pleiotropic mediator.[50] Especially after the finding of inducible NO synthase, which generates NO from arginine in macrophages, except human monocytes/macrophages, the possible involvement of NO in antimicrobial activity of macrophages has been extensively studied, because NO has a strong direct antimicrobial activity against a wide variety of microorganisms, similar to reactive oxygens, because of their strong affinity to iron-containing enzymes.[51] In fact, NO dependent antimicrobial activity of macrophages has been found against *Mycobacteria*,[52] *Legionella*,[53] *Cryptococcus*,[54] *Toxoplasma*,[55] *Francisella*,[56] and *Schistosoma*.[57] Therefore, NO mediated antimicrobial activity is now considered a reactive oxygen independent antimicrobial mechanism by macrophages. On the other hand, the participation of NO in autoimmune tissue destruction has been also recognized by recent studies. For example, the lysis of pancreatic islet cells by activated macrophages has been shown to be mediated by NO.[50] Thus, NO has a wide variety of biological activities and plays an important role in immunological and physiological regulations.

The effect of steroids on NO production has recently been examined using a variety of cells, including macrophages. Steroids, i.e., dexamethasone, hydrocortisone, and cortisole, inhibit induction but not activity of the Ca^{2+}-independent NO synthase *in vitro* in the macrophage cell line, J774, and other cells such as vascular endothelial cells,[58] fresh vascular tissue,[59] and the EMT-6 adenocarcinoma cells[60] after stimulation with LPS, either alone or in combination with IFNγ. It has also been observed in bacterial infection using mouse macrophages that induction of nitric oxide synthase mRNA by *Legionella pneumophila* infection *in vitro*, either alone or in combination with IFNγ, was significantly inhibited by treatment with hydrocortisone (Fig. 1). In addition, the *in vitro* inhibition of NO synthase induction by steroids was prevented by a partial agonist for steroid receptor,[58,61] indicating that this effect is specific and, therefore, related to the pharmacological and probably some of the physiological effects of steroids.[51] *In vivo* induction of NO synthase in liver, lung, and vascular tissue of animals after treatment with LPS was also prevented by dexamethasone and cortisole.[62] Thus, steroid treatment reduces NO production in cellular as well as tissue levels and may contribute to the reduced resistance of macrophages and thus is a host to infections.

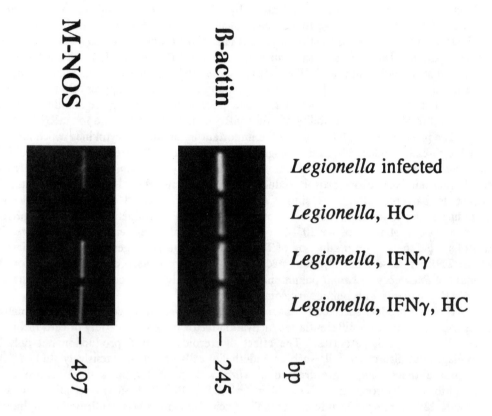

Fig. 1. Effect of hydrocortisone (HC) on the levels of nitric oxide synthase (MNOS) mRNA in *Legionella pneumophila* infected mouse macrophages. Macrophages (A/J mouse) were pretreated with 10^{-5}M of HC for 24 hrs and then primed with or without 20 U/ml of IFNγ for 24 hrs in the presence of HC. After treatment, macrophages were infected with *L. pneumophila* for 6 hrs. Infectivity ratio was 10 bacteria per macrophage. Total RNA was isolated and reverse transcription of RNA followed by the polymerase chain reaction (RT-PCR) was performed with mouse inducible nitric oxide synthase (M-NOS) and ß-actin specific primers. HC treatment decreased the levels of M-NOS mRNA induced by *L. pneumophila* in both IFNγ primed and non-primed macrophages.

4. Cytokine production

Cytokines produced by macrophages, such as interleukin 1 (IL-1), interleukin 6 (IL-6), and tumor necrosis factor (TNF), are important for the inflammatory response as well as for generating immunity to invading microbes. Therefore, it is assumed that disturbance of cytokine production by steroids may affect the final outcome of an infection. In this regard, the effect of steroids on IL-1 production has been examined. IL-1 is one of the cytokines released by macrophages and has an important role in immunity, since this cytokine binds to lymphocytes and augments T cell proliferation by effects on T cell growth factor production. Treatment of mouse[62a] or rat[63] peritoneal macrophages with 10^{-8}M of hydrocortisone inhibited the release of IL-1 in response to stimuli such as bacterial LPS or carrageenan. These *in vitro* effects are paralleled *in vivo*: the sera of mice injected with LPS contain IL-1, the level of which is decreased by steroid treatment.[64] The mechanism of inhibition has been examined using the human promonocytic cell line, U937, and it was found that steroids suppress IL-1ß synthesis by two distinct mechanisms, i.e., blocking transcription and decreasing the stability of IL-1ß mRNA, without affecting the stability of other mRNA such as ß-actin and *fos* mRNAs.[65,66]

The production of TNFα, one of the important inflammatory cytokines which initiates a cytokine cascade produced by macrophages, is also affected by the treatment with steroids *in vitro*[67-70] and *in vivo*.[70] Incubation of human blood monocytes *in vitro* with 10^{-8} to 10^{-6}M of steroids (dexamethasone or cortisol) reduced TNF production by 21-48% of the control in response to bacterial LPS.[69] A similar result was also obtained using porcine alveolar macrophages, i.e., dexamethasone treatment suppressed TNFα production by the macrophages by 86% at a concentration of 4×10^{-9}M.[68] Recent studies also demonstrated the suppressive effect of steroid therapy on production of TNF in alveolar macrophages from AIDS patients.[70] That is, TNFα release from alveolar macrophages from patients receiving steroids for the treatment of *Pneumocystis carinii* pneumonia was significantly less as compared to non-steroid treated AIDS patients with pulmonary complications.

IL-6 is a product secreted by macrophages in response to infection and inflammatory stimuli and may serve as a differentiation/activation factor for a wide variety of cells including B cells, T cells, and hepatocytes. The effect of steroids on IL-6 production not only by macrophages but also other cells such as endothelial cells has been extensively studied.[67,71-73] When mouse macrophages are treated *in vitro* with physiological and pharmacological concentrations of hydrocortisone, i.e., 10 ng/ml (2.8×10^{-8}M), IL-6 transcription levels in response to LPS were significantly decreased.[67] Such observation was confirmed using human peripheral blood monocytes. That is, the 50% effective dose of hydrocortisone on inhibition of IL-6 production from human monocytes was 2.0×10^{-7}M.[71] Such inhibitory activity of steroids on IL-6 production was observed not only for monocytes/macrophages but also endothelial cells[72,74] and fibroblast cells.[72] Study of the inhibition mechanism of steroids indicated that both LPS-stimulated and IL-1-stimulated IL-6 production was inhibited by steroid treatment at the transcription level and this effect was receptor mediated.

IL-8, neutrophil-activating peptide-1 (NAP-1), is secreted by activated mononuclear leukocytes, including monocytes, and plays a major role in neutrophil diapedesis through vascular endothelium and focal recruitment at inflamed sites. Therefore, the importance of IL-8 for defense against infections seems likely. However, knowledge concerning steroid effects on IL-8 production is still limited. Preincubation of peripheral blood mononuclear cells, which contain 16-28% monocytes, with 10^{-9}M of dexamethasone for 1 hr resulted in almost complete inhibition of zymosan induced IL-8 production.[75] Whether the steroid directly inhibited IL-8 production by certain immune cells is not clear due to the mixed cell culture used for analysis. However, inhibition of chemokine production by steroids, either directly or indirectly, may explain the poor migration of neutrophils to inflammatory sites in steroid treated patients.

It is apparent steroids inhibit cytokine production by monocytes/macrophages in a broad spectrum since IL-1, IL-6, TNF and probably IL-8, all of which are major pro-inflammatory cytokines produced by macrophages, are all suppressed. The molecular inhibition mechanism of cytokine production, including IL-1 and IL-6, has been examined and appears to involve reduced transcription of the message for these cytokines. However, it is still unclear whether a common inhibitory mechanism is involved in steroids induced inhibition of cytokine production, such as involvement of a common nuclear regulatory factor, because most of the inflammatory cytokines produced by macrophages are inhibited. Nevertheless, steroid induced inhibition of cytokine production by macrophages may contribute to the increased susceptibility to infections of patients who have enhanced steroid levels.

5. Differentiation

The reduced number of circulating monocytes, especially those bearing the Fc receptor, in the blood of a patient due to the administration of steroids has been widely recognized.[76-79] *In vivo* findings are supported by the *in vitro* study showing that hydrocortisone inhibits macrophage production in murine granulocyte-macrophage precursor cell (CFU-GM) cultures.[80] Inhibition of macrophage differentiation by steroids, demonstrated by use of the human monocytic cell line U937 to study the effect of cortisol on cellular differentiation to macrophages,[81] might also account for the reduced number of circulating monocytes.

B. NEUTROPHILS

Neutrophils play an important role in host defenses against invading microorganisms as a first line of defense, especially in early inflammatory responses to infection. Therefore, the inhibitory effects of a steroid on the number as well as the function of neutrophils may result in a more susceptible host to infections as compared to non-treated hosts. However, the effect of steroids on neutrophils appears to be different than the effects on other leukocytes, since a single dose of steroid induces a significant decrease of circulating monocytes, lymphocytes, eosinophils and basophils, but peripheral blood neutrophils increase two or three-fold rather than decrease.[82-87] The mechanism of such neutrophil increases is assumed to be due to an increase in the half-life of circulating neutrophils and the decrease in the size of the marginating neutrophil pool and/or increased output of neutrophils from bone marrow.[84,88]

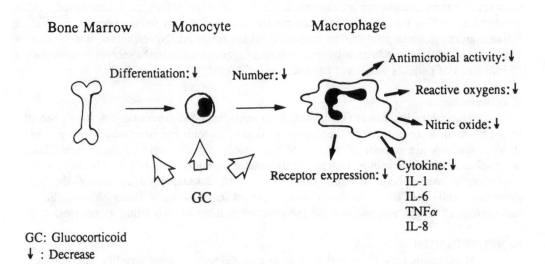

Fig. 2. Effect of steroid on macrophages.

Prevention of accumulation of neutrophils at a local inflammatory site by steroid treatment has also been reported by a number of groups.[89-91] However, the chemotactic activity of neutrophils obtained from steroid treated animals or humans is either enhanced, very weakly inhibited or unaffected,[92-95] except that neutrophils obtained from inflammatory exudates show reduced chemotactic activity.[96] Such data indicate that the poor accumulation of neutrophils in inflammatory sites may be mediated via inhibition of the local release of chemotactic factors rather than a direct effect of steroids on neutrophils. This hypothesis seems likely, because steroid treatment does not inhibit neutrophil accumulation induced by injection of C5a into human skin.[98] Furthermore, a recent study demonstrated that treatment of purified human neutrophils with dexamethasone (10^{-6}M) did not inhibit neutrophil chemotaxis in response to a range of concentrations of formyl peptides.[99]

In contrast to neutrophils, treatment of macrophages with steroids inhibit release of neutrophil chemotactic factor in response to LPS.[97] However, treatment of neutrophils *in vitro* with relatively high concentrations of steroids (2×10^{-5} to 2×10^{-3}M) decreases neutrophil functions including chemotaxis, respiratory burst, degranulation and production of arachidonic acid metabolites.[100-102] The relatively high concentrations of steroids required for a direct modulation of neutrophil functions are well beyond those required to saturate steroid receptors (Kd: 4.1×10^{-9}M for dexamethasone binding to neutrophils) and, therefore, such effects on neutrophil functions may not have relevance to *in vivo* effect of steroids on neutrophils.[99] Thus, steroid-induced inhibition of neutrophil migration may not be due to a direct action upon the migration cells but is most likely an indirect effect resulting from the inhibition of the release of an endogenous chemotactic factor.[97]

Surface receptors of neutrophils, especially receptors for the Fc portion of IgG, have been studied by several groups and three distinct types of IgG FcγR, i.e., FcγRI, FcγRII and FcγRIII, were identified on the neutrophils.[103-105] These receptors play a definite role in a number of neutrophil functions, including binding to and ingestion of IgG-coated particles[106] as well as IgG coated microbes. Treatment of neutrophils with IFNγ induced a 9- to 20-fold increase in the number of FcγRI sites per cell. However, when neutrophils were co-treated with dexamethasone (2×10^{-7}M) and IFNγ, induction of FcγRI by IFNγ was significantly inhibited, but the levels of FcγRI and FcγRII in resting neutrophils without activation by IFNγ were not modulated by dexamethasone.[107] The levels of CR3, the C3bi receptor which is a member of the leukocyte adherence glycoprotein family, integrin, on the neutrophils were also reduced by the treatment with steroids.[107] However, the reduction was slight and required a relatively long cultivation period, such as over night culture, for reduction. Nevertheless, modulation of certain types of receptor expression, especially inhibition of increased adhesion molecule expression by cytokine activation, may explain the modulating activity of steroids on the inflammatory response regarding mobilization of neutrophils to inflammatory sites.

MOLECULAR MECHANISMS

Human glucocorticoid receptor, one of the intracellular receptors, was cloned in 1985[108] and it is believed that the receptor has three main functional domains: DNA-binding, ligand-binding and immunogenic domains.[109] Such steroid receptors can be found in virtually all tissues and have a high affinity for steroids. However, during the resting condition, without binding of steroid, the receptors reside in the form of a hetero-oligomer with heat-shock protein 90 (HSP90) and immunophilin.[110] Once the free form of a steroid enters the cells by simple diffusion, since steroids are small lipophilic molecules, it can bind to the receptor complex and causes dissociation of HSP90 and immunophilin. The receptor-ligand then translocates into the nucleus through nuclear pores[111] and interacts with a select portion of DNA or DNA-binding proteins.

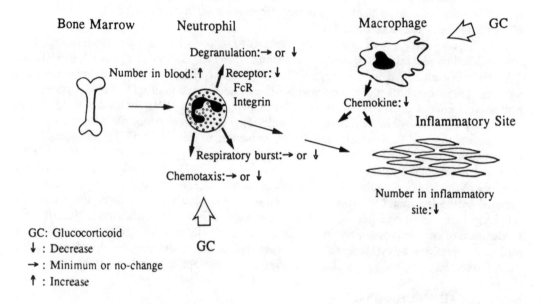

Fig. 3. Effect of steroid on neutrophils.

There are at least two possible mechanisms for gene regulation by the steroid-receptor complex. Direct interaction with glucocorticoid responsive elements (GRE) regulates the rate of transcription initiation from nearby promoters of the corresponding glucocorticoid-responsive genes.[112] Another possible regulation mechanism is the interaction with DNA-binding proteins associated with different regulatory elements of the DNA, such as glucocorticoid modulatory element-binding protein and CACCC-box-binding protein. Both of these transcription factors potentiate the modulatory effects of steroids after transcription of specific genes.[109] Thus, steroids regulate transcription of certain genes that code for proteins which augment steroid effects. This postulated mechanism is supported by the findings that many steroid actions require RNA and protein synthesis. The effect of a steroid is also mediated by post-transcriptional mechanisms such as regulation of RNA translation, protein synthesis, and protein secretion.[113]

Identification and characterization of steroid-regulated genes with potential anti-inflammatory and immunosuppressive activity have been studied by several groups. Induction of lipocortin, a phospholipase A_2 inhibitor, by steroids has been reported to be present in various tissues and macrophages and is supported by the recent study showing that dexamethasone treatment induces lipocortin mRNA in U937 cells, human monocytes and rat peritoneal exudate cells.[114] In contrast, steroid treatment suppresses cyclooxygenease mRNA levels, which probably is linked to inhibition of prostaglandin production by steroids.[115] Recently Helmberg *et al.* examined gene regulation by steroids using cDNA cloning and subtractive screening techniques to isolate and identify transcripts with steroid-regulated steady-state levels in the mouse macrophage-like cell line, P388D1.[116] Three hydrocortisone-regulated mRNA species, two of which cDNA species correspond to proteins (an endogenous retroviral protein and an enzyme in cholesterol biosynthesis) that modify *in vitro* immune responses, were isolated by the study.

Thus, the basic mechanisms of steroid effects on the cells at the molecular levels have been studied in detail, especially by cloning of the receptor. However, most of the anti-inflammatory and immunosuppressive effects of steroids regarding antimicrobial activity of phagocytic cells are still not clear at the molecular level, mainly because mechanisms of antimicrobial activity itself are not yet understood well.

CONCLUSION

Development of steroid therapy has had revolutionary therapeutic effects for many diseases, and it is widely recognized that this therapy constitutes a major success for modern medicine. However, such therapy with steroids is associated with serious side effects, since they have both beneficial and non-beneficial features which may be inseparable. The anti-inflammatory and immunosuppressive activity of steroids are considered essential for therapy, but also may result in an increased susceptibility of a patient to infections, especially if treatment is inappropriate. The spectrum of microorganisms causing infection during steroid therapy is not clear, although experimental animal studies have shown that steroid treatment is associated with infections caused by certain groups of microorganisms. An understanding of which microorganisms are mainly responsible for infection of steroid treated patients is essential, since infections caused by different microorganisms are often handled differently by the host defense system, which consists of cell types with different susceptibilities to steroids.

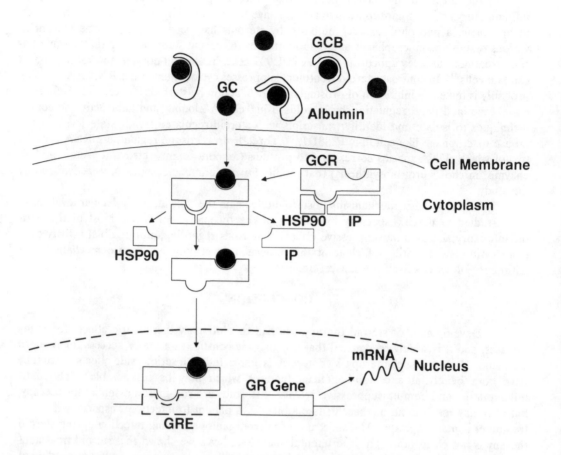

Fig. 4. Molecular mechanism of steroid action. GC, glucocorticoid; GCB, glucocorticoid binding globulin; GCR, glucocorticoid receptor; HSP90, heat shock protein 90; IP, immunophilin; GRE, glucocorticoid responsive element; GR Gene, glucocorticoid-responsive gene.

REFERENCES

1. **Dale, D. C., and Petersdorf, R. G.,** Corticosteroids and infectious diseases, *Med. Clin. North Am.,* 57, 1277, 1973.

2. **Ginzler, E., Diamond, H., Kaplan, D., Weiner, M., Schlesinger, M. and Sleznick, M.,** Computer analysis of factors influencing frequence of infection in systemic lupus erythematosis, *Arthritis Rheum.*, 21,37, 1978.

3. **Anderson, R. J., Schaffer, L. A., Olin, D. B. and Eickhoff, T. C.,** Infectious risk factors in the immunosuppressed host, *Am. J. Med.*, 54,453, 1973.

4. **Graham, S. and Tucker, W. S.,** Opportunistic infections in endogenous Cushing's syndrome, *Ann. Intern. Med.*, 101,334, 1984.

5. **Kovacs, J. A. and Masur, H.,** Are corticosteroids beneficial as adjunctive therapy for pneumocystis pneumonia in AIDS? *Ann. Intern. Med.*, 113,1, 1990.

6. **MacFadden, D. K., Edelson, J. D., Hyland, R. H.,** *et al.*, Corticosteroids as adjunctive therapy in treatment of *Pneumocystis carinii* pneumonia in patients with acquired immunodeficiency syndrome, *Lancet*, 1,1477, 1987.

7. **Schiff, M. J., Farber, B. F. and Kaplan, M. H.,** Steroids for *Pneumocystis carinii* pneumonia and respiratory failure in the acquired immunodeficiency syndrome - a reassessment, *Arch. Intern. Med.*, 150,1819, 1990.

8. **Bach, M. C., Howell, D. A., Valenti, A. J.,** *et al.*, Aphthous ulceration of the gastrointestinal tract in patients with the acquired immunodeficiency syndrome, *Ann. Intern. Med.*, 112,465, 1990.

9. **Perfect, J. R., Lang, S. D. R. and Durack, D. T.,** Chronic cryptococcal meningitis - a new experimental model in rabbits, *Am. J. Pathol.*, 101,177, 1980.

10. **Bennington, J. L., Haber, S. L. and Morgenstern, N. L.,** Increased susceptibility to cryptococcosis following steroid therapy, *Am. J. Pathol.*, 45,262, 1964.

11. **Bernstein, B., Flomenberg, P. and Letzer, D.,** Disseminated cryptococcal disease complicating steroid therapy for *Pneumocystis carinii* pneumonia in a patient with AIDS, *South Med. J.*, 87,537, 1994.

12. **Bozzette, S. A., Sattler, F. R., Chiu, J., Wu, A. W., Gluckstein, D., Kemper, C., Bartok, A., Niosi, J., Abramson, I., Coffman, J., Hughlett, C., Loya, R., Cassens, B., Akil, B., Meng, T-C., Boylen, C. T., Nielsen, D., Richman, D. D., Tilles, J. G., Leedom, J., McCutchan, J. A. and the California Collaborative Treatment Group,** A controlled trial of early adjunctive treatment with corticosteroids for *Pnemocystis carinii* pneumonia in the acquired immunodeficiency syndrome, *N. Engl. J. Med.*, 323,1451, 1990.

13. **Gustafson, T. L., Schaffner, W., Lavely, G. B., Stratton, C. W., Johnson, H. K. and Hutcheson, Jr., R.H.,** Invasive aspergillosis in renal transplant recipients: correlation with corticosteroid therapy, *J. Infect. Dis.*, 148,230, 1983.

14. **White, L. O.,** Germination of *Aspergillus fumigatus* conidia in the lungs of normal and cortisone-treated mice, *Sabouraudia*, 15,37, 1977.

15. **Sidransky, H.,** Experimental studies with aspergillosis, in *Opportunistic fungal infections: proceeding of the second international conference*, Chick, E. W., Balows, A., and Furcolow, M. L., Eds., Charles C. Thomas, Springfield, IL, 1975, p.165.

16. **Burry, H. C.,** Bacteriuria in rheumatoid arthritis, *Ann. Rheum. Dis.*, 32,208, 1973.

17. **Myerowitz, R. L., Medevios, A. A. and O'Brien, T. F.,** Bacterial infection in renal homotransplant recipients: a study of fifty-three bacteriemic episodes, *Am. J. Med.*, 53, 308, 1972.

18. Fauci, A. S., Dale, D. C. and Balow, J. E., Glucocorticosteroid therapy: mechanisms of action and clinical considerations, *Ann. Intern. Med.*, 84,304, 1976.

19. Cushing, H., The basophil adenomas of the pituitary body and their clinical manifestations (pituitary basophilism), *Bull. Johns Hopkins Hosp.*, 50,137, 1932.

20. Findling, J. W., Tyrell, J. B., Aron, D. C., Fitzgerald, P. A., Young, C. W. and Sohnle, P., G., Fungal infections in Cushing's syndrome [Letter], *Ann. Intern. Med.*, 95,392, 1981.

21. Plotz, C. M., Knowlton, A. I. and Ragan, C., The natural history of Cushing's syndrome, *Am. J. Med.*, 13,597, 1952.

22. Graham, B. S. and Tucker, W. S., Opportunistic infections in endogenous Cushing's syndrome, *Ann. Intern. Med.*, 101,334, 1984.

23. Baumann, G., Rappaport, G., Lemarchand-Beraud, T. and Felber, J. P., Free cortisol index: a rapid and simple estimation of free cortisol in human plasma, *J. Clin. Endocrinol. Metab.*, 40,462, 1975.

24. Peterson, R. E., Wyngaarden, S. L., Guerra, S. L., Brodie, B. B. and Bunim, B. B., The physiologic deposition and metabolic rate of hydrocortisone in man, *J. Clin. Invest.*, 34, 1779, 1955.

25. Webel, M., Ritts, R. E., Taswell, H. F., Donadio, Jr., J. V. and Woods, J. E., Cellular immunity after intravenous administration of methylprenisolone, *J. Lab. Clin. Med.*, 83, 383, 1974.

26. Plant, J. E., Higgs, G. A. and Easmon, C. S. F., Effects of antiinflammatory agents on chronic *Salmonella typhimurium* infection in a mouse model, *Infect. Immun.*, 42,71, 1983.

27. North, R. J., The action of cortisone acetate on cell-mediated immunity to infection: suppression of host cell proliferation and alteration of cellular composition of infective foci, *J. Exp. Med.*, 134,1485, 1971.

28. Bhatia, V. N. and Mohaptra, L. N., Experimental aspergillosis in mice. II. Enhanced susceptibility of the cortisone treated mice to infection with *Aspergillus fumigatus, Aspergillus flavus* and *Aspergillus niger, Mykosen,* 13,105, 1970.

29. Sandhu, D., Sandhu, R. S., Damodaran, V. N. and Randhawa, H. S., Effect of cortisone on bronchopulmonary aspergillosis in mice exposed to spores of various *Aspergillus* species, *Sabouraudia,* 8,32, 1970.

30. Sidransky, H. and Friedman, L., The effect of cortisone and antibiotic agents on experimental pulmonary aspergillosis, *Am. J. Pathol.*, 35,169, 1959.

31. White, L. O., Germination of *Aspergillus fumigatus* conidia in the lungs of normal and cortisone-treated mice, *Sabouraudia,* 15,37, 1977.

32. Boumpas, D. T., Chrousos, G. P., Wilder, R. L., Cupps, T. R. and Balow, J. E., Glucocorticoid therapy for immune-mediated diseases: basic and clinical correlates, *Ann. Intern. Med.*, 119,1198, 1993.

33. Taylor, C. E., Weiser, W. Y. and Bang, F. B., *In vitro* macrophage manifestation of resistance to mouse hepatitis virus, *J. Exp. Med.*, 153,732, 1981.

34. Rinehart, J. J., Balcerzak, S. P., Sagone, A. L. and LoBuglio, A. F., Effects of corticosteroids in human monocyte function, *J. Clin. Invest.*, 54,1337, 1974.

35. Webel, M. L., Ritts, Jr., R. E., Taswell, H. F., Donadio, Jr., J. V. and Woods, J. E., Cellular immunity after intravenous administration of methylprednisolone, *J. Lab. Clin. Med.*, 83,383, 1974.

36. **Peterson, R. E., Wyngaarden, J. B., Guerra, S. L., Brodie, B. B. and Bunim, J. J.,** The physiological disposition and metabolic rate of hydrocortisone in man, *J. Clin. Invest.,* 34, 1779, 1955.

37. **Stjernholm, R. L., Burns, C. P. and Hohnadel, S. H.,** Carbohydrate metabolism by leukocytes, *Enzyme (Basel),* 13,7, 1972.

38. **Zuvier, R. B., Hofftein, S. and Weissmann, G.,** Mechanisms of lysosomal enzyme release from human leukocytes. I. Effect of cyclic nucleotides and colchicine, *J. Cell Biol.,* 58,27, 1973.

39. **DeDuve, C., Wattiaux, R. and Wibo, M.,** Effects of fat-soluble compounds on lysosomes *in vitro, Biochem. Pharmacol.,* 9,97, 1962.

40. **Schaffner, A., and Schaffner, T.,** Glucocorticoid-induced impairment of macrophage antimicrobial activity: mechanisms and dependence on the state of activation, *Rev. Infect. Dis.,* 9,S620, 1987.

41. **Schaffner, A.,** Therapeutic concentrations of glucocorticoids suppress the antimicrobial activity of human macrophages without impairing their responsiveness to gamma interferon, *J. Clin. Invest.,* 76,1755, 1985.

42. **Van Zwet, T. L., Thompson, J., and Van Furth, R.,** Effect of glucocorticosteroids on the phagocytosis and intracellular killing by peritoneal macrophages, *Infect. Immun.,* 12,699, 1975.

43. **Nash, T. W., Libby, D. M. and Horwitz, M. A.,** Interaction between the Legionnaires' disease bacterium (*Legionella pneumophila*) and human alveolar macrophages. Influence of antibody, lymphokines, and hydrocortisone, *J. Clin. Invest.,* 74,771, 1984.

44. **Brummer, L., Hanson, L. H. and Stevens, D. A.,** Kinetics and requirements for activation of macrophages for fungicidal activity: effect of protein synthesis inhibitors and immunosuppressants on activation and fungicidal mechanism, *Cell. Immunol.,* 132,236, 1991.

45. **Yamamoto, Y., Friedman, H. and Klein, T.W.,** Nitric oxide has an immunoregulatory role other than antimicrobial activity in *Legionella pneumophila* infected macrophages, *Ann. N. Y. Acad. Sci.,* 730,342, 1994.

46. **Gangadharam, P. R. J. and Edwards III, C. K.,** Release of superoxide anion from resident and activated mouse peritoneal macrophages infected with *Mycobacterium intracellulare, Am. Rev. Respir. Dis.,* 130,834, 1984.

47. **Lehmeyer, J. E. and Johnston, Jr., R. B.,** Effect of antiinflammatory drugs and agents that elevate intracellular cyclic AMP on the release of toxic oxygen metabolites by phagocytes: studies in a model of tissue-bound IgG, *Clin. Immunol. Immunopathol.,* 9,482, 1978.

48. **Masur, H., Murray, H. W. and Jones, T. C.,** Effect of hydrocortisone on macrophage response to lymphokine, *Infect. Immun.,* 35,709, 1982.

49. **Nakagawara, A., DeSantis, N. M., Nogueira, N. and Nathan, C. F.,** Lymphokines enhance the capacity of human monocytes to secrete reactive oxygen intermediates, *J. Clin. Invest.,* 70,1042, 1982.

50. **Kolb, H. and Kolb-Bachofen, V.,** Nitric oxide: a pathogenic factor in autoimmunity, *Immunol. Today,* 13,157, 1992.

51.   **Moncada, S., Palmer, R. M. J. and Higgs, E. A.,** Nitric oxide: physiology, pathophysiology, and pharmacology, *Pharmacol. Rev.*, 43,109, 1991.
52.   **Flesch, I. E. A. and Kaufman, S. H. E.,** Mechanisms involved in mycobacterial growth inhibition by gamma interferon-activated bone marrow macrophages: role of reactive nitrogen intermediates, *Infect. Immun.*, 59,3213, 1991.
53.   **Summersgill, J. T., Powell, L. A., Buster, B. L., Miller, R. D. and Ramirez, J. A.,** Killing of *Legionella pneumophila* by nitric oxide in γ-interferon-activated macrophages, *J. Leuk. Biol.*, 52,625, 1992.
54.   **Granger, D. L., Hibbs, Jr., J. B., Perfect, J. R. and Durack, D. T.,** Specific amino acid (L-arginine) requirement for the microbiostatic activity of murine macrophages, *J. Clin. Invest.*, 81,1129, 1988.
55.   **Adams, L. B., Hibbs, J. B., Taintor, R. R. and Krahenbouhl, J. L.,** Microbiostatic effect of murine-activated macrophages for *Toxoplasma gondii*: role for synthesis of inorganic nitrogen oxides from L-arginine, *J. Immunol.*, 144,4338, 1990.
56.   **Anthony, L. S. D., Morrissey, P. J. and Nano, F. E.,** Growth inhibition of *Francisella turalensis* live vaccine strain by IFN-γ-activated macrophages is mediated by reactive nitrogen intermediates derived from L-arginine metabolism, *J. Immunol.*, 148,1829, 1992.
57.   **James, S. L. and Glaven, J.,** Macrophage cytotoxicity against shidtosomula of *Schistosoma mansoni* involves arginine-dependent production of reactive nitrogen intermediates, *J. Immunol.*, 143,4208, 1989.
58.   **Radomski, M. W., Palmer, R. M. J. and Moncada, S.,** Glucocorticoids inhibit the expression of an inducible, but not the constitutive, nitric oxide synthase in vascular endothelial cells, *Proc. Natl. Acad. Sci. USA,* 87, 10043, 1990.
59.   **Rees, D. D., Cellek, S., Palmer, R. M. J. and Moncada, S.,** Dexamethasone prevents the induction by endotoxin of a nitric oxide synthase and the associated effects on vascular tone. An insight into endotoxin shock, *Biochem. Biophys. Res. Commun.*, 173,541, 1990.
60.   **O'Connor, K. J. and Moncada, S.,** Glucocorticoids inhibit the induction of nitric oxide synthase and the related cell damage in adenocarcinoma cells, *Biochem. Biophys. Acta,* 1097,227, 1991.
61.   **Di Rosa, M., Radomski, M., Carnuccio, R. and Moncada, S.,** Glucocorticoids inhibit the induction of nitric oxide synthase in macrophages, *Biochem. Biophys. Res. Commun.*, 172,1246, 1990.
62.   **Knowles, R. G., Salter, M., Brooks, S. L. and Moncada, S.,** Anti-inflammatory glucocorticoids inhibit the induction by endotoxin of nitric oxide synthase in the lung, liver and aorta of the rat, *Biochem. Biophys. Res. Commun.*, 172,1042, 1990.
62a. **Snyder, D. S. and Unanue, E. R.,** Corticosteroids inhibit murine macrophage Ia expression and interleukin 1 production, *J. Immunol.*, 129,1803, 1982.
63.   **Stosic-Grujicic, S. and Simic, M. M.,** Modulation of interleukin 1 production by activated macrophages: *in vitro* action of hydrocortisone, colchicin, and cytochalasin B, *Cell. Immunol.*, 69,235, 1982.
64.   **Staruch, M. J. and Wood, D. D.,** Production of serum interleukin-1-like activity after treatment with dexamethasone, *J. Leuk. Biol.*, 37,193, 1985.

65. **Knudsen, P. J., Dinarello, C. A. and Strom, T. B.,** Glucocorticoids inhibit transcriptional and post-transcriptional expression of interleukin 1 in U937 cells, *J. Immunol.*, 139,4129, 1987.

66. **Lee, S. W., Tsou, A-P., Chan, H., Thomas, J., Petrie, K., Eugui, E. M. and Allison, A. C.,** Glucocorticoids selectively inhibit the transcription of the interleukin 1ß gene and decrease the stability of interleukin 1ß mRNA, *Proc. Natl. Acad. Sci. USA*, 85,1204, 1988.

67. **Doherty, G. M., Jensen, J. C., Buresh, C. M. and Norton, J. A.,** Hormonal regulation of inflammatory cell cytokine transcript and bioactivity production in response to endotoxin, *Cytokine*, 4,55, 1992.

68. **Dunham, D. M., Arkins, S., Edwards, C. K., Dantzer, R. and Kelley, K. W.,** Role of interferon-gamma in counteracting the suppressive effects of transforming growth factor-beta 2 and glucocorticoids on the production of tumor necrosis factor-alpha, *J. Leuk. Biol.*, 48,473, 1990.

69. **Waage, A. and Bakke, O.,** Glucocorticoids suppress the production of tumor necrosis factor by lipopolysaccharide-stimulated human monocytes, *Immunol.*, 63,299, 1988.

70. **Huang, Z. B. and Eden, E.,** Effect of corticosteroids on IL1ß and TNFα release by alveolar macrophages from patients with AIDS and *Pneumocystis carinii* pneumonia, *Chest*, 104,751, 1993.

71. **Breuninger, L. M., Dempsey, W. L., Uhl, J. and Murasko, D. M.,** Hydrocortisone regulation of interleukin-6 protein production by a purified population of human peripheral blood monocytes, *Clin. Immunol. Immunopathol.*, 69,205, 1993.

72. **Waage, A., Slupphaug, G. and Shalaby, R.,** Glucocoticoids inhibit the production of IL6 from monocytes, endothelial cells and fibroblasts, *Eur. J. Immunol.*, 20,2439, 1990.

73. **Zanker, B., Walz, G., Wieder, K. J. and Strom, T. B.,** Evidence that glucocorticosteroids block expression of the human interleukin-6 gene by accessory cells, *Transplantation*, 49,183, 1990.

74. **Hettmannsperger, U., Detmar, M., Owsianowski, M., Tenorio, S., Kammler, H. J. and Orfanos, C. E.,** Cytokine-stimulated human dermal microvascular endothelial cells produce interleukin 6: inhibition by hydrocortisone, dexamethasone, and calcitriol, *J. Invest. Dermatol.*, 99,531, 1992.

75. **Seitz, M., Dewald, B., Gerber, N. and Baggiolin, M.,** Enhanced production of neutrophil-activating peptide-1/interleukin-8 in rheumatoid arthritis, *J. Clin. Invest.*, 87,463, 1991.

76. **Wenck, U. and Speirs, R.,** The effect of cortisone on blood leukocytes and peripheral fluid cells of mice, *Acta Haemotol.*, 17,193, 1957.

77. **Thompson, J. and vanFurth, R.,** The effect of glucocorticosteroids on the kinetics of mononuclear phagocytes, *J. Exp. Med.*, 131,429, 1970.

78. **Rindhart, J. J., Sagone, A. I., Balcerzak, S. P., Ackerman, G. A. and LoBuglio, A. F.,** Effects of corticosteroid therapy on human monocyte function, *N. Engl. J. Med.*, 292,236, 1975.

79. **Melewicz, F. M., Zeiger, R. S., Mellon, M. H., O'Connor, R. D. and Spiegelberg, H. L.,** Increased peripheral blood monocytes with Fc receptors for IgE in patients with severe allergic disorders, *J. Immunol.*, 126,1592, 1975.

80. **Kaneko, S., Motomura, S. and Ibayashi, H.,** Differentiation of human bone marrow-derived fibroblastoid colony forming cells (CFU-F) and their roles in hemopoiesis *in vitro, Br. Haematol.*, 51,217, 1982.

81. **Baybutt, H. N. and Holsboer, F.,** Inhibition of macrophage differentiation and function by cortisol, *Endocrinol.*, 127,476, 1990.

82. **Saunders, R. H., and Adams, E.,** Changes in circulating leukocytes following the administration of adrenal cortex extract (ACE) and adrenocorticotropic hormone (ACTH) in infectious mononucleosis and chronic lymphatic leukemia, *Blood,* 5,732, 1950.

83. **Fauci, A. S. and Dale, D. C.,** The effect of *in vivo* hydrocortisone on subpopulations of human lymphocytes, *J. Clin. Invest.*, 53,240, 1974.

84. **Bishop, C. R., Athens, J. W., Boggs, D. R., Warner, H. R., Cartwright, G. E. and Wintrobe, M. M.,** Leukokinetic studies, XIII. A non-steady-state kinetic evaluation of the mechanism of cortisone-induced granulocytosis, *J. Clin. Invest.*, 47,249, 1968.

85. **Boseila, A. W. A.,** Hormonal influence on blood and tissue basophilic granulocytes, *Ann. N. Y. Acad. Sci.*, 103,394, 1963.

86. **Michael, N. and Whorton, C. M.,** Delay of the early inflammatory response by cortisone, *Proc. Soc. Exp. Biol. Med.*, 76,754, 1951.

87. **Saavedra-Delgado, A. M. P., Mathews, K. P., Pan, P. M., Kay, D. R. and Muilenberg, M. L.,** Dose response studies of the suppression of whole blood histamine and basophil counts by prednisone, *J. Allergy Clin. Immunol.*, 66,464, 1980.

88. **Mishler, J. M.,** The effects of corticosteroids on mobilization and function of neutrophils, *Exp. Hematol.*, 5, 15, 1977.

89. **Zweiman, B., Slott, R. I. and Atkins, P. C.,** Histological studies of human skin test responses to ragweed and compound 48/80, *J. Allergy Clin. Immunol.*, 58,57, 1976.

90. **Slott, R. I. and Zweiman, B.,** Histologic studies of human skin test responses to ragweed and compound 48/80, *J. Allergy Clin. Immunol.*, 55,232, 1975.

91. **Eidinger, D., Wilkinson, R. and Bose, B.,** A study of cellular responses in immune reactions utilizing the skin window technique, I. Immediate hypersensitivity reactions, *J. Allergy,* 35,77, 1964.

92. **Ackerman, N., Martinez, S., Thieme, T. and Mirkovich, A.,** Accumulation of rat polymorphonuclear leukocytes at an inflammatory site, *J. Pharmacol. Exp. Ther.*, 221,701, 1982.

93. **Clark, R. A. F., Gallin, J. I. and Fauci, A. S.,** Effects of *in vivo* prednisone on *in vitro* eosinophil and neutrophil adherence and chemotaxis, *Blood,* 53,633, 1979.

94. **Roth, J. A. and Kaeberle, M. L.,** Effects of *in vivo* dexamethasone administration on *in vitro* bovine polymorphonuclear leukocyte function, *Infect. Immun.*, 33,434, 1981.

95. **Stevenson, R. D.,** Effect of steroid therapy on *in vivo* polymorph migration, *Clin. Exp. Immunol.*, 23,285, 1976.

96. **Kurihara, A., Ohuchi, K. and Tsrufuji, S.,** Reduction by dexamethasone of chemotactic activity in inflammatory exudates, *Eur. J. Pharmacol.*, 101,11, 1984.

97. **Cunha, F. Q. and Ferreira, S. H.,** The release of a neutrophil chemotactic factor from peritoneal macrophages by endotoxin: inhibition by glucocorticoids, *Eur. J. Pharmacol.*, 129,65, 1986.

98. **Yancey, K. B., Hammer, C. H., Horvath, L., Renfer, L., Frank, M. M. and Lawley, T. J.,** Studies of human C5a as a mediator of inflammation in normal human skin, *J. Clin. Invest.,* 75,486, 1985.

99. **Schleimer, R. P., Freeland, H. S., Peters, S. P., Brown, K. E. and Derse, C. P.,** An assessment of the effect of glucocorticoids on degranulation, chemotaxis, binding to vascular endothelium and formation of leukotriene B$_4$ by purified human neutrophils, *J. Pharmacol. Exp. Ther.,* 250,598, 1989.

100. **Ketchel, M. M., Favour, C. B. and Strugis, S. H.,** The *in vitro* action of hydrocortisone on leukocyte migration, *J. Exp. Med.,* 107,211, 1958.

101. **Gallin, J. I., Durocher, J. R. and Kaplan, A. P.,** Interaction of leukocyte chemotactic factors with the cell surface. I. Chemotactic factor-induced changes in human granulocyte surface charge, *J. Clin. Invest.,* 55,967, 1975.

102. **Ward, P. A.,** The chemosuppression of chemotaxis, *J. Exp. Med.,* 124,209, 1960.

103. **Looney, R. J., Ryan, D. H., Takahashi, K., Fleit, H. B., Cohen, H. J., Abraham, G. N. and Anderson, C. L.,** Identification of a second class of IgG Fc receptors on human neutrophils, *J. Exp. Med.,* 163,286, 1986.

104. **Fleit, H. B., Wright, S. D. and Unkeless, J. C.,** Human neutrophil Fcγ receptor distribution and structure, *Proc. Natl. Acad. Sci. USA,* 79,3275, 1982.

105. **Perussia, B., Dayton, E. T., Lazarus, R., Fanning, V. and Trinchieri, G.,** Immune interferon induces the receptor for monomeric IgG1 on human monocytic and myeloid cells, *J. Exp. Med.,* 158,1092, 1983.

106. **Scribner, D. J. and Fahrney, D.,** Neutrophil receptors for IgG and complement: their roles in the attachment and ingestion phases of phagocytosis, *J. Immunol.,* 116,892, 1976.

107. **Petroni, K. C., Shen, L. and Guyre, P. M.,** Modulation of human polymorphonuclear leukocyte IgG Fc receptors and Fc receptor-mediated functions by IFN-γ and glucocorticoids, *J. Immunol.,* 140,3467, 1988.

108. **Hollenberg, S. M., Weinberger, C., Ong, E. S., Crelli, G., Oro, A., Lebo, R., Thompson, E. B., Rosenfeld, M. G. and Evans, R. M.,** Primary structure and expression of a functional human glucocorticoid receptor cDNA, *Nature,* 318,635, 1985.

109. **Boumpas, D. T., Chrousos, G. P., Wilder, R. L., Cupps, T. R. and Balow, J. E.,** Glucocorticoid therapy for immune-mediated diseases: basic and clinical correlates, *Ann. Inter. Med.,* 119,1198, 1993.

110. **Ku Tai, P. K., Albers, M. W., Chang, H., Faber, L. E., and Schreiber, S. L.,** Association of a 59 kd immunophilin with the glucocorticoid receptor complex, *Science,* 256,1315, 1992.

111. **Picard, D. and Yamamoto, K. R.,** Two signals mediate hormone-dependent nuclear localization of the glucocorticoid receptor, *EMBO J.,* 6,3333, 1987.

112. **Yamamoto, K. R.,** Steroid receptor regulated transcription of specific genes and gene networks, *Ann. Rev. Genet.,* 19,209, 1985.

113. **Boggaram, V., Smith, M. E. and Mendelson, C. R.,** Posttranscriptional regulation of surfactant protein: a messenger RNA in human fetal lung *in vitro* by glucocorticoids, *Mol. Endocrinol.,* 5,414, 1991.

114.   **Wallner, B. P., Mattaliano, R. J., Hession, C., Cate, R. L., Tizard, R., Sinclair, L. K., Foeller, C., Chow, E. P., Browning, J. L., Ramachandran, K. L. and Pepinsky, R. B.**, Cloning and expression of human lipocortin, a phospholipase A$_2$ inhibitor with potential anti-inflammatory activity, *Nature*, 320,77, 1986.

115.   **Bailey, J. M., Makheja, A. N., Pash, J. and Verma, M.**, Corticosteroids suppress cyclooxygenase messenger RNA levels and prostanoid synthesis in cultured vascular cells, *Biochem. Biophys. Res. Commun.*, 157,1159, 1988.

116.   **Helmberg, A., Fässler, R., Geley, S., Jöhrer, K., Kroemer, G., Böck, G. and Kofler, R.**, Glucocorticoid-regulated gene expression in the immune system. Analysis of glucocorticoid-regulated transcripts from the mouse macrophage-like cell line P388D1, *J. Immunol.*, 145,4332, 1990.

Chapter 10

PSYCHOLOGICAL STRESS AND UPPER RESPIRATORY ILLNESS

Dana H. Bovbjerg
Department of Neurology
Memorial Sloan-Kettering Cancer Center
New York, NY

Arthur A. Stone
Department of Psychiatry
State University of New York at Stony Brook
Stony Brook, NY

INTRODUCTION

Accumulating evidence from both naturalistic and experimental studies indicates that psychological stress can affect upper respiratory illness (URI). This literature has recently been reviewed in considerable detail by three separate research groups with complementary perspectives.[1-4] In their reviews, Boyce and Jemerin focused on infectious illness in children, emphasizing the role of individual differences in susceptibility.[1,2] Cohen and Williamson reviewed the broader literature on stress and disease in humans, with an emphasis on psychological factors.[3] Peterson and colleagues, on the other hand, reviewed both human and animal studies, with an emphasis on biological factors.[4]

Our purpose in this chapter is not to present another exhaustive review of the literature on psychological stress and URI, but rather to highlight some of the important issues for this area of research, as well as research conducted since those previous reviews. Throughout the chapter, our focus will be on human URI. We will usually focus our consideration of URI to the common cold, both because colds are the most common URI, and because colds have been the most frequent subject of studies on psychological stress in URI.

This chapter is divided into five sections. The first section provides a brief introduction to the various ways stress has been conceptualized in psychobiological and health studies. The second section provides an overview of upper respiratory disease, including etiology, pathophysiology, and immune defenses. The third section outlines the steps in the pathogenesis of URI where psychological stress could conceivably exert its influence, and possible mechanisms of such stress effects are discussed. The fourth section provides illustrative examples, organized by study design (cross-sectional, longitudinal, and viral exposure), of studies that have examined the association between stress and URIs. The fifth section provides a summary and recommendations for future research.

WHAT IS PSYCHOLOGICAL STRESS?

There has been considerable debate concerning the definition of psychological stress.

At least four different ways of conceptualizing stress have been discussed.[5-7] 1) The *environmental approach* defines stress as significant, objective changes in the environment. Some environmental definitions are based solely on the degree to which events require readjustment in one's lifestyle, whereas other definitions are based on the extent to which events are upsetting. Examples of this approach are the studies of major life events (e.g., death of spouse, change in employment, major purchase) and daily minor events (e.g., argument with spouse, problem at work, issues with children). 2) The *appraisal/emotion approach* defines stress as judgements of a situation as challenging or threatening (appraisal approach) or as the experience of negative emotions, such as anxiety or sadness (the emotion approach). This conceptualization hinges on psychological reactions to environmental events. 3) The *physiological approach* defines stress in terms of relatively stereotypic alterations in physiological processes. Examples of stress defined this way include activation of the hypothalamic-pituitary-adrenal axis (e.g., evident in increased plasma cortisol concentrations), or activation of the autonomic nervous system (e.g., evident in increased blood pressure). 4) The *integrative approach* defines stress as certain combinations of the above elements. An example of this approach is Lazarus' transactional model of stress and coping.[8] According to this model, stress is the result of a unique interaction of environmental change, specific appraisals, and coping efforts to modify emotions and/or problems. It is conceptualized as a temporally changing process (hence, transactional), which includes feedback from later processes that can, in turn, affect earlier processes.

A common characteristic of these conceptualizations of stress is that an external stimulus of some sort (a stressor) is always thought to be involved. It is also commonly presumed that stress has the potential to result in a negative health outcome.[7] In the review that follows, we have included studies that examined relations between URI and either some external stimulus (stressor) or some change in appraisal or affect. For the purpose of this review, defining stress as a physiological response seems, to us, conceptually too closely related to URI. We have limited our review to studies of clear environmental changes that are perceived as unpleasant or stressful. We have not included articles of more general psychological states (e.g., personality) which may have their own effects on URI.

UPPER RESPIRATORY INFECTION
Background

Nationwide annual surveys regularly indicate that URI is the single most common cause of physician visits and missed days of work among all acute medical conditions.[9] In 1990, Couch[9] estimated that the common cold and influenza accounted for 165 million significant illnesses, resulting in an average of 3.2 days of restricted activity and 1.6 days in bed; half of those illnesses received medical attention. It has been estimated that upper respiratory symptoms precipitate as many as 14% of all visits to physicians.[10] The common cold has been calculated to be responsible for 30 million lost days at school and work each year; to cost $3 billion for physician office visits; and to cost $2 billion for over-the-counter drugs.[11] Common colds, although not serious medical conditions, thus pose a major public health problem.[10]

The incidence of common colds is highest in infants and children, who suffer from four to eight colds each year; adults typically get two to five colds, except in households with children, where the incidence is higher.[12] Indeed, children's noses have been characterized as the chief reservoir for cold viruses,[10] with dissemination into the broader community occurring as infection moves from school to home, where parents become infected.

The viruses responsible for colds (see below) are transmitted by airborne droplets and aerosol (e.g., by sneezing), as well as through environmental contamination with virus-laden nasal secretions (e.g., by touching the eyes or nose with contaminated fingers).[12] For example, rhinovirus in nasal secretions of an infected individual has been demonstrated to be easily spread by the fingers to a variety of common household objects (e.g., doorknobs), where it can survive for several hours, serving as a source of infection.[13]

INFECTIOUS AGENTS

Rhinoviruses are the most common cause of the common cold.[13] Based on survey studies with samples collected for viral identification, rhinoviruses have been estimated to be responsible for 40 to 60% of all colds in adults.[9] The virus can either be identified in nasal secretions from the infected individual or by the presence of specific antiviral antibodies in serum or nasal secretions following infection.[13] More than a hundred different types of rhinoviruses (serotypes) have now been identified.[13] Coronaviruses (3 types), which are technically more difficult to identify, have been estimated to be responsible for another 20% of colds in adults and perhaps more in children.[9] Other viruses known to be associated with classic cold symptoms (e.g., rhinorrhea) include the influenza viruses (3 types), parainfluenza viruses (4 types), and respiratory syncytial virus; if not contained by host defense mechanisms, these viruses typically go on to cause more serious illness and associated systemic symptoms (e.g., fever).[9]

SYMPTOMS OF INFECTION

The multitude of different upper respiratory viruses, reviewed above, all typically cause a strikingly consistent constellation of symptoms including rhinorrhea, nasal obstruction, sneezing, pharyngeal discomfort, and cough, which are familiar to all of us as a common cold.[9,10,13,14] It should be noted, however, that for reasons that are as yet obscure, about a third of all verified rhinovirus infections have been found not to result in symptomatic illness (i.e., in clinical colds).[13] For many adults symptoms begin with a dry, "scratchy," or sore throat, soon followed by a watery nasal discharge, inflammation of the nasal mucosa, and sneezing.[12] Some systemic symptoms such as general malaise and myalgia may be evident, but fever is rare.[12] After one to three days, nasal obstruction is common, the nasal secretions thicken and are no longer clear.[12] In infants, fever and other systemic symptoms including anorexia, vomiting, and diarrhea are more commonly associated with colds.[12] Symptoms typically begin to decline within a few days and last for about a week, although in some cases they may linger for as long as a month.[15]

Confirming the commonality of symptoms across different infectious agents, a recent study conducted at the Common Cold Unit in England conducted detailed symptom assessments of viral shedding following exposure of healthy subjects to several different URI viruses.[16] Three different rhinoviruses, a respiratory syncytial virus, and a coronavirus were administered to quarantine volunteers, and symptoms were carefully monitored over the next several days. Although there were some modest differences in the rapidity with which symptoms developed, suggesting differences in the incubation period, by two to three days after the viral exposure, symptoms peaked. There were no major differences in the pattern of symptom development across the five viruses.

The ubiquity of this symptom constellation across different viruses suggests common pathogenic mechanisms (see below) and raises the possibility that these symptoms may confer an adaptive advantage to the infected individual. Ewald[17] has theorized that from an evolutionary perspective, symptoms of infection can be viewed as either: 1) adaptations of the pathogen to increase its reproductive success; 2) adaptations of the host to defend against the pathogen; 3) "side effects" of the infection that do not serve adaptive functions for pathogen or host. Which role is played by the symptoms of a common cold has yet to be determined.

ASSESSMENT OF ILLNESS

As might be expected from the discussion above, the literature includes two ways of assessing URIs: the presence of the symptom syndrome (clinical cold) or the presence of an infection, verified by isolation of virus from nasal secretions or by increases in specific antibody titers. As will be discussed below in the section on viral exposure studies, individuals can manifest an infection yet not demonstrate the clinical syndrome. Although one might demand that both infection and syndrome be present as a conservative definition of URI, this would virtually eliminate the study of URI in natural observation studies. The reason for this is that it is difficult to confirm viral infection when the particular virus is not known, which is the typical case in epidemiological studies of URI syndromes. Viral exposure studies can, on the other hand examine both clinical syndromes and infection (because the strain of virus is known and can be specifically tested for).

IMMUNE DEFENSES AND THE PATHOGENESIS OF
RHINOVIRUS INFECTION

Neither the pathogenesis of infection nor the relevant immune defenses is well understood for any of the more than 200 distinct viruses known to infect the human upper respiratory tract.[9] In the discussion that follows we will again focus on the rhinoviruses, the pathogenesis of which has been under increasing scrutiny in a series of experimental inoculation studies.[12]

Besides avoiding exposure to the rhinovirus, the best established initial immune defense against infection is the presence of protective levels of specific antibody for the virus as a result of previous exposure.[13] An early study by Hendley and colleagues[18] revealed an inverse relation between individuals' levels of serum neutralizing antibodies to the challenge rhinovirus and their subsequent susceptibility to experimental infection. More recent research, using a more sensitive ELISA assay, has continued to indicate the protective value of specific antibody to the challenge virus in either serum or nasal secretions prior to experimental inoculation.[19] Immunoglobulin in nasal secretions, predominantly secretory Immunoglobulin A (sIgA), but also including Immunoglobulin G (IgG), is thought to reduce the risk of infection by interfering with viral attachment to the epithelial surface.[20] It should be noted that the nasal mucosa has a wide range of nonspecific defense mechanisms that are also likely to play a preventative role; not the least of these is the outer layer of the mucous blanket itself, which by ciliary action transports particles to the posterior pharynx where they are swallowed.[21]

Although the protective role of secretory antibody at the mucosal surface is widely accepted, the relative importance of secretory and circulating antibody is still debated, with some investigators arguing that circulating antibody may be responsible for long lasting immunity.[22] It is also too early to rule out possible contributions of cellular defenses, whose role at mucosal surfaces has received little attention.[23]

Viral infection is, beyond a doubt, the critical initiating step in the pathogenesis of a clinical cold, but the chain of events leading from infection to the manifestation of symptoms has yet to be fully elucidated.[24] Several lines of evidence suggest that the virus itself is not directly responsible for symptoms. First, it is quite possible to be infected and have no symptoms. Typically, in both experimental studies and field studies, a third of the subjects with confirmed infections have no symptoms.[13] Second, histological studies with light and/or electron microscopy have found that rhinovirus infection, unlike influenza, does not cause damage to the nasal epithelium.[25] Third, although symptoms tend to be most severe when viral shedding is at its peak (e.g., day 2), shedding typically continues for several days after symptoms have resolved.[25] Fourth, accumulating evidence indicates that viral infection triggers the release of inflammatory mediators, which in turn cause the symptoms.[13]

Increased levels of kinins (e.g., bradykinin and lysylbradykin) have been found in the nasal secretions of symptomatic but not asymptomatic individuals, following both natural and experimental infection with rhinovirus.[26,27] Consistent with the possibility that kinins may play a role in symptomatology, provocation experiments have shown that nasal application of bradykinin causes rhinorrhea, nasal obstruction, and sore throat in volunteers.[28] In addition to their direct effects on vascular permeability, a major contributor to nasal secretions early in the course of a rhinovirus infection,[29] kinins are thought to affect local secretory responses and pain by stimulating nerve fibers in the nasal mucosa.[28,30] Although neural regulation of nasal secretions is complex and, as yet, poorly understood, both cholinergic and sympathetic nerves are thought to regulate the passive diffusion of plasma proteins, as well as active secretions (including sIgA) from serous and mucous glands.[21,31] Psychological influences could thus affect the secretion of kinins or the neural regulation of nasal secretions.

Unlike allergic rhinitis, no increases in histamine or prostaglandin D2 have been found in nasal secretions following rhinovirus infection, suggesting that mast cells may not play a role in the symptoms of a cold.[26,27] The severity of symptoms has been found to be correlated with an increase in the numbers of lymphocytes and neutrophils in nasal secretions.[25,32] Increased numbers of neutrophils have also been histologically documented in biopsies of the nasal mucosa of symptomatic individuals.[25] One possible explanation for these effects is suggested by a recent study indicating that nasal lavage fluids from symptomatic subjects contained significantly higher levels of interleukin-1 (IL-1) than asymptomatic, or sham-infected, subjects.[24] It is tempting to speculate that this IL-1 reflects the activation of local cellular defense mechanisms, but IL-1 is known to be secreted by a wide variety of nonlymphoid tissues, including nasal epithelial cells.[33,34] In any case, local secretion of IL-1 could upregulate a wide range of both local and systemic cell-mediated immune defenses.[33]

The mechanisms responsible for recovery from rhinovirus infection are not yet clear. A role for local and/or systemic neutralizing antibody cannot be ruled out, although in most studies increased levels of specific antibody have not been detected until the illness has resolved.[13,15] As with other viral infections, cellular immune defenses may play an important role in recovery, although local secretion of interferon may also be involved.[13] Although direct evidence for cell-mediated recovery mechanisms is lacking, there is some circumstantial support. As noted above, the number of lymphocytes and neutrophils in nasal lavages from infected subjects is significantly elevated within the first few days after experimental infection with RV25, raising the possibility that leukocytes in the nasal mucosa may play a role in resolving the infection.[32]

There is also some initial evidence consistent with the possibility that systemic cellular responses may play a role in resolving the infection. For example, three days after infection with the same rhinovirus serotype (RV25), Levandowski and colleagues[32] found decreases in peripheral blood lymphocyte numbers (including T cells, but not B cells); the decreases in cell numbers were inversely correlated with the number of days of viral shedding. Three days after infection with a different rhinovirus (Hanks), Hsia and colleagues[35,36] found no change in lymphocyte numbers, perhaps reflecting differences in the kinetics of infection which have been demonstrated across different viruses.[14,16] After five days, however, these investigators[36] found significant increases in lymphocyte numbers (including T cells, but not B cells). *In vitro* assessment of isolated peripheral blood mononuclear cells (PBMC) revealed increased levels of PHA-stimulated interleukin-2 and interferon production, which were inversely correlated with the duration of viral shedding. Confirming a previous report,[37] these investigators also found that the *in vitro* proliferative responses to the challenge virus were also increased following infection, as was natural killer (NK) cell activity.[36] *In vitro* proliferative responses to the challenge virus were also found to be increased following experimental infection (RV39) in another recent study,[38] but these investigators found reduced levels of NK cell activity in peripheral blood mononuclear cells collected seven days after infection.

All of these studies can be viewed as consistent with the possibility that cell mediated responses may play a role in recovery from rhinovirus infection, but none provides proof. The immune mechanisms responsible for recovery from rhinovirus infection remain to be determined. Although it is tempting to theorize that mucosal immune defenses must be more critical in resolving what is widely accepted as a localized infection,[13] systemic responses to rhinovirus may also play an important role. Indeed, there may be a false dichotomy in this thinking. Optimal recovery will likely depend on local and systemic responses working in concert.

HOW COULD PSYCHOLOGICAL STRESS AFFECT URIs?

There are several points in the pathway of pathogenesis where stress could affect URI, as is schematically shown in Figure 1. First, psychological stress may alter exposure to virus in some way. For example, people experiencing high stress levels may cope with the stress in ways that encourage exposure (Arrow #1). One method of coping with stress is called seeking social supports,[39] and it involves procuring practical advice and/or emotional support from others. This contact with others, some of whom may be shedding URI virus, could increase exposure to pathogens.

Lowered host prevention mechanisms to initial viral exposure is a strong possibility for mediating stress effects (Arrow #2). There is substantial research showing that psychological stress affects immune system function,[40] including many of the immune components previously discussed (e.g., sIgA). Stress may alter immunity by affecting behaviors such as sleep patterns, eating, consumption of alcohol medication use, and exercise.[41] It has also been shown that stress can affect immunity via direct neural connections and/or hormonal changes (reviewed in previous chapters).

Similar to host prevention mechanisms, the immune system is thought to play a role in controlling viral replication in the nasal mucosa (Arrow #3). Both cellular and humoral systems probably contribute to controlling viral replication and both are likely to be influenced by the same stress-related factors mentioned for host prevention mechanisms.

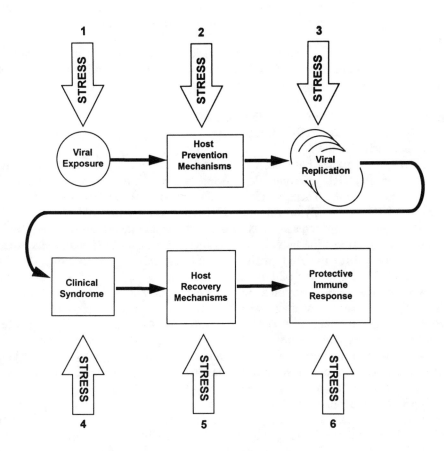

Fig 1. Possible ways that psychological stress could affect URIs

Clinical syndrome refers to the constellation of symptoms associated with URIs (mentioned above). It is important to note that URI does not always lead to clinical syndromes; as noted earlier, for rhinovirus only about two thirds of those infected (confirmed by viral shedding and/or raised antibody titers later) manifest symptoms. There is currently no explanation for this discrepancy, but clearly there must be individual differences in the physiological processes that affect symptom expression. Psychological stress may contribute to the individual differences by affecting the underlying processes responsible.

Recovery from a clinical syndrome usually takes several days. Stress could potentially influence the processes that are responsible for determining the duration of infection and symptoms (Arrow #5). In addition to physiological processes, stress could influence patients' use of medications.

Finally, after the experience with URI viruses, the immune system will be better poised to defend against future infection through both specific (e.g., secretory antibody response) and nonspecific mechanisms. Psychological stress has been shown, for example, to affect antibody response to vaccination.[42] It may be that psychological stress influences the magnitude of this secondary response.

Although there may be other routes by which psychological stress influences URI, these six routes are likely candidates. As will be evident from the discussion below, many of these routes have not yet received any research attention.

REPRESENTATIVE STUDIES EXAMINING THE ASSOCIATION BETWEEN STRESS AND URIs
Cross-sectional Field Studies

A number of studies have been conducted exploring associations between naturally occurring stressors and URI symptomatology. Some of these studies examined objective indices of URI infections as well. These studies provide support for the hypothesis that stress is associated with URI, yet clearly much research remains to be done.

Graham *et al.*[43] reported a survey of 2,618 children and their families in Australia. The survey examined the association between the mothers' stress and URI symptoms in children. Families were selected for the study if the proband (the child) fell into either the upper or lower quintile of the distribution of URI symptoms over the last year. Mothers' stress was assessed by a combination of scores on a major life event checklist, a daily hassles checklist, and an emotional distress scale. Mothers were labelled as stressed if their scores on all three measures were above the median. These procedures yielded a sample of 255 children with frequent URIs and 227 with infrequent URIs. Their average age was about 2.5 years. Children whose mothers were in the high stress group were four times as likely to be in the high URI group than those whose mothers were in the low stress group. Although other risk factors significantly contributed to a model developed by the investigators to predict URI group membership (e.g., chest illness in the child's first year of life was a strong predictor), the mothers' stress level was the second most powerful predictor of URIs. The investigators acknowledge the bidirectional interpretations that are consistent with this data, namely, that mothers of ill children may be stressed for that reason or that their stress causes increased illnesses. Regarding the latter hypothesis, it may be that a decrement in family hygiene due to the mother's stress or enhanced susceptibility of children could be operative.

A negative finding was shown in a study notable for its verification of infection. Clover *et al.*[44] studied 281 adults and children during an influenza outbreak. Sixty six subjects developed verified influenza: 35% of the children and 17% of the adults. A major life events inventory was administered yet showed no association with influenza status. However, a measure of family cohesion was negatively related to infection.

LONGITUDINAL FIELD STUDIES

Longitudinal studies have followed subjects on more than a single occasion. They have many of the same methodological weaknesses as cross-sectional field studies. The studies are correlational and, hence, cannot rule out third variable explanations for observed relations. However, the repeated measurements of both stress and URI allow for predictions of URIs from levels of stress reported *before* URIs are reported. Such prospective prediction allows stronger statements about causal relationships between stress and URI. Despite this strength, data from longitudinal studies have often not been analyzed in ways that capitalize on prospective prediction.

An excellent example of an early longitudinal field study is one that was conducted by Roghmann and Haggerty.[45] In this 1973 study, 512 families were monitored over a 28-day period by mothers. Using a health diary, mothers recorded significant events that were perceived as stressful (e.g., losses, arguments, financial problems), as well as health symptoms in each family member and his or her medical care utilization. Unlike many survey studies, these investigators utilized the longitudinal data in sophisticated ways, which served as a model for later studies of stress and health. Although the authors did not specifically analyze URIs, they state that most illnesses included fevers, coughs, headaches, and "colds." It is thus very likely that the vast majority of the illnesses analyzed in this study were URIs.

Considering all of the members of families studied, a total of 71,346 person-days were examined. On a day-by-day basis, the investigators found that 30% of the days could be characterized as stressful, but only 10% were found to be in the upper range of the stress score. A health complaint was reported by mothers about themselves on 25% of days, and for their youngest child on 17%. Same-day analyses showed that the probability of a mother having an illness doubled when stress was present on that day versus when it was not present; a 50% increase was observed for the youngest child. However, the most powerful analyses were based on the notion of stress and illness *episodes*. In accord with the commonsense notion that difficult times tend to stretch over multiple days and that the same is true for illnesses, the authors analyzed the correspondence between episodes of stress and illness. Importantly, they analyzed the data such that lagged associations would be evident: that is, stress episodes preceding illness episodes and vice versa. They found a 250% increase in illness episodes following stress episode onsets compared to what would be expected by chance, suggesting a causal relation.

In another of the earlier studies of URI in children, Boyce *et al.*[46] repeatedly assessed 58 children in a daycare setting. Children (average age 4 years) were observed on a daily basis on weekdays for an entire year, during which time all illnesses were assessed by a nurse practitioner. Additionally, biweekly nasopharyngeal cultures were taken; cultures were also taken at the start of each illness. At the end of the year, parents completed several questionnaires, including a major life events inventory for events in the child's life and a questionnaire about weekly family routines. Observational data yielded several measures of illness: frequency, average duration, average severity, and a composite measure (days of illness times average severity). When age, sex, race, family income, and family size were controlled, results indicated that the child's life event stress was associated with illness duration. More life stress was associated with longer illnesses. Interestingly, life event stress interacted with family routine such that those children with high stress and strong family routines had more severe illnesses than other children. It is important to note that stress was not associated with frequency of illness. In contrast to the previous study, these results are less convincing since prospective analyses (lagged relations) were not computed.

In the 1980s a number of studies examined the associations between minor symptomatology and two types of stress: major life events and daily events/hassles. At issue was not the nature of the association between stress and illness, but the then recent emergence of minor event and hassle checklists as a way of assessing stress. Therefore, the goal of these studies was to compare how well minor versus major events could predict symptomatology. Since the focus was not on the symptoms, it is often not clear exactly what sorts of symptoms respondents reported. Importantly, the concept of symptom episodes was not part of the methodology, making it difficult to know what a total symptom score meant (several individual days of symptoms or a few long episodes, etc.). Despite these shortcomings, several of these studies demonstrated that major life events and minor events both predicted symptom rates, with the edge going to minor events.[47,48]

Stone, Reed, and Neale[49] conducted a study capitalizing on the prospective associations of daily data collected in a longitudinal manner. Seventy nine community-dwelling, middle-aged, married males were studied for 84 consecutive days. An important feature of this study was the careful assessment of daily events. Husbands completed the daily event questionnaire about themselves each evening and their wives confirmed the reports of husbands' events (prior work had shown that this procedure improves event reporting). Another important design feature of this study was that questionnaires were completed on a daily basis and mailed to the investigators on the following day; this procedure has been shown to reduce the possibility that subjects complete multiple days at one sitting.[50]

This study following the conceptualization of URIs used by Roghmann and Haggerty,[45] isolating episodes of symptoms with symptom clusters consistent with a cold or flu (unlike Roghmann and Haggerty, however, no single-day episodes were allowed). The frequency of desirable and undesirable daily events was examined for several days prior to the onset of URI episodes. For control periods, days that did not precede URIs (matched for day of the week as well) were selected from the same subject. An increase in undesirable events and a decrease in desirable events were observed in the three to five day period prior to URIs relative to control days. Notably, there were no differences in event report, either desirable or undesirable, one or two days before URI onsets. This pattern of data makes it less likely that there was a reporting of subsyndromal symptoms, not recognized by subjects, since the most likely period for such an effect to be observed would be immediately before the URI onset.

Three additional studies have extended the Stone, Reed, and Neale[49] findings. Evans, Pitts, and Smith[51] replicated the dip in desirable events preceding URIs, using a similar methodology to that of Stone *et al.*[49] However, these investigators did not observe an increase in undesirable events prior to the onset of URIs. In an independent study, Evans and

Edgerton[52] again found a decrease in the frequency of desirable events before URIs, and this time they observed a trend for an increase in undesirable event frequency. Finally, Stone, Porter, and Neale[53] examined the same association in yet another longitudinal, daily diary study. No replication of either a dip in desirable events or a peak in undesirable events was observed. A comparison of the methods and analytic techniques employed in all four of these studies can be found in Stone, Porter, and Neale,[53] but suffice it to say that the bulk of the evidence is in favor of an effect of fewer desirable events and, less strongly, more undesirable events prior to URIs.

VIRAL EXPOSURE STUDIES

Naturalistic studies have the advantage of ecological validity[54] because subjects are exposed to "real" levels of stressors and naturally occurring exposures to pathogens. A disadvantage of these designs is that determining causal pathways is difficult because stressor and pathogen exposures are not controlled. Viral exposure studies can address these problems. These studies typically manipulate exposure to selected pathogens and can manipulate stressor levels as well, unlike field studies where the pathogen responsible for the URI symptoms is usually not identified (in part, because any of scores of viruses could be responsible). We should note, however, that given the expense of these types of studies, only a handful have been conducted, and these examined a very small selection of URI viruses.

One of the first studies exploring stress and experimentally-induced URI came from the Common Cold Unit (CCU) in England.[55] The CCU conducted a series of studies designed to understand the pathophysiology of the common cold by inoculating healthy individuals with live cold viruses. In 1977, Totman *et al.*[55] explored whether cognitive dissonance produced by a difficult decision paradigm was related to the incidence of infection or to cold symptoms in individuals exposed to either of two rhinoviruses. Although stress was not explicitly assessed, subjects who experienced decision making may be thought of as being stressed by the procedure (an interpretation advanced by the authors). After controlling for pre-existing antibody levels to the challenge viruses, "stressed" subjects had significantly higher levels of symptoms, but no differences in the incidence of infection or the amount of shed virus in nasal secretions.

A second study by Totman[56] from the CCU paradigm explicitly examined life event stress. In addition to the life event checklist, an interview fashioned after Brown's method[57] of objectively recording major experiences was used to create several stress indices. These included the SEI, assessing the time-adjusted impact of events, the SDI, assessing the total magnitude of life events, the TLI, assessing changes in "goal-directed" activities due to events, and the TCI, another measure of change in activities due to events. Measures of extroversion and neuroticism were also administered. After controlling for preantibody status prior to the experimental inoculation with the viruses, only the TLI was associated with total symptom score. Only the TCI was associated with amount of viral shedding. These relationships were complicated by the overlap between the stress measures associated with response to viral exposure (TLI and TCI) and extroversion, which had the strongest association with both symptoms and viral shedding. Regression analyses controlling for the overlap did, however, show that the TCI had an independent effect on shedding over that explained by extroversion.

More recently a landmark study by Cohen *et al*.[58] also at the CCU showed that levels of stress prior to inoculation with live virus were associated with susceptibility to and clinical syndromes of several URI viruses. In this study, 394 volunteers were exposed to one of several URI-inducing viruses. A psychological stress index was created by combining three separate stress scales that were administered upon entry to the CCU: a life events inventory, the Perceived Stress Scale, and a negative affect scale (for the past week). After controlling for pre-existing antibody levels to the viruses and for a variety of subject variables (e.g., age, sex, education, allergic status, weight, season), the association between the stress index and two outcomes was examined. Outcomes were: 1) the percentage of subjects who became infected with the experimental virus, as indicated by isolation of the shed virus or by increase in antibodies to the virus, and 2) the percentage of subjects who manifested clinical cold syndromes. For infection, there was a linear rise in proportion of subjects infected as stress levels increased. At the lowest stress levels, less than 75% of the subjects were infected, whereas at the highest stress levels, about 90% were infected. A parallel pattern emerged for the proportion of subjects with colds: at the lowest stress levels, under 30% had colds, while at the highest levels, about 45% had colds. This data provides some of the strongest evidence in support of the hypothesis that psychological stress affects URI.

A further analysis of the data from the Cohen, Tyrrell, and Smith[58] study explored relationships between individual components of the stress index and susceptibility to URI.[59] Analyses in this paper indicated a somewhat surprising set of associations. Only the life event measure was a significant predictor of clinical colds. In contrast, negative effect and perceived stress were significant predictors of infection.

A study by Stone and colleagues[60] essentially replicated the life event findings of the Cohen study in a smaller sample. Seventeen college undergraduates, all of whom had no pre-existing antibody titers to the experimental virus, were experimentally inoculated with a rhinovirus. At the outset of the study, subjects completed a life event inventory, resulting in a stress score reflecting the total number of events experienced in the last year, and a mood assessment. Because prior to entry into the study subjects were screened for having no antibody to the experimental virus, all subjects became infected after viral exposure (unlike the Cohen study where some individuals did have pre-existing antibodies to the experimental agents). Subjects were classified by whether or not they had a clinical cold: 12 of 17 (71%) did. Subjects who did not develop colds had fewer life events (2.6 vs. 7.3) than those who developed colds. Interestingly, when events were categorized according to subjects' perceptions, subjects without colds had fewer negative (nonsignificant) and more positive (significant) events. There were no differences in the groups of subjects in terms of negative or positive effect (an alternative way of conceptualizing stress).

PHYSIOLOGICAL MEDIATION OF PSYCHOLOGICAL STRESS AND URIs

With rare exception, the studies reviewed above have not directly examined possible mediating pathways that could be responsible for the associations between stress and URI. This comment should not be taken as criticism. It is appropriate that early investigators focus on establishing phenomenon before attempting to explain how it works. The validity of the demonstrated relations between psychological variables and URI is not compromised, after all, by our current lack of knowledge concerning the mediating mechanisms. On the other hand, consideration of possible mechanisms might help to explain some of the apparent anomalies in the present literature and suggest improved strategies for detecting psychological influences in future research. For example, are there plausible mechanism(s) that could account for the apparent selectivity of the effects of life events on colds due to rhinovirus infection? As reviewed above, differences in life events were not found to predict which individuals would become infected following experimental inoculation with rhinovirus, but did predict who would develop the symptoms of a cold.[59,60]

As we discussed above, there are several possible pathways that could link psychological variables with URI. The problem is knowing where to begin. We present one line of research exploring a possible pathway that may mediate the effects of psychological stress on URIs.

Some of the strongest evidence reviewed above supporting the association between stress and URIs was from prospective daily diary studies. There have been two studies which have examined a potential immunological mediator of stress and URIs, secretory IgA antibody, which was mentioned in the section on pathogenesis of URI.[62] We focus here only on the studies using specific sIgA antibody to a known antigen and bypass studies that have examined stress and nonspecific sIgA protein, because it is not clear that total sIgA protein is a meaningful index of immunological protection. It is more likely that specific antibody to a known antigen behaves in an analogous manner to URI pathogens.

The immunologic model used in both studies[61,61] involved repeated challenge with a (relatively) novel protein for participants, purified rabbit albumin. Every morning for a number of consecutive days, subjects ingested 100 mg of albumin. Specific antibody responses were monitored via nightly saliva samples obtained directly from the parotid gland with a Curby cup. An index of antibody activity was obtained by dividing sIgA activity to the albumin (obtained through ELISA) by overall levels of sIgA protein (obtained through RID), in order to control for salivary flow rates.

The stressors differed in the two studies: in the first it was affective states (negative and positive moods) and in the second it was the number of undesirable daily events (and the number of desirable daily events as well). In both studies, effects of stress on sIgA antibody to the albumin were demonstrated. In the first study,[61] thirty dental students were studied over an 8 week period, 3 times per week. Days with high negative mood had lower levels of specific antibody activity compared to days with low negative mood. Opposite findings emerged for days with high and low positive mood; namely, high positive mood days were associated with higher levels of specific antibody. (Notably, these associations were not observed for total sIgA protein levels.)

The second study examined 96 community-dwelling, married males who participated for 12 consecutive weeks.[62] On a daily basis they recorded undesirable and desirable events as well as mood; an evening saliva sample (as described above) was also taken to examine sIgA response to the ingested antigen. On days with relatively high numbers of undesirable events, lower levels of sIgA antibody were observed. Conversely, days with higher numbers of desirable events were associated with higher levels of antibody. In addition to concurrent daily analyses, lagged analyses, where event levels on one predicted antibody on a latter day, were also computed. Surprisingly, the effect of undesirable events was limited to the same day, whereas the effect of desirable events appeared to last for two subsequent days. Additional analyses explored the possibility that the effects of events on specific sIgA response were mediated through shifts in mood.[63] These analyses were consistent with the hypothesis: effects of both desirable and undesirable events were largely, although not exclusively, mediated through their effects on mood.

SUMMARY

This chapter has highlighted the substantial evidence from field studies and experimental viral exposure studies which suggest an association between psychological stress and URI. Overall, the majority of the studies found significant relations between stress and URI. The strength of the cross-sectional field studies is their epidemiological nature: large numbers of subjects were studied for an impressive number of days. Many of these studies focused on URIs in children, an important subject population given the high prevalence of URIs in this age group and their important role as a reservoir of infection. An additional strength of the prospective, longitudinal field studies is their attention to the timing of stress relative to URI. These studies generally show that stress precedes the onset of URI episodes. For example, several studies have found a significant increase in minor stressors a few days before the onset of URI symptoms. Of course, it is difficult to confirm infection in field studies. Exquisite control of viral exposure and intensive monitoring of symptoms (including objective measures such as mucosal weight) are strengths of the viral exposure studies. However, it must be noted that stress was not explicitly manipulated in these studies. The isolation of subjects that is required for these studies could be considered a stress-reducing intervention, as there may be fewer minor stressors compared to those experienced in the outside world. From the viewpoint of the psychological independent variable, these viral inoculation studies are not true experimental designs. As with the field studies, the relations between psychological stress and URI have been correlational. Nevertheless, the viral exposure studies represent the strongest evidence that stress can affect infection as well as symptoms of infection. It is particularly noteworthy that two independent studies have raised the possibility that the *symptoms* of URI may be affected by different types of psychological stress than the *incidence* of infection; these apparently selective effects can be best investigated in further viral inoculation studies.

We outlined several ways that psychological stress could affect URI, including influences on viral exposure, initial host susceptibility, viral replication, the clinical syndrome, recovery mechanisms, and the enduring protective immune response. Although there are many possible mechanisms by which stress could affect URIs, few have been examined. This focus is quite appropriate from the standpoint of public health issues. Regardless of the underlying biological mechanisms, significant effects of stress on the symptom syndrome in infected individuals could have a major impact on societal costs. Initial evidence indicates that stress may play a role in determining which infected individuals develop symptoms. Recall that nearly a third of those infected do not develop the clinical syndrome of a cold. These symptom-free individuals are unlikely to seek medical attention or to purchase medications. From a public health perspective, psychological effects on the incidence of infection are also important, of course. In addition to increasing the number of individuals with symptoms (who would then seek treatment), increased rates of infection due to stress could increase the pool of contagious individuals, thus furthering the spread of the virus throughout the population.

Although one can cautiously conclude from the literature that there is an association between stress and URI, a number of issues remain to be investigated before we can fully understand the relations. Five of these were highlighted in the review by Cohen and Williamson.[3] (1) *Little is known about how the timing of the stressor and viral exposure influence the development of URI.* Some evidence suggests that stress occurring after viral exposure may increase susceptibility, whereas mixed findings (increased and decreased susceptibility) have been reported when stress occurs prior to viral exposure. Temporal relations clearly go uncontrolled in naturalistic studies, and are more difficult to determine, as no information is available about the timing of subjects' exposure to virus. Such temporal issues could be addressed in viral exposure trials. (2) *Little is known about differential effects of acute (shorter-term), chronic (longer-term), and repetitive stressors.* The health psychology literature suggests that chronic stress may have particularly pernicious effects on health, so distinctions among the duration and patterns of stress should be researched for their effects on URI. However, additional conceptual work may be needed to define stressors according to a chronicity dimension. (3) *True experimental studies of stress exposure should be conducted (namely, where subjects are randomly assigned to stressor conditions).* Although viral exposure studies appear to be experimental (because they have a high degree of experimenter control over the situation), as mentioned above, they generally are not because the independent variable, stress, is not manipulated. A number of experimental stress procedures have been developed and used extensively in cardiovascular reactivity and psychoendocrinology research. These procedures could easily be adapted for use in conjunction with viral exposure trials, and would increase confidence in the causal processes suggested by observational studies. (4) *Little is known about the biological pathways that may be responsible for the association between stress and URI.* We have outlined several points in the process from viral exposure through resulting protective memory response after the infection where stress could affect URI. As we noted above, very few of these have been researched. (5) *Biological mechanisms underlying associations between psychological stress and URI should be investigated in viral exposure studies.* Control over extraneous factors (e.g., smoking, alcohol consumption, exercise) that may affect URI can be controlled or at least adequately assessed in such studies, which should aid in the detection of mechanisms. (6) *The results obtained from the experimental, viral inoculation studies should be used as sources of hypotheses that can be tested in real world settings.* Although experimental infection studies lend themselves to the investigation of psychobiological mechanisms involved in URI, it is critical to establish their relevance outside the laboratory. As with studies of cardiovascular reactivity, it is important to demonstrate the clinical significance of laboratory results. Field studies will also be necessary to determine the relative importance of psychobiological relations in comparison to other risk factors in the community. (7) *The impact of psychosocial interventions designed to capitalize on results obtained from the experimental, viral inoculation studies should be assessed in real world settings.* The effectiveness of psychosocial interventions in ameliorating a number of health problems (e.g., cardiovascular disease, cancer) is increasingly being investigated. If an appropriate psychosocial intervention were found to result in reduced stress and reduced URI, one could make a compelling case for causal relationships in a field study. Such intervention studies would also be informative with regard to the public health significance of psychosocial effects in URI.

REFERENCES

1. **Boyce, W. and Jemerin, J.,** Psychobiological differences in childhood stress response. I. Patterns of illness and susceptibility, *Developmental and Behavioral Pediatrics*, 11, 86, 1990.

2. **Jemerin, J. and Boyce, W.,** Psychobiological differences in childhood stress response. II. Cardiovascular markers of vulnerability, *Developmental and Behavioral Pediatrics*, 11,140, 1990.

3. **Cohen, S. and Williamson, G.,** Stress and infectious disease in humans', *Psychological Bull.*, 109,54, 1991.

4. **Peterson, P., Chao, C., Molitor, T. W., Murtaugh, M., Strgar, F. and Sharp, B.,** Stress and pathogenesis of infectious disease, *Reviews of Inf. Dis.*, 13,710, 1991.

5. **Weiner, H.,** *Perturbing the organism. The biology of stressful experience,* Chicago, The University of Chicago Press, 1992.

6. **Cohen, S., Doyle, W., Skoner, D., Fireman, P., Gwaltney, J. Jr. and Newsom, J.,** State and trait negative affect as predictors of objective and subjective symptoms of respiratory viral infections, *J. Personality and Soc. Psychol.*, 68,159, 1995.

7. **Stone, A.,** Measures of affective response, in S. Cohen, R. Kessler & L. Gordon, Eds., *Measuring stress: A guide for health and social scientists* [cfs], Oxford, Oxford University Press., 1995, pp.148-171.

8. **Lazarus, R. and Folkman, S.,** *Stress, appraisal and coping,* Springer, New York, 1984.

9. **Couch, R.,** Respiratory diseases, in *Antiviral agents and viral diseases of man*, G. Galasso, R. Whitley and T. Merigan, Eds., Raven Press, New York, 1990, pp.327-372.

10. **Lowenstein, S. and Parrino, T.** Management of the common cold, *Adv. in Int. Med.*, 32,207, 1987.

11. **Springer, T.,** Stalking the cold trail, *The New Republic*, October 29, 17, 1990.

12. **Hall, C. and McBride, J.** Upper respiratory tract infections: The common cold, pharyngitis, croup, bacterial tracheitis and epiglottits, in *Respiratory infections: Diagnosis and management*, Third edition, J. Pennington, Ed., Raven Press, New York, 1994, pp.101-124.

13. **Gwaltney, J. Jr.,** Rhinoviruses, in *Viral infections of humans*, Third edition, A. Evans, Ed., Plenum Medical Book, New York, 1989, pp.593-615.

14. **Jackson, G., Dowling, H., Spiesman, I. and Boand, A.,** Transmission of the common cold to volunteers under controlled conditions, *A.M.A. Arch. Int. Med.*, 101,267, 1958.

15. **Couch, R.,** Rhinoviruses, in *Virology*, B. Fields, D. Knipe, R. Chanock, J. Melnick and B. Roizman, Eds., Raven Press, New York, 1985, pp.795-816.

16. **Tyrrell, D., Cohen, S. and Schlarb, J.,** Signs and symptoms in common colds, *Epidemiol. and Infect.*, 111,143, 1993.

17. **Ewald, P.,** Evolutionary biology and the treatment of signs and symptoms of infectious disease, *J. Theoretical Biol.*, 86,169, 1980.

18. **Hendley, J., Edmondson, W. and Gwaltney, J. Jr.,** Relation between naturally acquired immunity and infectivity of two rhinoviruses in volunteers, *J. Infect. Dis.*, 125,243, 1972. 1973, June 28

19. **Barclay, W., al-Nakib, W., Higgins, P. and Tyrrell, D.,** The time course of the humoral immune responses to rhinovirus infection, *Epidemiol. Infect.*, 103,659, 1989.
20. **Brandtzaeg, P.,** Humoral immune response patterns of human mucosae: Introduction and relation to bacterial respiratory tract infections, *J. Infect. Dis.*, 165,S167, 1992.
21. **Kaliner, M.,** Human nasal host defense and sinusitis, *J. Aller. and Clin. Immunol.*, 90,424, 1992.
22. **Tyrrell, D.,** A view from the common cold unit, *Antiviral Res.*, 18,105, 1992.
23. **Bienenstock, J., Croitoru, K., Ernst, P. and Stanisz, A.,** Nerves and neuropeptides in the regulation of mucosal immunity, *Adv. in Exper. Med. and Biol.*, 257,19, 1989.
24. **Proud, D., Gwaltney, J. Jr., Hendley, J., Dinarello, C., Gillis, S. and Schleimer, R.,** Increased levels of interleukin-1 are detected in nasal secretions of volunteers during experimental rhinovirus colds, *J. Infect. Dis.*, 169,1007, 1994.
25. **Turner, R.,** The role of neutrophils in the pathogenesis of rhinovirus infections, *Pediat. Infect. Dis. J.*, 9,832, 1990.
26. **Naclerio, R., Proud, D., Lichtenstein, L., Kagey-Sobotka, A., Hendley, J. and Gwaltney, J. M.,** Kinins are generated during experimental rhinovirus colds, *J. Inf. Dis.*, 157,133, 1988.
27. **Proud, D., Naclerio, R., Gwaltney, J. Jr. and Hendley, J.,** Kinins are generated in nasal secretions during natural rhinovirus colds, *J. Infect. Dis.*, 161,120, 1990.
28. **Proud, D. and Kaplan, A.,** Kinin formation: Mechanisms and role in inflammatory disorders, *Ann. of Rev. Immunol.*, 6,49, 1988.
29. **Igarashi, Y., Skoner, D., Doyle, W., White, M., Fireman, P. and Kaliner, M.,** Analysis of nasal secretions during experimental rhinovirus upper respiratory infections, *J. Aller. and Clin. Immunol.*, 92,722, 1993.
30. **Baraniuk, J., Lundgren, J., Mizoguchi, H., Peden, D., Gawin, A., Merida, M., Shelhamer, J. and Kaliner, M.,** Bradykinin and respiratory mucous membranes, *Amer. Rev. Resp. Dis.*, 141,706, 1990.
31. **Raphael, G., Baraniuk, J. and Kaliner, M.,** How and why the nose runs, *J. Aller. and Clin. Immunol.*, 87,457, 1991.
32. **Levandowski, R., Weaver, C. and Jackson, G.,** Nasal-secretion leukocyte populations determined by flow cytometry during acute rhinovirus infection, *J. Med. Virol.*, 25,423, 1988.
33. **di Giovine, F. and Duff, G.,** Interleukin 1: The first interleukin, *Immunol. Today*, 11,13, 1990.
34. **Kenney, J., Baker, C., Welch, M. and Altman, L.,** Synthesis of interleukin-1 alpha, interleukin-6, and interleukin-8 by cultured human nasal epithelial cells, *J. Aller. and Clin. Immunol.*, 93,1060, 1994.
35. **Hsia, J., Sztein, M., Naylor, P., Simon, G., Goldstein, A. and Hayden, F.,** Modulation of thymosin alpha-1 and thymosin beta-4 levels and peripheral blood mononuclear cell subsets during experimental rhinovirus colds, *Lymphokine Res.*, 8,383, 1989.
36. **Hsia, J., Goldstein, A., Simon, G., Sztein, M. and Hayden, F.,** Peripheral blood mononuclear cell interleukin-2 and interferon-V production, cytotoxicity, and antigen-stimulated blastogenesis during experimental rhinovirus infection, *J. Infect. Dis.*, 162,591, 1990.

37. **Levandowski, R., Pachucki, C. and Rubenis, M.,** Specific mononuclear cell response to rhinovirus, *J. Infect. Dis.*, 148,1125, 1983.
38. **Skoner, D., Whiteside, T., Wilson, J., Doyle, W., Herberman, B. and Fireman, P.,** Effect of rhinovirus 39 infection on cellular immune parameters in allergic and nonallergic subjects, *J. Aller. and Clin. Immunol.*, 92,732, 1993.
39. **Cohen, S. and Wills, T. A.,** Stress, social support and the buffering hypothesis, *Psychol. Bull.*, 98,310, 1985.
40. **Ader, R., Felten, D. and Cohen, N.,** *Psychoneuroimmunology*, 2nd Edition, San Diego: Academic Press, 1991.
41. **Kiecolt-Glaser, J. and Glaser, R.,** Methodological issues in behavioral immunology research with humans, *Brain, Behav., and Immunity*, 2,67, 1988.
42. **Stone, A. and Bovbjerg, D.,** Stress and humoral immunity: A review of the human studies, *Adv. in Neuroimmunol.*, 4,49, 1994.
43. **Graham, N., Woodward, A., Ryan, P. and Douglas, R.,** Acute respiratory illness in Adelaide children. II., The relationship of maternal stress, social supports and family functioning, *Internatl. J. Epidemiol.*, 19,937, 1990.
44. **Clover, R. T., Becker, L., Crawford, S. and Ramsey, C.,** Family functioning and stress as predictors of influenza B infection, *J. Fam. Prac.*, 28,535, 1989.
45. **Roghmann, K. and Haggerty, R.,** Daily stress, illness, and use of health services in young families, *Pediat. Res.*, pp. 520-526, 1973.
46. **Boyce, W., Jensen, E., Cassel, J., Collier, A., Smith, A. and Ramey, C.,** Influence of life events and family routines on childhood respiratory tract illness, *Pediatrics*, 60,609, 1977.
47. **Kanner, A. D., Coyne, J. C., Schaefer, C. and Lazarus, R.,** Comparison of two modes of stress measurement: Daily hassles and uplifts vs. major life events, *J. Behav. Med.*, 4,1, 1981.
48. **Stone, A. Jandorf, L., and Neale, J.,** Trigger or aggravators of symptoms? *Social Science and Med.*, 22,1015, 1986.
49. **Stone, A., Reed, B. and Neale, J.,** Changes in daily events frequency precede episodes of physical symptoms, *J. Human Stress,* 13,70, 1987.
50. **Stone, A., Kessler, R. and Haythornthwaite, J.,** Measuring daily events and experiences: decisions for the researcher, *J. Personality*, 59,575, 1991.
51. **Evans, P. D., Pitts, M. K. and Smith, K.,** Minor infection, minor life events, and the four day desirability dip, *J. Psychosomatic Res.*, 32,533, 1988.
52. **Evans, P. and Edgerton, N.,** Life events and mood as predictors of the common cold, *Brit. J. Med. Psychol.*, 64,35, 1991.
53. **Stone, A., Porter, L. and Neale, J.,** Daily events and mood prior to the onset of respiratory illness episodes: A nonreplication of the 3-5 day "desirability dip," *Brit. J. Psychol.*, 66,383, 1993.
54. **Brunswick, E.,** *Systematic and representative design of psychological experiments.* Berkeley and Los Angeles: University of California Press, 1949.
55. **Totman, R., Reed, S. and Craig, J.,** Cognitive dissonance, stress and virus-induced common colds, *J. Psychosomatic Res.*, 21,55, 1977.
56. **Totman, R., Kiff, J., Reed, S. and Craig, J.,** Predicting experimental colds in volunteers from different measures of recent life stress, *J. Psychosomatic Res.*, 24,155, 1980.

57. **Brown, G. W. and Harris, T.,** *Social origins of depression: A study of psychiatric disorder in women,* Wiley, New York, 1978.
58. **Cohen, S., Tyrrell, D. and Smith, A.,** Psychological stress and susceptibility to the common cold, *New Engl. J. Med.,* 325,606, 1991.
59. **Cohen, S., Tyrrell, D. and Smith, A.,** Negative life events, perceived stress, negative affect and susceptibility to the common cold, *J. Personality and Soc. Psychol.,* 64,131, 1993.
60. **Stone, A., Bovbjerg, D., Neale, J., Napoli, A., Valdimarsdottir, H., Cox, D., Hayden, F., and Gwaltney, J. Jr.,** Development of common cold symptoms following experimental rhinovirus infection is related to prior stressful life events, *Behav. Med.,* 18,115, 1992.
61. **Stone, A., Neale, J., Cox, D., Napoli, A., Valdimarsdottir, H. and Kennedy-Moore, E.,** Daily events are associated with a secretory response to an oral antigen in humans, *Hlth. Psychol.,* 13,440, 1994.
62. **Stone, A., Cox, D., Valdimarsdottir, H., Jandorf, L. and Neale, J.,** Evidence that secretory IgA antibody is associated with daily mood, *J. Personal and Soc. Psychol.,* 52,988, 1987.
63. **Stone, A., Marco, C., Cruise, C., Cox, D. and Neale, J.,** Are stress-induced immunological changes mediated by mood? A closer look at how both desirable and undesirable daily events influence sIgA antibody, State University of New York at Stony Brook, 1995 (submitted).

Chapter 11

THYMIC HORMONES, VIRAL INFECTIONS AND PSYCHONEUROIMMUNOLOGY

N. Trainin
Department of Cell Biology
The Weizmann Institute of Science
Rehovot, Israel

M. Pecht
Department of Organic Chemistry
The Weizmann Institute of Science
Rehovot, Israel

Y. Burstein
Department of Microbiology and Immunology
Faculty of Health Sciences
Ben Gurion University of the Negev
Beer-Sheva, Israel

B. Rager-Zisman
Department of Microbiology and Immunology
The Weizmann Institute of Science
Rehovot, Israel

THE THYMIC HORMONE: THF-γ2 AS A MODULATOR OF CELL MEDIATED IMMUNE RESPONSES

The thymus gland plays a key role in the development of the immune system during embryonic and early life in mammals. It is well established that neonatal thymectomy induces peripheral lymphopenia accompanied by general atrophy of the lymphoid tissues and damage of the immune response.[74] In addition, more recent evidence suggested that the thymus also plays an active role in the maintenance of the immune system. Precursor cells migrate from the bone marrow during the neonatal period throughout adulthood to carry out this function.[65]

Reconstitution of the resultant profound deficit in humoral and cell-mediated immunity by thymus grafts in cell impermeable diffusion chambers[32] or by injection of thymic extracts[71] provided convincing evidence for the existence of thymic hormones.

One of these hormones, Thymic Humoral Factor THF-γ2 isolated in our laboratory, has been purified to homogeneity from calf thymus dialysate, chemically characterized and prepared by chemical synthesis.[11] The synthetic product contains essentially all the biological properties of the natural hormone. THF-γ2 is an octapeptide with a molecular weight of 918

Daltons and has the following amino acid sequence: LEU-GLU-ASP-GLY-PRO-LYS-PHE-LEU. This sequence lacks homology to the published sequence of any other characterized thymic hormone, such as serum thymic factor,[48] thymopentin and thymopoetin[2,18] or the thymosins.[33] THF-γ2 participates in T-cell differentiation and proliferation in all three lymphoid cell compartments: the bone marrow, the thymus gland and the peripheral lymphoid system. Some of these functions have been shown only *in vitro* while others are manifested also *in vivo* thus hinting at its physiological role.[75a]

Treatment of murine bone marrow cells with THF-γ2 *in vitro* was found to increase their capacity to proliferate into granulocyte-macrophage colonies (GM-CFC) in the presence of suboptimal concentrations of colony stimulating factor (CSF). THF-γ2 was neither able to replace CSF as an inducer nor did it induce IL-6 activity in bone marrow cells. Serial injections of THF-γ2 into NTx mice repaired the damage caused to myeloid progenitors, and in parallel increased the percentage of Thy1+ cells in their spleens.[47] A similar effect of THF-γ2 on the proliferative capacity of bone marrow cells was found also *in vitro* in normal and diseased humans.[3] THF-γ2 augmented multiple T-lymphocyte functions in mice, including the proliferative responses to T-cell lectins (Figure 1), mixed lymphocyte reaction (Figure 2), helper effect in antibody production,[45] and cytotoxic responses.[46]

THF-γ2 was found to increase the ConA induced IL-2 production by spleen cells from normal mice[11] and also raised the levels of IL-2 in experimental models of immune impairment such as those induced by neonatal thymectomy[11] and by chemotherapy in murine plasmacytoma.[45,46]

Figure 3 summarizes the restorative effect of THF-γ2 on the immune competence of neonatally thymectomized mice.

Human umbilical cord blood lymphocytes (UCBL) responded to treatment with THF-γ2 by an increased percentage of CD4+ and CD8+ cells and by elevation in the PHA-induced IL-2 production.[6]

It was also found that a single injection of THF-γ2 into immune deficient aging mice led to a rise in the frequency of mitogen responsive T-cell population in their thymus and spleen as well as in IL-2 secretion and helper activity in anti-TNP-SRBC response of primed mice.[19] The above results led us to suggest that THF-γ2 has a modulatory function in stimulating the production and differentiation of T-cells and in regulating the normal balance of the T-cell subpopulations. Preliminary data indicate that UCBL incubated *in vitro* in the presence of THF-γ2 manifest elevated production of CSF, following PHA stimulation, in a dose-dependent manner. On the other hand, IL-6 and TNFα levels in the supernatants of the PHA-activated cells were not affected.

Taken together, it seems that THF-γ2 selectively modulates those lymphokines secreted by type 1 helper T-cells (Th1).

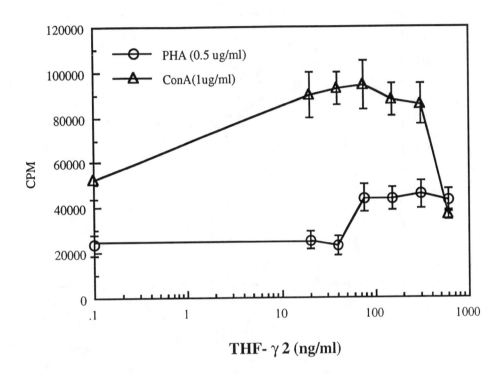

Fig 1. *In vitro* dose-response effect of THF-γ2 on normal Balb/c mitogenic proliferation. Spleen cells from 6-8 wk old Balb/c or C57BL/6 female mice were preincubated at 6×10^5/well (Falcon flat-bottomed 96-well plates) with various doses of THF-γ2, in 150 μl RPMI-1640 (with penicillin 100 U/ml, streptomycin 100 μg/ml and 2 mM glutamine), at 5% CO_2, 37°C for 18 hr. Phytohemagglutinin (PHA) or Concanavalin A (ConA) were then added in 50 μl RPMI supplemented with FCS (2.5% final concentration). Cultures were incubated for 48 hr and pulsed with 1 μCi ^{3}H-thymidine for the last 4 hr. The plates were then harvested on glass-filter paper and counted in a scintillation counter. Results were expressed as counts per minute (cpm) mean + SD of 6 replicates. CPM of cells in medium: 301 ± 160 of cells with THF-γ2 (300 ng/ml) without added mitogen: 288 ± 62.

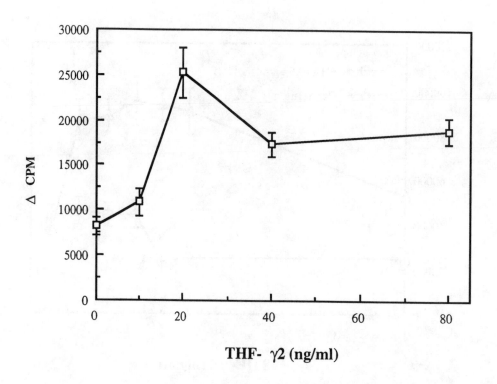

Fig 2.   *In vitro* dose-response effect of THF-γ2 on normal Balb/c mixed lymphocyte culture (MLC).
Spleen or thymus cells from 6-8 wk old Balb/c female mice were preincubated at 1×10^6/well (Falcon flat-bottomed 96-well plates) with various doses of THF-γ2 in 150 μl RPMI-1640 (containing penicillin 100 U/ml, streptomycin 100 μg/ml and 2 mM glutamine) at 5% CO_2, 37°C for 1 hr. Spleen cells from 2-3 month old C57BL/6 female mice, irradiated at 1500 Rad (Co source), were added at 0.5×10^6/well in a final volume of 200 μl. Cultures were incubated for 3 days and pulsed with 1 μCi/well of ³H-thymidine during the final 18 hr. The cultures were then processed as described in the legend to Fig 1. The results are calculated as cpm (mean ± SD) of 6 replicates of each allogeneic and syngeneic response and expressed as Δcpm (Allogeneic minus Syngeneic).

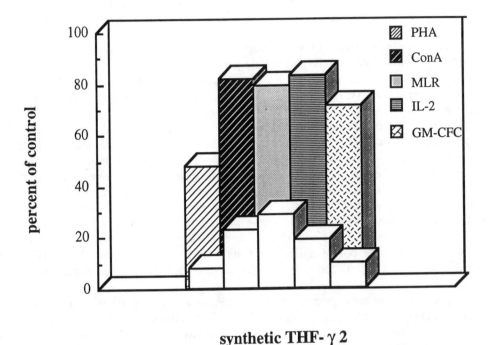

synthetic THF-γ 2

Fig 3. Restoration of the immune competence of neonatally thymectomized mice by systemic treatment with THF-γ2. Four-week old neonatally thymectomized (NTx) Balb/c mice were injected (i.p.) daily with THF-γ2 (4-400 ng/Kg) or saline (total 14 injections). Spleen cells, pooled from four mice, were incubated in quadruplicates in microcultures with PHA (2 μg/ml) or ConA (4 μg/ml) or with irradiated spleen cells of C57BL mice, at 5% CO_2, 37°C for 72 hr. Cultures were harvested after a 4 hr or 18 hr pulse (mitogenic or allogeneic response, respectively) with ^{3}H-thymidine and counted in a β-scintillation counter. IL-2 activity in ConA conditioned medium was tested on CTLL as described.[11] Methyl cellulose clonal cell culture for myeloid progenitor cells from bone marrow was done as previously described.[47] Results are expressed as percentage relative to control cultures of intact age-matched mice. Plain columns represent results obtained with NTx mice injected with saline, in the respective assays.

INTERACTION BETWEEN VIRUSES
AND THE IMMUNE SYSTEM

Immune defense against virus infections involves both nonspecific and antigen specific components. Studies on interactions between viruses and the immune system revealed that a large repertoire of specific mechanisms protects the host against infection, diminishes viral spread and promotes recovery. The relevance of any particular mechanism hinges on the nature of the virus in question and on the way the viral antigens or virions are encountered, i.e. the phase of the infection.[38,39]

The specific response to viral antigens is almost entirely T-cell dependent, thus susceptibility to viral infection is particularly associated with T cell dysfunction. T helper (CD4+) cells are pivotal in regulating cell mediated immune responses including that of specific cytotoxic T lymphocytes (CTL), natural killer cells (NK), macrophages, antibody mediated cytotoxicity (ADCC) and the production of antibody by B lymphocytes. The CD4+ cell population is composed of two distinct subsets, T helper 1 (Th1) and T helper 2 (Th2). Th1 cells produce IL-2 and IFNγ and execute cell mediated immune responses, whereas Th2 produce IL-4, IL-5, IL-6 and IL-10 and assist in antibody production for humoral immunity.[41,59]

CD4+ lymphocytes, both helper T cells of the type Th1, and Class II MHC restricted CTL, secrete a similar range of cytokines as MHC class I restricted CD8+ CTL. Antigenic stimulation of CD4+ T cells results in the release of a variety of cytokines for which an antiviral role has been demonstrated. These include IFN α/β and γ, which inhibit viral synthesis, and activate NK cells and macrophages.[49] They produce tumor necrosis factor a (TNFoc), which exhibits potent antiviral activity and can act synergistically to IFNγ.[54] In addition, transforming growth factor β (TGFβ) has also been shown to exert anti-viral activity *in vitro*.[68]

Recovery from most primary viral infections is largely dependent on classical class I restricted, CD8+ cytotoxic lymphocytes (CTL).[8] These effectors identify infected cells via receptors that recognize viral peptides only in association with cell surface class I MHC molecules.[70] MHC class I determinants are involved in the presentation of endogenously synthesized viral antigens, while those of class II are involved in the presentation of endogenously fragmented proteins.[35] It is thought that lysis of virus infected cells is effective at controlling infection only if it occurs before virus assembly in the infected cells. Otherwise, cell lysis will promote the release and subsequent spread of mature infectious virus from infected cells and tissues.[76] Therefore, these mechanisms are maximally effective against those viruses such as herpes and influenza, which mature at the cell membrane and insert viral proteins into host cells membrane long before maturation of progeny virions. Although there is circumstantial support for direct cytolysis it is still not clear how the virus infection is halted by CTL. Recent data suggest that CD8+ T cells, which have long been regarded simply as cytotoxic cells, produce different combinations of cytokines which regulate the immune response. The cytokines IFNγ, and perhaps also TNFα, are likely candidates.[28,36]

It has been recently suggested that in addition to recognition and lysis of virus infected cells, another major function of CD8+ T cells, following class I MHC restricted recognition of virus infected target cells, is to focus the antiviral cytokines, such as IFN-γ and TNFα, at the site of virus replication which exert directly and locally their anti-viral activity. For example, it has been shown that the protective bystander effect mediated by soluble CD4+ and CD8+ T cell dependent factors against vaccinia virus was obtained only in localized organs such as the central nervous system.[30]

T cells (CD4+) are also required for most anti-viral antibody responses, which may serve to restrict spread of infection by viral neutralization and limit the extent of viremia which in turn prevents viral dissemination and affords protection against re-infection.[38]

In the process of evolution, vertebrates have developed various mechanisms of self defense, while the invading organisms have found ways to perturb or evade these mechanisms especially the specific reactions of antibody and T cells. Viruses which have developed successful strategies to evade the immune system produce latent or persistent infections.[34] During a persistent infection, there is frequently a reduction in virally encoded surface proteins,[66] which provides a selective advantage for the persisting virus. Such infected cells fail to activate lymphocytes with natural killer activity and they become resistant to immunologic assault by humoral or cell mediated effector mechanisms. EBV replication for instance is suppressed by virus specific CTLs and the virus enters a state in which B cells and cells of the salivary gland epithelium are stably infected. Only a few viral genes are expressed in B cells that are poor sources for the virus specific peptides which are essential for targeting the latently infected cells for killing by cytotoxic cells.[34]

Viral replication can be modulated by antibodies to the viral glycoproteins, which helps to initiate and maintain the persistent state.[64] Many viruses are capable of decreasing the expression of MHC class I genes in virus infected cells, thereby avoiding recognition by CTLs and favoring viral persistence.[10,52,66]

A common viral stratagem is the suppression of host effector immune responses.[38,59] Prominent examples are measles virus, cytomegalovirus (CMV) and other herpes viruses and HIV. With the increasing use of anticancer aggressive chemotherapy, immunosuppressive therapy and radiotherapy, long term immunosuppression became a common clinical phenomenon. As a result, reactivation of chronic virus infection in these patients is seen; in addition these patients become targets to viruses which normally cause a mild illness. Thus, fulminant measles pneumonia and encephalitis or incapacitating adenovirus pneumonitis are common challenges in the pediatric oncology units, while CMV is a major threat to transplant patients.[26,56,58]

An important indirect effect of viruses on the immune system is through the generation of cytokines which enhance class I expression. Overexpression of class I glycoproteins provides restriction elements for cellular proteins, thus recognized by autoreactive CTL. Even more important in the generation of virus induced autoimmunity are the class II MHC antigens. An acute virus infection usually causes some tissue damage eliciting an inflammatory reaction with an influx of lymphokine secreting activated T cells. As a result somatic cells are induced to express class II MHC antigens, thus forming a functional unit capable of presenting "self" antigens, and eliciting a large scale autoimmune response.[63]

To summarize, numerous viruses can perturb immune functions and regulation during an acute or persistent infection. Interest is now focusing on a possible viral etiology through these indirect mechanisms for disorders of the nervous system and for which no cause is known. Immunomodulation of the afflicted host may provide new means of controlling this mode of pathogenic events.

THERAPEUTIC EFFECTS OF THF-γ2 AND OTHER
THYMIC HORMONES IN VIRAL INFECTIONS

As mentioned above, cytotoxic T-cells develop in response to viral infection as specific effector/cytotoxic cells; helper T-cells are essential for the development of an optimal anti-viral response. The first evidence of a direct antiviral effect of THF was reported.[50] It was found that a significant percentage of mice infected by a potentially lethal dose of Sendai virus were salvaged by concomitant administration of THF, as compared to infected untreated controls.

Cytomegalovirus (CMV), a ubiquitous herpes virus responsible for congenital birth defects, mononucleosis, herpes and interstitial pneumonitis,[55,67,75] is also one of the main causes of mortality in AIDS patients. Murine cytomegalovirus (MCMV) has been used as an experimental model for the human virus, since acute infection by both viruses results in depressed immune functions. The immune impairment in infected mice is expressed by a decreased response to the T-cell mitogens PHA and ConA and reduction in ConA-induced IL-2 secretion. Systemic treatment of MCMV infected mice with THF resulted in a reconstitution of the mitogenic responses and IL-2 secretion.[29] As in the infection of humans with CMV, immunosuppression is a prerequisite for the development of a fatal murine CMV infection in adult immunocompetent mice.[57] Adult Balb/c mice resistant to murine CMV become highly susceptible following immunosuppression by cyclophosphamide.

Passive transfers of MCMV immune spleen cells prevented to some extent the development of a fatal disease in the recipient mice. Injections of MCMV-immune-donor mice with THF-γ2 considerably enhanced the therapeutic potential of virus-specific immune cells.[52] These experiments provided direct evidence for the antiviral capacity of THF-γ2 through its immunomodulatory effect on immune T-cells. In a more detailed analysis it was found that the efficacy of passively transferred immune cells is dose dependent and that both CD4+ and CD8+ T-cell subsets are essential in the THF-γ2 mediated restorative effect. It was also confirmed that the passive transfer of MCMV-immune spleen cells did not involve viral DNA, using PCR amplification with a 27 bp DNA oligonucleotide primer selected from MCMV glycoprotein B gene. Analysis of the donor mice after THF-γ2 treatment showed increased levels of MCMV neutralizing antibodies while enhancement of NK activity was transient and lasted only during the early stage of the infection.[52]

In studies performed in humans, THF increased *in vitro* proliferation of human peripheral blood lymphocytes in response to a challenge with Varicella Zoster viral antigen.[73] In addition, THF was capable to reconstitute *in vitro* cell mediated immunity in lymphocytes of immune deficient Subacute Sclerosing Panencephalitis (SSPE) patients.[27] The beneficial results of THF therapy were first described in a group of 27 patients suffering from primary or secondary disseminated viral infections. In the patients with severe infection of the herpes viruses, including herpes zoster and varicella, THF treatment resulted in a rapid decline in fever and regression of the herpetiform and varicella lesions and in an increase in T-cell populations.[25] Randomized controlled trials with THF were performed in African children (Red Cross Memorial Hospital, Cape Town, South Africa) suffering from severe complicated acute measles. The treatment resulted in objective benefits as evidenced by improvement in the ESR and the fall in C-reactive proteins. Fewer secondary complications were observed and the incidence of secondary herpetic infections was reduced. The THF treated children presented an increased CD4/CD8 ratio and a greater lymphoproliferative response to PHA.[5] A controlled trial was also performed in a small group of asymptomatic homosexual males with reduced CD4 cells and other lymphocytic defects. Following THF treatment, a significant increase was observed in total lymphocyte counts and in the mean relative increment of CD3 and CD4 cells compared to the placebo group.[24] In a pilot clinical study with synthetic THF-γ2, a group of patients suffering from various lymphoproliferative disorders, in remission following chemotherapy, resulted in an enhancement in the T-cell subsets and their function.[25]

In addition to THF-γ2, three other chemically characterized thymic hormones with immune-modulating properties[72] have been used in treatment of viral infections.

In a double-blind placebo-controlled clinical study in elderly men, thymosin α1 augmented the anti-influenza antibody response when administered concomitantly with influenza vaccine.[20] Recently thymosin α1 has been tested in the woodchuck model of hepadnavirus.[16] In a pilot trial, thymosin α1, administered to patients with chronic hepatitis B, improved some parameters of the disease including level of serum aminotransferase, peripheral blood lymphocytes, and CD3 and CD4 counts. Furthermore hepatitis B DNA was cleared from serum.[42]

Thymopentin (TP-5), a synthetic pentapeptide derived from the active site of thymopoetin,[18] was used to treat children suffering from relapsing herpes and recurrent infections of the respiratory tract. A progressive increase of CD3+ and CD8+ cells was observed during the treatment.[37] Vaccination against hepatitis B with thymopentin in addition to viral DNA vaccine led to a significant increase in the number of seroconvertng uremic patients.[15] HlV-infected patients suffering from persistent generalized lymphadenopathy have manifested a more stable immunological picture and improvement in subjective symptoms following treatment with thymopentin.[13] A similar improvement in CD4 and blastogenesis to PWM was found in HIV positive drug addicts treated with thymopentin.[31] When thymopentin was administered to HIV-infected subjects, an increase was observed in the percentage of CD4 cells in asymptomatic patients but no such effect was obtained in symptomatic patients or any change in the disease progression.[12] In a model of suckling mice infected with reovirus type 2, and suffering from thymic atrophy, TP-5 enhanced antibody production to SRBC and LPS.[44] Thymic serum factor (FTS) increased or decreased NK activity in cancer patients and in mice, depending on the dose administered, on strain sensitivity and whether the thymus was present.[4] Viruses such as Friend leukemia virus, influenza and reovirus type-2 were found to decrease the level of FTS in the serum, concomitantly with a decrease in cellular immune functions.[14,44,69] Thymic hormones are presently tested in clinical trials against a variety of viruses including HIV, aiming at their introduction to the clinics.

THYMIC HORMONES IN THE NEUROENDOCRINE AXIS

Stress and emotional distress may influence the immune system via the central nervous system and/or endocrine mediation.[1]

When the effect of sound stress on the bone marrow-thymus axis was studied by using the *in vitro* migration assay, a significant reduction was found in the percentage of migrating bone marrow cells towards thymus supernatants separated by NUCLEPORE membrane.[9] The effect of emotional stress due to the loss of a loved object such as an unborn child on some parameters of cell mediated immunity was studied in women who had undergone an abortion. The percentage of T-cells and the response of peripheral blood lymphocytes to the mitogens

PHA and ConA of women accepting or not accepting the loss were compared. It was found that non acceptance of the loss of the fetus, as evaluated by rating scale of depression, guilt feelings and anxiety, was correlated with a significant reduction in the proliferative response to both PHA and ConA while the percent of T-cells remained constant. The most marked immunologic effect was found in those women who had the highest score in the depression scale.[43]

The central nervous system seems to modulate the activity of the immune system by the pituitary-adrenal axis.[21] The activated immune system may on the other hand modulate the central nervous system most probably by secreted molecules and thus close this controlled circuit.[22] Thymosin $\alpha 1$ and $\beta 4$ were found to stimulate hypothalamic pituitary-gonadal and adrenal axis *in vitro*, respectively.[23] Thymopoietin, the biological source of thymopentin, was originally identified by its inhibitory effect on neuromuscular transmission in rats[17] and was found to bind specifically to nicotinic acetylcholine receptor of torpedo fish and human.[40] Electrophysiological experiments have shown changes in the electrical activity of single neurons in the endocrine hypothalamus during an immune response.[7] Such changes may be recorded during the whole period of antibody production in conscious rats.[60] Intracerebroventricular injections of THF-$\gamma 2$ decreased preoptic area/anterior hypothalamus multi-unit electrical activity, increased EEG synchronization,[61] and decreased plasma levels of corticosterone.[62]

In conclusion, the availability of thymic hormones with a defined structure opens the way for a more detailed study of their participation in the afferent and/or efferent pathway of the neuro-endocrine system. Presently, THF-$\gamma 2$ is submitted to extensive double-blind clinical trials in AIDS patients, at different stages of thymus disease, with the aim to establish its therapeutic value.

REFERENCES

1. **Ader, R.,** Ed. *Psychoneuroimmunology,* Academic Press, New York, 1981.
2. **Audhya, T., Schlesinger, D. H. and Goldstein, G.,** Complete amino acid sequences of bovine thymopoietins I,II and III: closely homologous peptides, *Biochemistry,* 20,6195, 1981.
3. **Barak, Y., Hahn, T., Pecht, M., Karov, Y., Berrebi, A., Zaizov, R., Stark, B., Buchner, V., Burstein, Y. and Trainin, N.,** Thymic humoral factor-$\gamma 2$, an immunoregulatory peptide, enhances human hematopoietic progenitor cell growth, *Exp. Hematol.,* 20,173, 1992.
4. **Bardos, P., and Bach J. F.,** Modulation of mouse natural killer cells activity by the serum thymic factor, *Scan. J. Immunol.,* 16,321, 1982.
5. **Beatty, D. W., Handzel, Z. T., Pecht, M., Ryder, C. R., Hughes, J., McCabe, K. and Trainin, N.,** A controlled trial of treatment of acquired immunodeficiency in severe measles with thymic humoral factor, *Clin. Exp. Immunol.,* 56,479, 1984
6. **Ben-Hur, H., Pecht. M., Netzer, L., Borenstein, R., Blickman, I., Burstein, Y. and Trainin, N.,** Immune modulation exerted by thymic humoral factor THF-$\gamma 2$ on T-cell

subsets and IL-2 production of umbilical cord blood lymphocytes, *Immunopharmacol. and Immunotoxicol.*, 12,123, 1990.

7. **Besedovsky, H. O., Sorkin. E., Felix, D. and Haas, H.,** Hypothalamic changes during the immune response, *Europ. J. Immunol.*, 7,325, 1977.

8. **Bloom, B.R. and Rager-Zisman, B.,** Cell-mediated immunity in virus infections, in *Viral Immunology and Immunopathology*, A.L. Notkins, Ed., Academic Press, New York, 1975, pp. 113-136.

9. **Bomberger, C. E. and Haar, J. L.,** Effect of sound stress on the migration of prethymic stem cells, *Ann. N.Y.Acad Sci.*, 540,700, 1988.

10. **Boshkov, L.K., Macen, J. L. and McFadden, G.,** Virus-induced loss of class I MHC antigens from the surface of cells infected with myxoma virus and malignant rabbit fibroma virus, *J. Immunol.*, 146,881, 1992.

11. **Burstein, Y., Buchner, V., Pecht, M. and Trainin, N.,** THF-γ2: Purification and amino acid sequence of immunoregulatory peptide from calf thymus, *Biochemistry*, 27,4066, 1988.

12. **Conant, M. A., Calabrese, L. H., Thompson, S. E., Poiesz, B. J., Rasheed, S., Hirsch, R. L., Meyerson, L. A., Kremer, A.B., Wang, C. C. and Goldstein, G.,** Maintenance of CD4+ cells by thymopentin in asymptomatic HIV-infected subjects: results of a double-blind, placebo-controlled study, *AIDS*, 6,1335, 1992.

13. **Costigliola, P., Ricchi, E., Colangeli, V., Pintori, C., Boni, P. and Chiodo, F.,** Thymo-pentin (TP-5) therapy during lymphadenopathy syndrome (LAS/ARC): preliminary report, *J. Exp. Pathol.*, 3,705, 1987.

14. **Del Gobbo, V., Calio-Villani, N., Garaci, E. and Calio, R.,** PR8 influenza virus infection impairs serum thymic activity levels and thymus-derived immune functions in mice, paper presented at *Boll. 1st. Sieroter*, Milan, Italy, 1985, 64, pp. 207-215.

15. **Ervo, R., Faletti, P., Magni, S. and Cavatorta, F.,** Evaluation of treatments for the vaccination against hepatitis B+ thymopentine, *Nephron*, 61,371, 1992.

16. **Gerin, J. L., Korba, B. E., Cote, P. J. and Tennant, B. C.,** A preliminary report of a controlled study of thymosin α-l in the woodchuck model of hepadnavirus, *Adv. Exp. Med. Biol.*, 312,121, 1992.

17. **Goldstein, G.,** Isolation of bovine thymin: a polypeptide hormone of the thymus, *Nature*, 247,11, 1974.

18. **Goldstein, G., Sheid, M. R., Boyse, E. A., Schlesinger, D. H. and van Wauwe, J.,** A synthetic pentapeptide with biological activity characteristic of the thymic hormone thymopoietin, *Science*, 204,1309, 1979.

19. **Gozo, C., Frasca, D. and Doria, G.,** Effect of synthetic thymic humoral factor (THF-γ2) on T-cell activities in immunodeficient aging mice, *Clin. Exp. Immunol.*, 87,346, 1992.

20. **Gravenstein, S., Duthie, E. H., Miller, B. A., Roecker, E., Drinka, P., Prathipati, K. and Ershler, W. B.,** Augmentation of influenza antibody response in elderly men by thymosin alpha one. A double blind placebo-controlled clinical study, *J. Am. Geriatr. Soc.*, 37(1),1, 1989.

21. **Hall, N. R. and Goldstein, A.L.,** Endocrine regulation of host immunity, in *Immune Modulation Agents and their Mechanisms.*, Fenichel, R. L. and Chirigos, M. A., Eds., Marcel Dekker, New York, 1984, pp. 533-563.

22. **Hall, N. R. S. and O'Grady, M. P.**, Regulation of pituitary peptides by the immune system, *Bioassays,* 11,141, 1989.

23. **Hall, N. R., MacGillis, J. P., Spangelo, B. L., Palaszynski, E., Moody, T. W. and Goldstein, A. L.**, Evidence for a neuroendocrine-thymus axis mediated by thymosin polypeptidase, *Dev. Immunol.,* 17,653, 1982.

24. **Handzel, Z. T., Berner. Y., Segal, O., Burstein, Y., Buchner, V., Pecht, M., Levin, S., Burstein, R., Milcham, R., Bentwich, Z., Ben-Ishai, Z. and Trainin, N.**, Immunoreconstitution of T-cell impairments in asymptomatic male homosexuals by thymic humoral factor (THF), *J. Immunopharmacol.,* 9,166, 1987.

25. **Handzel, Z. T., Burstein, Y., Buchner, V., Pecht, M. and Trainin, N.**, Immunomodulation of T-cell deficiency in humans by thymic humoral factor: from crude extract to synthetic THF-γ2, *J. Biological Response Modifiers,* 9,269, 1990.

26. **Handzel, Z. T., Burstein, Y. and Rager-Zisman, B.**, The effects of thymic humoral factor (THF) on viral infections in humans, *EOS,* 2,68, 1985.

27. **Handzel, Z. T., Gadot, N., Idar, D., Schlesinger, M., Kahana, E., Dagan, R., Levin, S. and Trainin, N.**, Cell mediated immunity and effects of thymic humoral factor in 15 patients with SSPE, *Brain and Devel.,* 5,29, 1983.

28. **Karupiah, G., Blanden, R. V. and Ramshaw, I. A.**, Interferon gamma is involved in recovery of a thymic nude mice from recombinant vaccinia virus/interleukin 2 infection, *J. Exp. Med.,* 172,1495, 1990.

29. **Katorza, E., Pecht, M., Apte, R. N., Benharroch, D., Burstein, Y., Trainin, N. and Rager-Zisman, B.**, Restoration of immunological responses by THF, a thymic hormone, in mice infected with murine cytomegalovirus (MCMV), *Clin. Exp. Immuunol.,* 70,268, 1987.

30. **Kundig, T.M., Hengartner, H. and Zinkernagel, R.M.**, T-cell dependent IFN γ exerts an antiviral effect in the central nervous system but not in peripheral solid organs, *J. Immunol.,* 150,2316, 1993.

31. **Lazzarin, A., Barcellini, W., Uberti-Foppa, C., Borghi, M. O., Franzetti, R., Cinque, P. and Moroni, M.**, Experiences with immunomodulant in HIV infections, *Acta Haematologica,* suppl. 1,84, 1987.

32. **Levey, R.H., Trainin, N. and Law, L. W.**, Evidence for function of thymic tissue in diffusion chambers implanted in neonatally thymectomized mice, *J. Nat. Cancer Inst.,* 31,199, 1963.

33. **Low, T. L. K. and Goldstein, A. L.**, The chemistry and biology of thymosin. II. Amino acid sequence analysis of thymosin α1 and polypeptide β1, *J. Biol. Chem.,* 254,987, 1979.

34. **Marrack, P. and Kappler, J.**, Subversion of the immune system by pathogens, *Cell,* 76,323, 1994.

35. **Martz, E. and Gample, S. R.**, How do CTL control virus infection? Evidence for prelytic halt of herpes simplex, *Viral Immunol.,* 5,81, 1992.

36. **Martz, E. and Howell, D. M.**, CTL: virus control cells first and cytolytic cells second? DNA fragmentation, apoptosis, and the prelytic halt hypothesis, *Immunol. Today,* 10,79, 1989.

37. **Melaranci, C. and Gammaria, P.**, Variations of the immunological parameters and the clinical response in 25 children treated with thymopentin, *Pediatr. Med. Chur.,* 13(6),609, 1991.

38. **Mims, C. A. and White, D. O.,**. *Viral Pathogenesis and Immunology,* Blackwell Scientific Publications, London, 1984, pp.39-158.

39. **Mims, C. A., Playfair, J. H. L., Roitt, I. M., Wakelin. D. and Williams, R.,** *Medical Microbiology,* Mosby, Europe Ltd, England, 11-14.6, 1993.

40. **Morel, E., Vernet-der-Garabedian, B., Raimond, F., Audhya, T. K., Goldstein, G. and Bach, J. F.,** Thymopoietin: a marker of the human nicotinic acetylcholine receptor, *Ann. N. Y. Acad. Sci.,* 540,298, 1988.

41. **Mosmann, T. R. and Coffman, R. L.,** Th1 and Th2 cells: different patterns of lymphokine secretion lead to different functional properties, *Ann. Rev. Immunol.,* 7,145, 1989.

42. **Mutchnick, M. G., Appleman, H. D., Chung, H. T., Aragona, E., Gupta, T. P., Cummings, G. D., Waggoner, J. G., Hoffnagle, J. H. and Shafritz, D. A.,** Thymosin treatment of chronic hepatitis B: a placebo-controlled pilot trial, *Hepatolog,* 14(3),409, 1991.

43. **Naor, S., Assael, M., Pecht, M., Trainin, N. and Samuel, D.,** Correlation between emotional reaction to loss of unborn child and lymphocyte response to mitogenic stimulation in women, *Isr. J. Psychiatry Relat. Sci.,* 20,231, 1983.

44. **Onodera, T., Taniguchi, T., Tsuda, T., Yoshihara, K., Shimizu, S., Sato, M., Awaya, A. and Hayashi, T.,** Thymic atrophy in type 2 retrovirus infected mice: immunosuppression and effects of thymic hormone, Thymic atrophy caused by reo-2, *Thymus,* 18,95, 1991.

45. **Ophir, R., Pecht, M., Rashid, G., Halperin, D., Lourie, S., Burstein,Y., Ben-Efraim, S. and Trainin, N.,** A synthetic thymic hormone, THF-γ2 repairs immunodeficiency of mice cured from plasmacytoma by melphalan, *Intern. J.Cancer,* 45,1190, 1990a.

46. **Ophir, R., Pecht, M., Relyveld, E. H., Burstein, Y., Ben-Efraim, S. and Trainin, N.,** THF-γ2, a synthetic thymus hormone increases effectiveness of combined chemotherapy and immunotherapy against RPC-5 murine plasmacytoma, *Int. J. Immunopharmac.,* 19,751 1990b.

47. **Pecht, M., Lourie, S., Burstein, Y., Zipori, D. and Trainin, N.,** Potentiation of myeloid colony formation in bone marrow of intact and neonatally thymectomized mice by the thymic hormone THF-γ2, *Exp. Hematology,* 21,277, 1993.

48. **Pleau, J. M., Dardenne, M., Blouquite, Y. and Bach, J. F.,** Structural study of circulating thymic factor. A peptide isolated from pig serum. II. Amino acid sequence, *J. Biol. Chem.,* 252,8045, 1977.

49. **Rager-Zisman, B. and Bloom, B. R.,** Natural killer (NK) cells in resistance to virus infected cells, in *Seminars in Immunopathology,* Miescher, P., Ed., Springer-Verlag, Heidelberg, Germany, 1982, pp.397-414

50. **Rager-Zisman, B., Harish, Z., Rotter, V., Yakir, Y. and Trainin, N.,** Treatment of mice infected with Sendai virus with THF, a thymic hormone, in *Advances in Allergology and Immunology,* Oehbling, A. *et al.,* Eds., Pergamon Press, Oxford and New York, 1980, pp. 25-31.

51. **Rager-Zisman, B., Ju, G., Rajan, T. V. and Bloom, B. R.,** Decreased expression of H-2 antigens following acute measles virus infection, *Cellular Immunol.,* 59,319, 1981.

52. **Rager-Zisman, B., Segev, Y., Blagerman, S., Palmon, A., Tel-Or, S., Pecht, M., Trainin, N. and Burstein, Y.,** Thymic humoral factor, THF-γ2, enhances immunotherapy of

murine cytomegalovirus (MCMV) infection by both CD4+ and CD8+ immune T cells, *Immunol. Lett.,* 39,23, 1994.

53.   **Rager-Zisman, B., Zuckerman, F., Benharroch, D., Pecht, M., Burstein, Y. and Trainin, N.,** Therapy of a fatal MCMV infection with THF-γ2 treated immune spleen cells, *Clin. and Exper. Immunol.,* 79,246, 1990.

54.   **Ramsay, A. J., Ruby, J. and Ramshaw. I.,** A case for cytokines as effector molecules in the resolution of virus infection, *Immunology Today,* 14,155, 1993.

55.   **Rapp, F.,** The biology of cytomegalovirus, in *The Herpes Viruses, Vol.* 2, Roizman, B. Ed., Plenum Publishing, New York, 1983, p. 1.

56.   **Reboul, F., Donaldson, S. S. and Kaplan, H. S.,** Herpes-zoster and varicella infections in children with Hodgkin's disease, *Cancer,* 41,95, 1978.

57.   **Reddehase, M. J., Weiland, F., Munch, K., Jonjic, S., Luske, A. and Koszinowski, U. H.,** Interstitial murine cytomegalovirus pneumonia after irradiation. Characterization of cells that limit viral replication during established infection of the lungs, *J. Virol.,* 55,264, 1985.

58.   **Reusser, P., Fisher. L. D., Buchler, C. D., Thomas, E. D. and Meyers, J. D.,** Cytomegalovirus infection after autologous bone marrow transplantation: occurrence of cytomegalovirus disease and effect on engraftment, *Blood,* 75,1888, 1990.

59.   **Rook, G.,** Immunity to viruses, bacteria and fungi, in *Immunology,* 3rd Ed. Roitt, I. M., Brostoff, J. and Male, D. Eds., Mosby, Europe Limited, England, 1993, pp. 15.1-15.22.

60.   **Saphier, D., Abramsky, O., Mor, G. and Ovadia, H.,** A neurophysiological correlate of an immune response, *Ann. N. Y. Acad. Sci.,* 496,354, 1987

61.   **Saphier, D., Kidron, D., Ovadia, H., Trainin, N., Pecht, M., Burstein, Y. and Abramsky, O.,** Neurophysiological challenges in the brain following central administration of immunomodulatory factors, *Isr. J. Med. Sci.,* 24,261, 1988.

62.   **Saphier, D. Ovadia, H. and Abramsky, O.,** Neural responses to antigenic challenges and immunomodulatory factors, *Yale J. Biol. Med.,* 63,109,1990.

63.   **Schattner, A. and Rager-Zisman, B.,** Virus induced autoimmunity, *Rev. Inf Dis.,* 12,204, 1990.

64.   **Schneider-Schaulies, S., Segev, Y., Liebert, U.G., Rager-Zisman, B., Wolfson, M., Isakov, N., Koschel, K. and ter Meulen, V.,** Antibody dependent transcriptional regulation of measles virus in persistently infected neuronal cells, *J. Virol.,* 66,5534, 1992.

65.   **Scolay, R., Smith, J. and Stauffer, V.,** Dynamics of early T cells: prothymocyte migration and proliferation in the adult mouse thymus, *Immunol. Rev.,* 91,129, 1986.

66.   **Southern, P. and Oldstone, M. B. A.,** Medical consequences of persistent viral infection, *New Engl. J. Med.,* 314,359, 1986.

67.   **Shanley, J, D., Pesanti, E. L. and Nugent, K. M.,** The pathogenesis of pneumonitis due to murine cytomegalovirus, *J. Infect. Dis.,* 146,388, 1982.

68.   **Su, C. H., Kimberley, A., Leite, M., Braun, L. and Biron, C. L.,** A role for transforming growth factor β1 in regulating natural killer cells and T lymphocytes proliferative responses during acute infection with lymphocytic choriomeningitis virus, *J. Immunol.,* 147,2717, 1991.

69.   **Tonietti, G., Rossi, G. B., Del Gobbo, V., Accinni, L., Ranucci, A., Titti, F., Premrov, M. G. and Garaci, E.,** Effects of *in vivo* Friend leukemia virus infection on levels of serum thymic factors and on selected T-cell functions in mice, *Cancer Res.,* 43,4355, 1983.

70. **Townsend, A. R., Rothbard, J., Gotch, F. M., Bahadur, G., Wraith, D. and McMichael, A. J.,** The epitopes of influenza nucleoprotein recognized by cytotoxic T lymphocytes can be defined with short synthetic peptides, *Cell*, 44,959, 1986.

71. **Trainin, N. and Linker-Israeli, M.,** Restoration of immunologic reactivity of thymectomized mice by calf thymus extracts, *Cancer Res.*, 27,309, 1967.

72. **Trainin, N., Pecht, M. and Handzel, Z. T.,** Thymic hormones, *Immunol. Today*, 4,16, 1983.

73. **Trainin, N., Rotter, V., Yakir, Y., Levey, R., Handzel, Z. T., Shohat, B. and Zaizov, R.,** Biochemical and biological properties of THF in animal and human models, in *Subcellular Factors in Immunity*, H. Friedman, Ed., *Annals of the New York Academy of Science*, 332, 1979, pp.9-22.

74. **Trainin, N.,** Thymic hormones and the immune response, *Physiol.Reviews*, 54,272, 1974.

75. **Yow, M. D.,** Congenital cytomegalovirus disease: a now problem, *J. Infect. Dis.*, 159,163, 1989.

75a. **Trainin, N.,** Prospects of AIDS therapy by thymic humoral factor, a thymic hormone, *Natl. Immun. Cell Growth Regul.*, 9,155, 1990.

76. **Zinkernagel, R. M. and Althage, A..** Antiviral protection by virus immune cytotoxic T cells: infected virus progeny is assembled, *J. Exp. Med.*, 145,644, 1977.

Chapter 12

HERPES VIRUS INFECTIONS AND PSYCHONEUROIMMUNOLOGY

Susan Kennedy
Department of Psychology
Denison University
Granville, OH

DEDICATION

The author wishes to dedicate this chapter to the memory of Dr. David S. Tuber, psychologist, colleague, and friend, who passed away on March 12, 1995. His spirit and uncompromising enthusiasm for science will never be forgotten, and will always remain an inspiration.

INTRODUCTION

Within the growing literature relating stressful psychological events in humans to changes in endocrine and immune function, there is a wealth of data examining stress-related increases in susceptibility to developing primary herpes virus infections,[1] as well as in the role of psychological factors in predicting recurrent herpes virus infection or reactivation of latent virus.[2,3]

The human herpes viruses are unique, in that following primary infection, an individual will remain infected for life.[4] Although poorly understood, it is believed that replication of the latent virus is normally inhibited by cellular immune mechanisms, and that reactivation represents the inability of cellular immune control mechanisms to keep the virus latent within host cells.[4] This hypothesis is supported by the well-documented finding that patients undergoing immunosuppressive therapy, or patients with immunosuppressive illnesses (e.g., AIDS), show characteristic elevations in serum antibody titers to several herpes viruses (often resulting in disease),[5] and that these antibody titers return to normal levels when cellular immune competency is restored (e.g., when immunosuppressive therapy is terminated).[6] Moreover, in individuals with a healthy cellular immune response, reactivation of latent herpes viruses and accompanying elevations in serum antibody titer level can occur periodically, oftentimes in the absence of clinical symptoms.[5]

This chapter presents recent literature examining the relationship between psychological factors and herpes virus infections in humans, with an emphasis on reactivation of latent viruses and recurrent infection. The first section of the chapter reviews data relating psychological factors to primary herpes virus infection, recurrent infections and reactivation of latent virus. The second section relates the implications of these findings to disease susceptibility and overall health. Although stress-related changes in herpes virus infection have been studied with virtually all of the herpes viruses (including cytomegalovirus and Varicella-zoster virus), this chapter focuses on three human herpes viruses, Epstein-Barr Virus (EBV), herpes simplex Type-1 (herpes labialis; oral herpes; HSV-1) and herpes simplex Type-2 virus (genital herpes; HSV-2).

PSYCHOLOGICAL VARIABLES RELATING TO PRIMARY INFECTION, RECURRENT INFECTION, AND VIRUS REACTIVATION

There are a number of studies that have examined the role of psychological factors in the development of primary herpes virus infections, and in the reactivation of latent herpes virus. Some representative studies are discussed below.

A. EPSTEIN-BARR VIRUS

In an early study, Kasl and colleagues[1] examined psychological "risk factors" associated with the development of a primary infection with EBV, the presumptive agent for infectious mononucleosis. In this prospective study, a group of cadets entering the United States Military Academy who were seronegative for EBV antibody upon arrival were studied over the course of four years. Of the original group of cadets, one-fifth eventually seroconverted, and 25% of those seroconverting exhibited clinical symptoms of infectious mononucleosis. Kasl, *et al.* identified three "risk factors" that were predictive of those individuals most likely to seroconvert and subsequently develop clinical disease. These included poor academic performance, high motivation, and having a father who was an overachiever.

More recently, the relationship between coping style and reactivation of latent EBV has been examined in healthy undergraduates by Esterling and his colleagues.[7,8] In one study, for example, students who engaged in disclosing emotional information in a laboratory task had lower antibody titers to latent EBV than students who did not engage in this task. Moreover, students with repressive coping styles were found to have higher antibody levels to EBV, suggesting poorer cellular control over the latent virus. In a follow-up study, students with middle- or high-anxiety levels on the Taylor Manifest Anxiety Scale were found to have higher antibody levels than those students with low levels of anxiety. Elevated antibody titers were also found in those students scoring high on a measure of defensiveness.

Glaser and his colleagues[9-14] have studied stress-related immune changes in medical students during academic examinations, as well as during a baseline period one month prior to exams. In these studies, students completed a battery of psychological questionnaires assessing anxiety, depression, loneliness, and other psychological variables; blood samples were also collected for immune analysis. During examination periods, changes in several immune measures were found, including decreases in the ability of lymphocytes to respond to mitogens, decreases in the percentages of lymphocyte subsets, decreases in the ability of natural killer (NK) cells to lyse target cells, and decreases in the receptor for interleukin-2 (IL-2). Moreover, examination stress resulted in *increases* in antibody titer for EBV, indicating poorer cellular immune control over keeping the virus latent.

In one study,[10] changes in the antibody titer for EBV VCA (viral capsid antigen) were compared in medical students reporting high levels of loneliness to those who reported low levels of loneliness. Figure 1 depicts changes in antibody titer to EBV in "high loneliness" and "low loneliness" students across three sample points: the first sample was taken at baseline (one month prior to examinations), the second was obtained during examinations, and the third was taken upon the students' return from summer vacation. Seroepidemiological studies have indicated that the usual EBV VCA antibody titer in healthy adults is about 1:80,[23] and that over 90% of the population is seropositive for the virus. Students who self-reported high levels of loneliness had significantly higher antibody titers to EBV VCA at all three sample points than did those reporting low loneliness, despite the fact that titer levels had returned to near-normal levels upon the students' return from summer vacation. Importantly, the elevated antibody titers observed during stressful periods were specific for latent EBV, and were not reflective of all viruses, such as poliovirus type 2, a virus for which most students were seropositive, but for which titers remained unchanged over the same sample points.

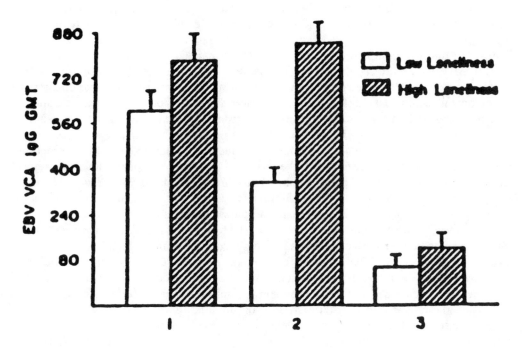

Fig 1. Changes in antibody titer to EBV VCA in "low"- vs "high"-loneliness medical students across three sample points. The first sample was taken at baseline, the second during examinations, and the third upon the students' return following summer vacation. (Taken from Glaser *et al.*, 1985a,[10] reprinted with permission of the publisher).

In subsequent work, Kiecolt-Glaser and her colleagues[16-18] studied stress-related changes in antibody titer to EBV in other stressed populations, including separated/divorced women and men, and spousal caregivers of Alzheimer's Disease patients.

The disruption of a marriage is reported to be one of the most stressful events experienced by an individual, particularly if the separation or divorce is not initiated by the individual.[19,20] It is well-documented, for example, that separated/divorced individuals visit physicians more frequently than do married persons, and have higher rates of acute and chronic illnesses that may limit their daily activities.[21] Given these differences in health-related behaviors, Kiecolt-Glaser *et al.* hypothesized that separated/divorced individuals might also be immunologically impaired. In one study, women who had been separated or divorced within the last year were significantly impaired on several immune parameters, including lower percentages of NK cells, lower percentages of helper T-lymphocytes and lower blastogenic responses of lymphocytes to mitogens when compared to a group of well-matched (i.e., for age, socioeconomic status, length of marriage, etc.) married women. Moreover, separated/divorced women showed average antibody titers to EBV that were 3.5 times higher than those of married comparison subjects, suggesting poorer cellular immunity in keeping the virus in its latent state.

In a follow-up study examining marital disruption in men,[17] differences in EBV titers were also found between separated/divorced and married subjects. In this study, men in the separated/divorced group reported higher levels of distress and loneliness, and reported more illnesses in the two months prior to testing than did married comparison subjects. Antibody titers to EBV VCA were significantly higher, as well, in the separated/divorced group of men. Within the separated/divorced group, however, those men that initiated the separation reported better overall health, less distress, and had *lower* EBV antibody titers than noninitiators; moreover, married men reporting less marital satisfaction had *higher* EBV titers relative to other married subjects, indicating poorer immune control over virus latency.

Taken together, the data on loneliness in medical students and marital disruption on reactivation of latent EBV suggest that the quality of one's interpersonal relationships may be a critical determining factor of one's immune status. In support of this hypothesis, epidemiological evidence suggests that social relationships may be as much of a risk factor for morbidity and mortality as hypertension, smoking, obesity and physical exercise.[22]

Given the relationship between psychosocial factors and changes in immune status, one might speculate that greater impairments in stress-associated immune function would be found in older adults who are already somewhat immunocompromised due to aging.[23,24] Although it is believed that the *numbers* of immune cells do not change with age, it is generally agreed that the functional response of the immune system is significantly impaired in older adults.[14,15] To investigate the effects of a long-term stressor on immune function in an elderly population, Kiecolt-Glaser and her colleagues[18] studied spouses of patients with Alzheimer's Disease who were providing care for their chronically-ill partner. Caregivers were seen initially (intake) and again 13 months later. At the time of the study, caregivers had a mean age of 67 years, and had been caring for their spouse for several years. Antibody titers to EBV VCA were assessed for the caregiver group at both time points, as well as for a group of well-matched non-caregiver comparison subjects. In the comparison subjects, no change was found between titers at intake and at follow-up, although average titer values were well above normal (approximately 1:530 at both time points). In the caregiver group, however, titers were significantly higher at follow-up relative to intake, and compared to the non-caregiver group (average titer values for caregivers were 1:900 at follow-up). In addition to differences in EBV antibody titers, caregivers reported significantly more days of illness than non-caregivers, and were also more depressed than comparison subjects. Importantly, those caregivers reporting lower levels of social support at intake on the Social Support Interview[25] were those most likely to show impaired immune function at follow-up. When one considers the tremendous time demands placed on caregivers and the consequent limitations on the caregiver's social activities and social contacts, it is not surprising to find possible health-related immune changes in these individuals. In light of these findings, the importance of healthy social relationships may be of even greater significance with regard to health status in individuals who are already immunocompromised, such as the elderly. In the final section of this chapter, specific possible health implications of reactivated virus are discussed.

B. HERPES SIMPLEX VIRUSES

1. Herpes Simplex Virus Type-2

Genital herpes (HSV-2) has been described as one of the fastest spreading viruses in the United States.[26] Typically, the virus is transmitted from one individual to another by sexual contact, after which it travels along nerve pathways and enters and remains latent in the sacral ganglia. Reactivation of the virus in an infected individual may occur several times throughout a year,[23,27] although the precise factors responsible for reactivation have not been clearly delineated.[29]

Individuals with genital herpes infections have often been reported to experience significant emotional trauma, particularly with respect to their interpersonal relationships.[30,31] For example, fear and anxiety associated with infecting a sexual partner may limit close interpersonal relationships in these patients.[32] In view of the psychological impact that genital herpes often has on infected individuals, several studies have explored the relationship between psychological variables and herpes recurrences in subjects previously infected with the virus. A representative sample of these studies is presented below.

Taylor[33] retrospectively examined the relationship between stressful events and herpes recurrences, and found that those individuals with more frequent herpes recurrences reported higher stress levels than did those with less frequent recurrences.[1] Differences were found between genital herpes sufferers and controls in terms of frequency of recent stressful events. In a subsequent study, Watson[34] found additional support linking stress with genital herpes recurrences. In this study, volunteers were asked to complete questionnaires assessing locus of control, as well as social relationships and life experiences. In addition to finding an association between recent stressful life events and genital herpes recurrences, it was also found that those subjects with an external locus of control experienced significantly more recurrences than those with an internal locus of control, and that social support was directly and negatively correlated with the number of recurrences.

In order to investigate the relationship between life stress, mood and lymphocyte subsets with the recurrence of genital herpes, Kemeny and her colleagues[35] studied a group of 36 subjects, all of whom had experienced several recurrences of HSV-2 during the six months preceding the study. Nineteen of these subjects were used for immunological analysis. Subjects reporting high levels of life stress were found to have lower percentages of helper T-lymphocytes (CD4+) and suppressor/cytotoxic T-lymphocytes (CD8+); moreover, those individuals with high levels of depressive mood were found to have lower percentages of CD8+ cells. In terms of herpes recurrence, depressive mood over the six month study was found to be predictive of subsequent infection, but only in those subjects not reporting symptoms of other infections. Importantly, the relationship between depressive mood and herpes recurrence was independent of changes in health-related behaviors, such as amount of sleep, alcohol consumption, or exercise level. Collectively, these data suggest the possibility that under normal conditions, CD8+ cells may be critical in maintaining virus latency, and that decreased percentages of these cells (perhaps as a consequence of depressed mood) may increase the likelihood of virus reactivation in previously-infected individuals.[35]

Auerbach and his colleagues[32] have investigated the relationship between recurrent genital herpes and several psychological factors, including emotional dysfunction, stress, style of coping and levels of social support in a group of 66 men and women. Subjects completed a battery of questionnaires, including the Symptom Checklist (SCL),[36] Life Experiences Survey,[37] and Rotter's[33] inventory measuring locus of control. When compared to a control group, subjects scored significantly higher on emotional dysfunction as assessed by the SCL; moreover, this measure was associated with the frequency and pain of herpes recurrences. Duration of the lesion was associated with higher levels of reported life stress, as well as the tendency not to employ cognitive coping strategies (as well as the use of wishful thinking as a coping mechanism) to deal with the stress of herpes. Similarly, patients with more frequent recurrences of herpes lesions scored higher on a measure of external locus of control, and tended not to employ cognitive coping strategies to deal with herpes-associated stress. Although social support was found to be unrelated to herpes symptomatology in this study, it *was* found to be inversely related to levels of psychopathology in the sample. Thus, subjects reporting low social support levels scored significantly higher on several of the SCL subscales, including depression and GSI (General Severity Index). These latter data offer additional evidence regarding the importance of healthy social relationships in maintaining virus latency.[13-16] In view of these data, it is somewhat of an irony that individuals infected with HSV-2 often experience difficulty in their interpersonal relationships, including intimate relationships, as well as relationships with friends and family.[30] In fact, difficulties with interpersonal relationships has been reported by a sample of HSV-2 patients as a primary stressor related to disease recurrence.[30]

The role of social support has been the focus of several other studies examining stress-related activation of herpes virus. For example, VanderPlate, Aral and Magder[39] examined herpes-specific social support and virus reactivation, and found that social support was inversely related to stress and the number of lesion recurrences, but only when support levels were low. No relationship between stress and herpes recurrences was found when social support levels were high, or when *general* levels of social support were considered.

Despite the data relating psychological factors to genital herpes recurrences, there are a number of studies that have failed to find significant correlations between these variables. For example, not all studies have found social support to be a "positive modulator" of herpes recurrence. Hoon *et al.*[40] found that social support *increased* the likelihood of lesion recurrence in a sample of over 140 individuals, possibly due to the additional stress created by feelings of pressure to maintain social commitments. Psychological stress, although not specifically related to herpes recurrence, may predispose individuals to general illness, which might result in herpes recurrence.[40] Similarly, in a prospective study of over 60 patients with genital herpes, no relationship was found between daily stress and herpes recurrence.[33] These latter data are in accordance with previous reports which failed to find evidence of a direct association between recurrent genital herpes lesions and self-reported psychological stress,[2] despite reports from patients and physicians that stress is often a precipitating factor in the appearance of genital lesions.[41] However, methodological weaknesses, including lack of certain statistical controls, make these studies difficult to interpret.[42]

2. Herpes Simplex Virus Type-1

Although it is estimated that up to 80 percent of the population is antibody positive for HSV-1, only about half of infected individuals experience recurrent oral lesions.[43] Following primary infection, the virus remains latent within the trigeminal ganglia, but may be reactivated, during which time the virus migrates from the trigeminal along peripheral nerve fibers, and may result in an oral lesion. In addition to the literature on HSV-2 and its relationship to psychological variables, stress-related activation of herpes labialis (HSV-1) has also been examined in several populations.

In one study,[44] several stated measures, including anxiety, stressful life events, frustrations and daily hassles were found to be significantly increased in the week prior to the recurrence of oral lesions in a sample of 18 subjects who had experienced previous episodes of HSV-1. The stressful life events that were ranked by subjects as most associated with the onset of the lesion included problems with personal relationships, including problems with a spouse/significant other, friends or family member, and job-related stress, such as a change in job, or problems with co-workers.

Using a prospective design, Schmidt and his colleagues[45] examined stress associated changes in HSV-1 recurrences and cellular immunity in a subject with a history of recurrent oral lesions, as well as in an individual who was antibody-positive for HSV-1, but who did not manifest lesions. Each subject completed weekly questionnaires for the duration of the 32-week study, and had weekly blood samples taken for quantitative analysis of lymphocyte subsets. For both subjects, an inverse relationship between reported stress level and percentages of helper/inducer T lymphocytes was found; for the subject with a history of HSV-1, however, the numbers of these cells were at their lowest levels in the beginning stages of lesion recurrence. These data concur with several earlier reports documenting changes in immunological parameters during recurrence of HSV-2,[46] and with reports by patients reporting the appearance of oral lesions shortly after experiencing a stressful situation.[44]

The role of "commonplace" stressors and more chronic stressors in reactivation of HSV-1 has been studied by Glaser and Kiecolt-Glaser and their colleagues.[9-14] In these studies, medical students completed questionnaires at several time points throughout the academic year that corresponded to exam (stressful) time points, and to baseline time points approximately one month prior to examinations. Although no overt symptoms of oral lesions were found, significant changes in antibody titers to HSV-1 were reported in the sample during examinations, with titers at their lowest levels upon the students' return from summer vacation.[10] Similarly, in a study by Kiecolt-Glaser et al.[17] antibody titers to HSV-1 were approximately ten times higher in a sample of separated/divorced men relative to a group of married comparison subjects, suggesting poorer cellular immune control over virus latency.

There are, however, several conflicting reports in the literature regarding psychological factors and reactivation of HSV-1. Ship and his associates,[47] for example, examined emotion and personality profiles of 108 HSV-1 positive subjects and 108 comparison subjects who were antibody negative for the virus. No significant differences between the two groups were found on emotional measures, or on personality characteristics, suggesting that virus reactivation was not associated with certain predisposing traits of an individual.

Likewise, Luborsky[2] found no significant relationship between daily self reported mood scores and physician-diagnosed oral herpes recurrences over three months in a sample of 16 nursing students, although students reported that recurrent lesions were almost always associated with stressful experiences. These latter data present an important methodological consideration; namely, that subjects' retrospective reports of stress may not be accurate indicators of parallel physical symptoms.

PSYCHOLOGICAL FACTORS AND HERPES VIRUS INFECTION: IMPLICATIONS FOR HEALTH

This chapter has presented studies that examine the relationship between psychological variables and changes in susceptbility to herpes virus infections, and reactivation of latent virus. Although the precise mechanism of virus reactivation is not well understood, it is believed that the cellular immune response is critical in maintaining virus latency, and that virus reactivation may result from a suppressed cellular immune response due to psychological stress.

Although a direct causal relationship between psychological factors and illness has not yet been firmly established in humans, one might speculate on possible health implications of psychologically-mediated immune changes with regard to reactivation of latent herpes viruses. For example, although most commonly associated with cold sores, HSV-1 infections may also produce more serious conditions, including encephalitis;[47] HSV-1 infections may also result in death.[47]In addition to being the presumptive agent for infectious mononucleosis, EBV is also associated with African Burkitt's lymphoma and nasopharyngeal carcinoma;[48] moreover, infections with EBV may result in subsequent neuropathies, including Guillain-Barre' syndrome and Bell's Palsy.[49]

Given the well-documented health risks associated with herpes virus infections and the evidence linking psychological and psychosocial factors to changes in virus latency, it is plausible that stress may increase one's susceptibility to illness by down-regulating certain components of the immune system, including cellular immune mechanisms that are critical in maintaining virus latency. Those who might be at a particularly high risk for developing illness include individuals who are already immunocompromised, such as the elderly, AIDS patients, and patients receiving chemotherapy.[50] Conceivably, psychological stress might also affect the time course and/or severity of illness via its effects on immune cells,[50] although more research is needed before this can be determined with certainty.

In light of the reported negative effects that psychological stress may have on herpes recurrence, several investigators have examined the possible ameliorative effects of various intervention techniques on herpes virus reactivation. In one study, for example, serum antibody titers to HSV-1 and self reported distress levels were significantly lower, and natural killer (NK) cell activity was significantly higher in a group of geriatric patients given relaxation training three times per week for a month.[51] Titers remained at low levels when assessed at a follow-up point one month following cessation of the intervention, although NK cell activity and self-reported distress were no different from baseline levels at this point. In another study, patients with recurrent genital herpes lesions randomly assigned to a psychosocial intervention group that included relaxation training, imagery techniques and stress management subsequently reported fewer herpes recurrences, as well as lower levels of stress, relative to a social support intervention group, and a waiting list control group.[52] Similar improvements in genital lesion recurrence have also been found using EMG biofeedback.[53] Taken collectively, these studies offer important additions to the literature on stress-related changes in herpes virus latency in humans, and offer promise for treating herpes infections.

In summary, several studies have found significant correlations between psychosocial factors and reactivation of herpes viruses, including EBV, HSV-1 and HSV-2. Future studies are needed that will more clearly elucidate the possible health changes associated with virus reactivation, and potential techniques that are effective in managing herpes infections.

REFERENCES

1. Kasl, S.V., Evans, A.S., and Neiderman, J.C., Psychosocial risk factors in the development of infectious mononucleosis, *Psychosom. Med.*, 41,445, 1979.

2. Luborsky, L., Mintz, J., Brightman, U.J., and Katcher, A.H., Herpes simplex and moods: A longitudinal study, *J. Psychosom. Res.*, 20,543, 1976.

3. Kiecolt-Glaser, J.K., Ricker, D., Messick, G., Speicher, C.E. and Glaser, R., Urinary cortisol, cellular immunocompetency and loneliness in psychiatric inpatients, *Psychosom. Med.*, 46,15, 1984b.

4. Glaser, R. and Gotlieb-Stematsky, T., *Human Herpesvirus Infections: Clinical Aspects*, Marcel Dekker, New York, 1982.

5. Henle, W. and Henle, G., Epstein-Barr virus and blood transfusions, in *Infection, Immunity and Blood Transfusion*, Dodd, R.Y. and Barker, L.F., Eds., A.R. Liss, New York, 1985.

6. Sekizawa, T., Openshaw, H., Wohlenberg, C., and Notkins, A.L., Latency of herpes simplex virus in absence of neutralizing antibody: Model for reactivation, *Science*, 210, 1026, 1980.

7. Esterling, B.A., Antoni, M.H., Kumar, M., and Schneiderman, N., Emotional repression, stress disclosure responses and Epstein-Barr viral capsid antigen titers, *Psychosom. Med.*, 52,397, 1990.

8. Esterling, B.A., Antoni, M.H., Kumar, M., and Schneiderman, N., Defensiveness, trait anxiety, and Epstein-Barr viral capsid antigen antibody titers in healthy college students, *Health Psych.*, 12,132, 1993.

9. Glaser, R., Kennedy, S., Lafuse, W.P., Bonneau, R.H., Speicher, C.E., and Kiecolt-Glaser, J.K., Psychological stress-induced modulation of IL-2 receptor gene expression and IL-2 production in peripheral blood leukocytes, *Arch. Gen. Psychiatry*, 47,707,1990.

10.   **Glaser, R., Kiecolt-Glaser, J.K., Speicher, C.E., and Holliday, J.E.,** Stress, loneliness, and changes in herpesvirus latency, *J. Behav. Med.,* 8,249, 1985a.

11.   **Glaser, R., Kiecolt-Glaser, J.K., Bonneau, R.H., Malarkey, W., Kennedy, S., and Hughes, J.,** Stress-induced modulation of the immune response to recombinant hepatitis B vaccine, *Psychosom. Med.,* 54,22, 1992.

12.   **Glaser, R., Kiecolt-Glaser, J.K., Stout, J.C., Tarr, K.L., Speicher, C.E., and Holliday, J.E.,** Stress-related impairments in cellular immunity in medical students, *Psychiatry Res.,* 16,233, 1985c.

13.   **Glaser, R., Rice, J., Speicher, C.E., Stout, J.C., and Kiecolt-Glaser, J.K.,** Stress depresses interferon production by leukocytes concomitant with a decrease in NK cell activity, *Behav. Neurosci.,* 100,675, 1986b.

14.   **Kiecolt-Glaser, J.K., Garner, W., Speicher, C., Penn, G.M., Holliday, J.E., and Glaser, R.,** Psychosocial modifiers of immunocompetence in medical students, *Psychosom. Med.,* 46,7, 1984.

15.   **Henle, W., and Henle, G.,** Epstein-Barr virus and infectious mononucleosis, in *Human Herpesvirus Infections: Clinical Aspects,* Glaser, R. and Gotleib-Stematsky, T., Eds., Marcel Dekker, New York, 1982, p.151.

16.   **Kiecolt-Glaser, J.K., Fisher, L.D., Ogrocki, P., Stout, J.C., Speicher, C.E., and Glaser, R.,** Marital quality, marital disruption, and immune function, *Psychosom. Med.,* 49,13, 1987a.

17.   **Kiecolt-Glaser, J.K., Kennedy, S., Malkoff, S., Fisher, L., Speicher, C.E., and Glaser, R.,** Marital discord and immunity in males, *Psychosom. Med.,* 50,213, 1988.

18.   **Kiecolt-Glaser, J.K., Dura, J.R., Speicher, C.E., Trask, O.J., and Glaser, R.,** Spousal caregivers of dementia victims: Longitudinal changes in immunity and health, *Psychosom. Med.,* 53,345, 1991.

19.   **Bloom, B.L., Asher, S.J., and White, S.W.,** Marital disruption as a stressor: A review and analysis, *Psychol. Bull.,* 85,867, 1978.

20.   **Pettit, E.F., and Bloom, B.L.,** Whose decision was it? The effects of initiator status on adjustment to marital disruption, *J. Marriage Fam.,* 46,587, 1984.

21.   **Verbrugge, L.M.,** Marital status and health, *J. Marriage Fam.,* 41,267, 1979.

22.   **House, J.S., Landis, K.R., and Umberson, D.,** Social relationships and health, *Science,* 241,540, 1988.

23.   **Murasko, D.M., Weiner, P., and Kaye, D.,** Association of lack of mitogen-induced lymphocyte proliferation with increased mortality in the elderly, *Aging Immunol. Infect. Dis.,* 1,1, 1988.

24.   **Wayne, S.J., Rhyne, R.L., Garry, P.J., and Goodwin, J.S.,** Cell mediated immunity as a predictor of morbidity and mortality in subjects over sixty, *J. Gerontol. Med. Sci.,* 45, M45, 1990.

25.   **Fiore, J., Becker, J., and Coppel, D.B.,** Social network interactions: A buffer or a stress? *Am. J. Community Psychol.,* 11,423, 1983.

26.   Genital herpes in U.S. called an epidemic, *Richmond Times Dispatch,* 27,13,1982.

27.   **Corey, L., Adams, E.G., Brown, Z.A., and Holmes, K.K.,** Genital herpes simplex virus infections: Clinical manifestations, course, and implications, *Internal Med.,* 98,958,1983.

28. **Knox, S., Corey, L., Blough, H. and Lerner, A.**, Historical findings in subjects from a high socioeconomic group who have genital infections with herpes simplex virus, *Sexually Trans. Diseases*, 9,15, 1982.

29. **Corey, L.**, First-episode, recurrent, and asymptomatic herpes simplex infections, *J. Amer Acad. Dermatol.*, 18,169, 1988.

30. **Keller, M.L., Jadack, R.A., and Mims, L.F.**, Perceived stressors and coping responses in persons with recurrent genital herpes, *Res. Nursing Health*, 14,421, 1991.

31. **VanderPlate, C., and Aral, S.O.**, Psychosocial aspects of genital herpes virus infection, *Health Psych.*, 6,57, 1987.

32. **Silver, P.S., Auerback, S.M., Vishniavsky, N., and Kaplowitz, L.G.**, Psychological factors in recurrent genital herpes infection: Stress, coping style, social support, emotional dysfunction, and symptom recurrence, *J. Psychosom. Res.*, 30,163, 1986.

33. **Taylor, B.J.**, The psychological and behavioral effects of genital herpes in women: High recurrers vs. low recurrers, *Diss. Abstr. Int.*, 39,2529B, 1978.

34. **Watson, D.**, The relationship of genital herpes and life stress as moderated by locus of control and social support, Unpublished manuscript, University of Southern California, 1983.

35. **Kemeny, M.E., Cohen, F., Zegans, L.S., and Conant, M.A.**, Psychological and immunological predictors of genital herpes recurrence, *Psychosom. Med.*, 51,195, 1989.

36. **Derogatis, L.B.**, *The SCL-90 Revised Manual 1*, Baltimore, Johns Hopkins University School of Medicine, 1977.

37. **Sarason, I.G., Johnson, J.G., and Siegel, J.M.**, Assessing the impact of life changes: Development of the life experience survey, *J. Consult. Clin. Psych.*, 46,932, 1978.

38. **Rotter, J.B.**, Generalized expectancies for internal versus external control of reinforcement, *Psychol. Mon.*, 80,609, 1966.

39. **VanderPlate, C., Aral, S.O., and Magder, L.**, The relationship among genital herpes simplex virus, stress, and social support, *Health Psych.*, 7,159, 1988.

40. **Hoon, E.F., Hoon, P.W., Rand, K.H., Johnson, J., Hall, N.R., and Edwards, N.B.**, A psycho-behavioral model of genital herpes recurrence, *J. Psychosom. Res.*, 35,25,1991.

41. **Hamilton, R.**, *The Herpes Book*, Boston: Houghton Mifflin, 1980.

42. **Longo, D., and Koehn, K.**, Psychosocial factors and recurrent genital herpes: A review of prediction and psychiatric treatment studies, *Intl. J. Psychiat. in Med.*, 23,99,1993.

43. **Nahmias, A., Dowdle, W.R., and Schinazi, R. F.**, Eds., *The Human Herpes Viruses: An Interdisciplinary Perspective*, New York: ElsevierNorthHolland, 143, 1981.

44. **Schmidt, D.D., Zyzanski, S., Ellner, J., Kumar, M.L., and Arno, J.**, Stress as a precipitating factor in subjects with recurrent herpes labialis, *J. Fam. Practice*, 20,359, 1985.

45. **Schmidt, D.D., Schmidt, P.M., Crabtree, B.F., Hyun, J., Anderson, P., and Smith, C.**, The temporal relationship of psychosocial stress to cellular immunity and herpes labialis recurrences, *Fam. Med.*, 23,594, 1991.

46. **Sheridan, J., Donnerberg., A., Aurelian, L., and Elpern, D.**, Immunity to herpes simplex virus type 2. IV. Impaired lymphokine production during recrudescence correlates with an imbalance in T lymphocyte subsets, *J. Immunol.*, 129,326, 1982.

47. **Ship, I., Brightman, V.J., and Laster, I.**, The patient with herpes labialis: A study of two population samples, *J. Amer. Dent. Assoc.*, 75,645, 1967.

48. **Adam, E.,** Herpes simplex virus infections, in *Human Herpesvirus Infections: Clinical Aspects*, Glaser, R. and Gotlieb-Stematsky, T., Eds., Marcel Dekker, New York, 1982, 1-56.

49. **Tuckwiller, L.S., and Glaser, R.,** Epstein-Barr virus and nasopharyngeal carcinoma, in *Comparative Respiratory Tract Carginogenesis*, ReznickSchuller, H.M., Ed., CRC Press, Boca Raton, 1983, pp.171-185.

50. **Gotlieb-Stematsky, T., and Glaser, R.,** Association of Epstein-Barr virus with neurologic diseases, in *Human Herpesvirus Infections: Clinical Aspects*, Glaser, R. and Gotlieb--Stematsky, T., Eds., Marcel Dekker, New York, 1982, pp.169-204.

51. **Kiecolt-Glaser, J.K., and Glaser, R.,** Psychosocial influences on herpesvirus latency, in *Viruses. Immunity and Mental Disorders*, Kurstak, E., Lipowski, Z.J., and Morozov, P.V., Eds., Plenum Medical, New York, 1987, pp.403-411.

52. **Kiecolt-Glaser, J.K., Glaser, R., Williger, D., Stout, J., Messick, G., Sheppart, S., Ricker, D., Romisher, S.C., Briner, W., Bonnell, G., and Donnerberg, R.,** Psychosocial enhancement of immunocompetence in a geriatric population, *Health Psych.*, 4,25, 1985a.

53. **Longo, D.J., Clum, G.A., and Yaeger, N.J.,** Psychosocial treatment for recurrent genital herpes, *J. Consult. Clin. Psych.*, 56,61, 1988.

54. **VanderPlate, C., and Kerrick, G.,** Stress reduction treatment of severe recurrent genital herpes virus, *Biofeedback Self Regul.*, 10,181, 1985.

Chapter 13

CERVICAL NEOPLASIA, HUMAN PAPILLOMA VIRUS AND PSYCHONEUROIMMUNOLOGY

Michael H. Antoni
Department of Psychology
University of Miami
Coral Gables, FL

Karl Goodkin
Department of Psychiatry
University of Miami School of Medicine
Miami, FL

INTRODUCTION

Cervical carcinoma is the fifth most common cancer in the world and the second major cause of cancer-related death in women, preceded only by breast cancer. It is diagnosed in more than 420,000 women annually across the world[1] and by some indications is actually increasing among women less than 35 years of age in developed countries.[2] Cervical intraepithelial neoplasia (CIN)--a set of conditions associated with pre-clinical neoplastic changes-- is defined by three levels: CIN I, II, III. Mild (CIN I) and moderate levels (CIN II) involve undifferentiated cells limited to less than 75% of the cervical thickness while severe dysplasia (CIN III) involves undifferentiated cells permeating from 75% to the entire cervical epithelium. Consistent evidence for viral initiation of CIN has been found for Human Papilloma Virus (HPV) infection. The number of cases of cervical dysplasia and carcinoma *in situ* has rapidly increased over the past two decades. Oncologists attribute this increase primarily to the increased incidence of HPV, types 16 and 18. Several studies have demonstrated that condyloma of the cervix (HPV lesions) often co-exists with dysplasia. These lesions, many of which contain HPV subtype 16 or 18, have at least a 30% incidence of malignant transformation. However, most work suggests that HPV infection alone is insufficient to induce dysplastic promotion in an immunocompetent host. Here, cofactors such as cigarette smoking, co-infection with other viruses (e.g., herpesviruses), and other phenomena associated with immunosuppression are likely to be necessary. Recent work has identified abnormalities in cellular immunity accompanying advancing levels of cervical neoplasia. Cervical neoplasia may be an especially pertinent pathophysiological phenomenon for which to identify psychoneuorimmunologic (PNI) influences because (a) promotion from viral infection to early CIN, and finally to invasive carcinoma appears to be associated with immunosuppressed states on the one hand and psychosocial stressors and coping strategies and resources on the other, and (b) these psychosocial factors have been associated with decrements in cellular immunity in several different populations.

CERVICAL NEOPLASIA AND HUMAN
PAPILLOMAVIRUS INFECTION

There has been a good deal of evidence suggesting that specific types of HPV infection may initiate or promote pre-cancerous squamous intraepithelial lesions (SILs) also referred to as varying levels of CIN. These cervical changes are commonly associated with the transformation to invasive carcinoma of the cervix, though the epidemiologic evidence for this association is not entirely conclusive.[3] It is noteworthy that the incidence of HPV-associated genital warts grew by a factor of four from 1966-1981, and that this trend was especially striking among African American women.[4] Given that HPV is known to be the most prevalent sexually transmitted disease (STD)[5] this would place cervical carcinoma as one of the few recognized cancers associated with a sexually transmitted virus. Sexual behaviors have been known for several years to be associated with the incidence of this cancer as evidenced by its association with early age at first intercourse and increasing numbers of sexual partners.[6,7] Moreover, transmission (of infectious agents including HPV) from male sexual partners has also been implicated.[8,9] Genital-genital contact is by far the most common transmission route for HPV and some studies of women with condylomata or CIN showed identical HPV types in the penile lesions of their male partners in 77% of cases.[10,11]

Because cervical carcinoma is one of the few human cancers believed to be linked to a viral infection, much work has been dedicated in recent years to identifying which specific types of HPV carry a malignant potential. HPV DNA has been found in approximately 90% of invasive squamous cell carcinomas of the cervix, a figure that varies considerably as a function of the assessment procedure employed (e.g., the most sensitive techniques such as polymerase chain reaction [PCR] have yielded prevalence rates as high as 100% in both invasive cervical carcinoma and carcinoma *in situ*).[12] Among the more than 60 types of HPV identified to date, at least 10 of them--types 16, 18, 31, 33, 35, 45, 51, 52, 56, and 58--have been associated with moderate dysplasia (CIN II), severe dysplasia (CIN III), primary invasive cervical carcinomas, and metastatic disease.[4,13,14]

HPV 16 is the most frequently identified type, present in 40-50% of pre-malignant cervical lesions, and in 50-70% of cervical carcinomas.[12] Recent work suggests that HPV 16 and 18 genomes increase systematically with progressive levels of CIN: in only 15% of CIN I lesions, up to 75% of CIN III lesions, and 92% of microinvasive carcinomas.[15] One study reported the results of 2627 women recruited into eight studies of HPV and cervical neoplasia during the period 1982-1989.[16] Using specific probes for 15 HPV subtypes they examined specimens from 153 cancers, 261 high-grade SILs (CIN II or III), and 377 low-grade SILs (CIN I or no CIN) and found HPV DNA in 79% of the overall samples, as compared to 6.4% (101 of 1566) of normal subjects. "Low-risk" HPV subtypes--6, 11, 42, 43, 44--were present in 20.2% of low-grade SILs but absent in all high-grade SILs and cancer samples. Intermediate risk HPV subtypes--including 31, 33, 35--were detected in 23.8% of high-grade SILs and 10.5% of cancers. HPV 16 was associated with 47.1% of high-grade SILs and cancers and HPV 18, 45, and 56 were found in 26.8% of cancers but only 6.5% of high-grade SILs.

Overall, the presence of an oncogenic HPV subtype yielded relative risks ranging from 65.1-235.7 for high-grade SILs and 31.1-296.1 for invasive cancer. One group theorized that integration of HPV results in disinhibited transcription of the E6 and E7 genes.[17] Although the evidence is inconclusive at present, some recent empirical studies do suggest a role for HPV 16/18--and possibly, types 31, 33, and 35--in cervical carcinogenesis, but an unlikely contribution for HPV 6/11.[18-22] For types 16 and 18, it is believed that viral DNA is integrated into the cellular chromosomes as opposed to in benign changes (e.g., CIN I) where HPV 16 and 18 DNA remains extrachromosomal. Thus, despite the fact that CIN II and CIN III (including carcinoma *in situ*, CIS) form a morphologic continuum that merges with CIN I, the more advanced grades of dysplasia display molecular differences that separate them from CIN I. In general, the virologic data shows a dichotomous pattern rather than a continuous distribution across CIN grades--whereas CIN I is associated with several benign HPV subtypes (e.g., 6 and 11), CIN II and III consist largely of HPV 16 as well as types 31, 33, 35 and 18, the latter one primarily seen in invasive carcinoma.[23] Thus some have designated women with HPV types 16,18,31,33,or 35 as being at high risk for cervical cancer and those with types 6 or 11 as representing a low-risk group.

PUTATIVE PSYCHOSOCIAL CO-FACTORS
FOR CERVICAL CARCINOMA

Many women who develop cervical neoplastic changes are young, African American, low socioeconomic status (SES) women - a group known to experience multiple chronic psychosocial stressors.[24] These include chronic financial hassles, drug and alcohol dependency, poor social networks, medical problems, and a general feeling of helplessness and powerlessness. These stressors may interact to aggravate a perceived loss of control over environmental stimuli. Over the past 11 years we have studied the relationship of psychosocial stressors, coping strategies, and social support in CIN and squamous cell carcinoma of the cervix in three separate samples made up of 208 women living in the greater Miami metropolitan area. This work was conducted in collaboration between the Departments of Obstetrics and Gynecology, Psychology, and Psychiatry at the University of Miami School of Medicine. We assessed a theroretically-derived set of psychosocial factors in gynecology clinic patients who were undergoing a colposcopy and directed cervical biopsy following determination of an abnormal Pap smear. All psychosocial assessments were made before subjects were notified of the results of their pathology report; thus, their self-reports on life stressors, coping strategies, and social support measures are presumably not confounded by knowledge of or psychological reaction to diagnosis. After controlling for known behavioral risk factors for cervical carcinoma, we found that across three different samples, a set of psychosocial factors appeared to be consistently related to higher levels of CIN. These included experiences of negatively rated stressful life events, the use of passive and emotionally non-expressive coping strategies to deal with stressors, an attitude of pessimism and helplessness , and social isolation.[25-28] We also found that the statistical association between these psychosocial variables and stage of CIN were strengthened by adding information on the women's perceived control over the negative life events that they had experienced[27] and the degree to which they experienced adequate social support.[29] We now review some of the results of this series of studies.

A. STUDY 1

Our first sample included 73 predominantly African American, married and high school-educated women ultimately diagnosed with CIN I, II, III, or stage I cervical squamous cell carcinoma using a control group of gynecology in-patients with leiomyomata (uterine fibroids - a benign tumor) awaiting total abdominal hysterectomy with bilateral salpingo - oophorectomy.[25] This control group was viewed as a stringent and relevant one given the fact that they were experiencing the stress of hospitalization and associated distress, yet had a benign medical condition related to atypical growth in the same organ. Subjects in these five groups showed no differences in age, Pap smear frequency/year, age at first coitus or other risk factors for cervical squamous cell carcinoma. A greater amount of negative life events experienced in the six months prior to testing was associated with a greater level of CIN; in fact, the CIN III and cancer stage I subjects reported approximately twice the number of negative events as compared to the other groups. Analysis of individual differences in coping styles indicated that subjects high in pessimism, social alienation, somatic anxiety, life threat reactivity (an index of autonomic responsivity to a given level of stress) and self-expectations showed the largest stressor association with level of cervical atypia.

B. STUDY 2A

To further test our hypothesis that psychosocial factors might interact with viral infections associated with CIN presence and severity we modified our design. This study[26] involved an entirely outpatient sample -- the control group were women ultimately diagnosed as positive for HPV but negative for CIN. This design removed the possible bias of hospitalization while at the same time providing a comparison of women who shared two critical factors: (a) possessing a viral infection associated with CIN, and (b) experiencing a shared psychosocial stressor--recent notification of an abnormal Pap smear. In this sample of 75 women there were no differences among diagnostic groups in age, Pap smear frequency, race or marital status. Structured interviews indicated that the cancer Stage I women had greater marital dissatisfaction and the highest proportion of cases reporting the use of passive/helplessness strategies to cope with stressors. Cancer subjects scored higher in dispositional pessimism, hopelessness and somatic anxiety with CIN III subjects showing the next highest elevations. A summative index of these maladaptive attitudes was positively correlated with level of cervical cellular atypia. Moreover, cancer stage I subjects scored higher in passive, conforming, and repressive coping styles as compared to HPV and CIN I subjects. Regression analyses showed that the interaction of a passive, conforming and repressive coping style with pessimism, hopelessness and somatic anxiety predicted 15% of the variance in level of cervical cellular atypia.[26] Generally, results indicated that a susceptible group of women --those who were characterized as passive, pessimistic, repressive, conforming, avoiding, and somatically anxious-- showed higher levels of cervical cellular atypia while, in contrast, a resilient group -- those who were characterized as more optimistic and used more active coping styles-- showed little or no association with level of cervical cellular atypia.

C. STUDY 2B

In a separate paper we conducted a more highly differentiated analysis of stressful life event factors in a subset of HPV and CIN subjects (no cancer subjects).[27] Findings indicated that the perceived controllability of negative life events experienced over the preceding year accounted for 21.6% the variance in level of CIN with lower controllability scores associated with greater levels of CIN. The controllability was also moderated by dispositional pessimism and an inhibited coping style in this sample.

D. STUDY 3

We conducted a third study of 60 women involving the same paradigm that we used previously with the addition of a situational coping assessment and several life style (control) measures.[28] Here we measured subjects' coping responses to a specific single stressor -- notification of an abnormal Pap smear -- and focused on the analysis of differences between HPV-only subjects and those with CIN I, II and III (no cancer subjects). The majority of subjects were African American and never married and there were no age, race, religion, education or marital status differences among diagnostic groups. Moreover, we found no group differences in self-reported diet, sleep, exercise frequency, smoking, alcohol or recreational substance use, prescribed medication use, or sexual history ; moreover, none of these variables were significantly correlated with the level of CIN. Perceived controllability, predictability, and duration of negatively rated life events were also assessed among diagnostic groups. Perceived controllability over all negatively related life event stressors occurring within the past year was found to differ among groups. A post-hoc comparison between the HPV infection control group and the combined CIN (I - III) group mean controllability scores was highly significant with the CIN groups reporting significantly less control over negatively-rated life events. As in our previous study,[27] the diagnostic groups did not significantly differ with regard to perceived predictability, or duration of negative life events. When the life event predictor variables (perceived impact, and perceived controllability) were entered singly along with the impact-by-controllability interaction term, a significant overall regression equation resulted. These predictors combined to predict 16.7% of the variance in CIN level, though only perceived impact and the impact-by-controllability interaction made significant contributions. Both correlational and polynomial trend analyses replicated our first study,[25] as we found a linear trend for increasing negatively rated stressful life events over the previous six months with increasing level of CIN. The findings regarding controllability of life events are similar in some ways to study 2B wherein 21.6% of the variance in CIN grade was predicted by stressor controllability.

Interestingly, the ethnic profile of the 208 women that we have studied to date seems to impact the pattern of psychosocial variable - CIN progression relationships that emerge. We have reasoned that host-environment transactional processes[30] may differ as a function of ethnicity. For instance, social support utilization and the consistency of coping responses may vary among these groups with implications for self-efficacy when encountering environmental challenges.[31] It was, however, noteworthy that in both a mostly African American sample[28] and a largely non-Hispanic White sample,[27] the strongest independent predictor of CIN level was the women's perceived control over negatively-rated major life events. Thus, a low sense of personal control over the environment (as reflected in elevated reports of uncontrollable negatively-rated life events) seems to be related to advancing levels of CIN across the populations that we have studied. It should be mentioned that in each of these samples, our psychosocial model accounted for approximately 25% of the variance in CIN grade - a significant portion of the remaining variance is undoubtably due to which HPV type that these women were infected with.   However, not all women with these viral infections go on to develop cervical neoplasia. We have reasoned that the relationships observed between psychosocial factors (such as uncontrollable negative life events and passive, repressive coping styles) and level of CIN may be mediated by changes in immune system status.[32]

IMMUNOMODULATORY CO-FACTORS
IN CERVICAL CARCINOMA

Most work suggests that HPV infection alone may be insufficient to induce dysplastic changes in an immunocompetent host. Here, cofactors such as cigarette smoking[33,34] and co-infection with herpesviruses and other pathogens[35] that may contribute to immunosuppression are possibly sufficient.[36,37] A recent meta-analysis of studies relating smoking to cervical cancer found that after adjusting for age and number of sexual partners that smoking conferred a weighted odds ratio of 1.42 (95% Confidence Interval: 1.33 -1.51).[38] Women with HPV 16/18 who smoke appear to have a 5 -10 fold increased risk of cervical cancer over their non-smoking counterparts[34] and this may be attributable to the suppressive effects of smoking on local immunity.[39] Also, passive smoking (>3 hrs/day) increased risk among smokers (2.96:1) and non-smokers (3.43:1).[40] Although this may be due to other partner factors, one study has shown the passive smoking effect after adjusting for sexual partners of the male (though smoking practices of the women themselves were not measured here).[41] Significantly more smokers than non-smokers have koilocytic cells in their cervical biopsies (93.5%), suggesting a synergism between HPV infection and smoking. This may be attributable to the suppressive effects of smoking on local immunity, systemic immunity, or by absorption of carcinogenic components from the blood by the cervix.

There is a well known increased incidence of cervical carcinoma among immunosuppressed populations (e.g., renal transplant recipients).[42,43] Moreover, cervical cancer increased 50-fold among other patients receiving similar immunosuppressive therapy.[44] In one of these studies 80% of women had decreased T-helper/suppressor (T4/T8) cell ratios and T cell mitogen responses[44] - similar to but less profound than those seen in HIV-1 infection. There may also be a greater likelihood of malignant transformation of HPV to CIN among immunosuppressed women[45] and progression from HPV to high-grade SIL and cervical carcinoma may be at least partially dependent upon depression in specific T-cell subpopulations (e.g., decreased T4/T8 ratio).[46] Natural killer (NK) cells, which have spontaneous cytotoxic effects against virally-infected cells, have been shown to be present in the subepithelial stroma of women with HPV infections and CIN,[47] where they may play a role in preventing stromal invasion. It is also known, however, that NK cell cytotoxicity (NKCC) is reduced in all stages of cervical carcinoma.[46,47] Nevertheless, there has been no systematic work investigating the degree to which NKCC is predictive of those HPV-infected women who go on to develop SILs, or those women with high-grade SILs who go on to develop cervical carcinoma and metastasis, despite the fact that a reasonable body of evidence exists relating NKCC to breast cancer metastasis.[50] It seems reasonable to propose that individual differences in cell-mediated immunity (CD4/CD8 ratio and NKCC) might differentiate those women with particular HPVs who do and do not progress to a high-grade SIL or invasive carcinoma.[32]

The fact that cervical cancer develops in many immunosuppressed hosts suggests that immunological escape may be of primary importance. Immunosuppression may be crucial for CIN promotion. That immunosuppression increases with the severity of CIN suggests that CIN may be monitored by the immune system for some time before the system is finally overcome or eluded (i.e., immunological escape). It is unlikely that cervical neoplasia fails to show adequate antigenicity for immune response during CIN since initiating viruses (e.g., HPV) have antigens which probably insert themselves near tumor antigens resulting in a carrier-hapten relationship and increased antigenicity. Since adequate immune system challenge is apparent in CIN, some depression of cellular immune functioning via insufficient production of cytokines (e.g., migration inhibitory factor, MIF) may be necessary to create the conditions for immunological escape allowing CIN promotion.

Other factors traditionally viewed as risk factors for cervical carcinoma include age at first coitus, age at first pregnancy, African American ethnicity, family history, low SES, years of education, number of sexual partners, number of pregnancies, incidence of STDs, non-barrier contraceptive methods, circumcision of partner, marital discord, divorce, bereavement, sexual hygiene, douching practices, poor nutrition, and Pap smear frequency.[51-54] Exposure to concomitant, multiple co-factors of progression such as smoking and STDs other than HPV does account for a significant proportion of the variance of CIN progression over time. For instance, one study found that more smokers than non-smokers present with CIN, especially CIN III, controlling for the incidence of miscarriage or abortion, pregnancy, prevalence of genital warts, use of oral contraceptive or barrier methods, and cytomegalovirus seropositivity.[53] Given the high prevalence of smoking, low SES, multiple partners, inadequate contraceptive methods, and STDs in African American women are critical to assess these variables as possible confounds in PNI research with this population. Other lifestyle factors that favor immunosuppression (i.e., sleep deprivation, alcohol and substance use, caffeine intake, inadequate nutritional status, lack of exercise) and use of prescribed medications that can adversely affect immune function[55] or other processes more directly associated with carcinogenesis such as DNA repair ability[56] may comprise the set of sufficient criteria that complete the formula.

THE PSYCHOSOCIAL MODEL, CERVICAL CANCER AND THE IMMUNE SYSTEM

Taking together all of the information just reviewed, we have proposed that the relationships observed between CIN and psychosocial factors (such as negatively-rated stressful life events and their controllability; passive, non-expressive coping strategies; and social support/isolation) in our prior work may be mediated by changes in immune system status that we have observed in association with these same psychosocial factors in HIV-1 infected gay men.[57] Such a mediational pathway is supported by the following. First, virally-associated cancers--such as cervical carcinoma- are more likely to undergo immunosurveillance than tumors initiated or promoted by chemical carcinogens, for example, and thus their clinical course may be more dependent on the immunocompetency of the host than is the case in cancers associated with chemical carcinogens.[58,59] Second, as noted previously, immunosuppressed transplant patients and HIV-infected women show a several-fold increase in risk for CIN and cervical cancer over their sociodemographically-equivalent counterparts. Third, uncontrollable life events, passive and emotionally non-expressive coping strategies, and social isolation have been related to impaired immunologic status [60-64] and poorer tumor outcomes.[65-71] Fourth, we have related these same psychosocial model variables with impairments in immunologic status[72-73] and likelihood of disease progression[74] in HIV-infected individuals. However, it should be pointed out that such relationships have not always been found in HIV-1 infected individuals.[75] Finally, we have found that changes in emotional expression, coping strategies, and social support occurring during the course of psychosocial group intervention predict improvements in NKCC and immunologic surveillance of latent herpesvirus infections in HIV-infected[76] and healthy[77] individuals.

HIV-1 INFECTION AND CERVICAL NEOPLASIA

Another disease that is impacting women is acquired immune deficiency syndrome (AIDS). Women comprise the fastest growing AIDS risk group and currently account for over 11% of known AIDS cases in the U.S. Among women aged 15-44, the death rate due to HIV-1 infection has more than quadrupled since 1985.[78] Although African-Americans account for only 19% of the women in this country, they make up 53% of all female cases of AIDS.[79]

Women with AIDS tend to be significantly younger than non-homosexual male cases, with an especially high incidence in the 20-29 year old age group.[80] Analysis of CDC female AIDS cases greater than 13 years of age from 1981-1988 showed that 85% were of reproductive age (15-44 years old) and 52% were African American.[81] The age-adjusted death rate for HIV/AIDS in African American women increased from 4.4/100,000 in 1986 to 10.3/100,000 in 1988.[78] In striking comparison, the 1988 death rate for HIV/AIDS in African American women 15-44 years of age was nine times the death rate for non-Hispanic White women in this age range.[78] African American females have a cumulative AIDS incidence 12 times that of White females in some areas. In New Jersey from 1981-1988 the relative risk (RR) of AIDS was 12.4 in Black vs. White women.[78] It is apparent that the greatest increases in female AIDS cases will occur in geographical regions characterized by high densities of substance abusers, and young, low SES minority populations.

Survival time (from the time of AIDS diagnosis to death) for women is typically much shorter than for men and this disparity appears to be magnified in ethnic minority women.[82] This disparity was believed to have been predominantly due to the failure of physicians to diagnose women with AIDS and pre-AIDS changes, as less was known about the gender-specific clinical manifestations in HIV-infected women. One such manifestation that is drawing considerable attention is cervical carcinoma and CIN or SIL. This set of clinical conditions is particularly relevant because of growing evidence that high-grade SILs and invasive cervical carcinoma are relatively common in HIV-infected African American women. Moreover, the incidence of SIL and cancer is remarkably high in those women who are co-infected with both HIV-1 and specific HPVs.

Because the development of cervical cancer and high-grade SILs has been linked to immunocompromised states and/or factors associated with impairments in the immune system it is imperative to examine the contribution of these factors to the development of cervical carcinoma in women who are significantly and chronically immunocompromised. Such is the case in women who have been infected with HIV-1. A recent report noted that 19% of 84 patients (< 50yrs. old) presenting with invasive cervical carcinoma were found to be HIV-1 seropositive (HIV+), including a 16-year old with stage IIIb disease.[83] Of the HIV+ women

14/16 were symptomatic as defined by CDC criteria. All but one HIV+ patient had squamous cell carcinomas. The HIV+ women had significantly more advanced disease as a group: only 6% of the HIV+ women had early surgical-pathologic stage disease as compared to 40% of the HIV- women. Regarding clinical stage, half of the HIV+ patients showed stage III-IV disease, while only 19% of their HIV- counterparts did so. Among patients for whom histologic confirmation of lymph node tissue was available, 5/8 (63%) of HIV+ patients were positive for metastatic disease as compared to 11/31 (35%) of HIV- patients. Following therapy all of the HIV+ patients who presented with advanced disease developed a recurrence as compared with 17/35 (49%) of HIV- women. The mean interval to recurrence was 2.3 months in these HIV+ patients. Of the original 16 HIV+ patients enrolled in the study 11 have died (mean survival time = 9.2 months) of whom nine died of cervical cancer, and two from other AIDS-related conditions. As expected HIV+ patients had significantly lower CD4 cell counts ($M = 360$ cmm) as compared to HIV- women ($M = 830$ cmm). While all HIV+ women with CD4 cell counts < 500/cmm had poor outcomes, all four patients with counts above 500 cmm had a prolonged course, and two continue to be free of disease at the close of the study.

These findings are also in line with case reports of rapidly progressive squamous cell cervical cancer in HIV+ women.[84,85] Finally, these immunologic findings point out that whereas the CD4 counts of HIV+ women, on average ($M = 360$ cmm), did not meet the World Health Organization (WHO) immunologic criterion for an AIDS diagnosis (CD4 < 200/cmm), individual differences in patients' immunologic status appeared to predict their response to treatment and thus may help in determining the aggressiveness of treatment required in these cases. There are also a growing number of published studies documenting associations between HPV infections and CIN in HIV+ women.[86-89] To the extent that promotion of HPV infections to CIN is dependent upon the host's immunologic status, one would expect a greater incidence and severity of CIN in women co-infected with HIV-1 and HPV, though the HPV type might provide the critical prognostic information. Of particular epidemiologic importance are findings that HIV+ women show an elevated prevalence of HPV infections.[88-90] Importantly, Vermund et al.[88] found that 70% of symptomatic HIV+ women were HPV infected compared with 22% of HIV- women, and that 52% of those with both viruses were SILs as compared to 18% of those with only one of the viruses, and 9% of those with neither virus. The symptomatic HIV+ women had the strongest association between HPV and SIL (odds ratio = 12) and the risk for SIL was highest among young women from ethnic minority groups.[88]

There is also preliminary evidence that the HPV-SIL association is mediated by immunologic status in HIV-1 infected women. One study[89] compared cytologic smears among 111 HIV+ women, 76 HIV- injection drug users (IDU), and 526 outpatient female controls and found the following trends: (a) 41% of HIV+ women had CIN compared to 9% of IDUs and 4% of controls; (b) HPV infections were four times more prevalent in the HIV+ vs. controls (similar to the Vermund et al. findings);[88] (c) the frequency and severity of CIN was associated with lower CD4 counts, and poorer blastogenic responses to phytohemagglutinin (PHA), pokeweed mitogen (PWM) and tetanus toxoid. Moreover, HPV-associated genital warts regressed with increased CD4 counts following zidovudine therapy in this study suggesting a pathophysiologic link between immunologic status, and HPV-associated CIN and HIV disease progression. Thus, individual differences in the degree of immunosuppression may increase the risk of an HIV+ woman developing HPV-related SIL or cervical carcinoma. It is likely that consideration of both immunomodulatory behaviors and HPV subtype will be critical considerations in the management of HIV+ women in coming years. Lymphomas and Kaposi's sarcoma comprise traditional opportunistic cancers in AIDS. It seems reasonable that cervical cancer is considered as an AIDS-defining criterion in women. Moreover, the identification of factors that are predictive of the development of high-grade SILs may identify candidates for imminent progression to cervical cancer on the one hand or another AIDS-defining event on the other.

CLINICAL DISEASE IN HPV AND HIV-1 CO-INFECTED WOMEN

Although it is impossible to predict at the present time, even in healthy people, which pre-cancerous lesions will progress, regress, or persist, generally, the more severe the grade of SIL, the greater the likelihood of progression and the shorter transit time to cervical carcinoma. For example, the early natural history studies of cervical dysplasia suggested that severe dysplasia progressed to invasive carcinoma twice as frequently as milder levels (40% to 20%)[91] and that as many as 50% of CIN I (by the terminology prior to the Bethesda System) and 70% of CIN III lesions progressed to invasive carcinoma if left untreated long enough.[92] Rates of progression, however, are not uniform and it is extremely difficult for clinicians or policy makers to predict the outcome for any specific patient, especially when prognosis is based solely upon observations of cervical morphology.

In HIV+ women there is evidence that once a woman advances to a high-grade SIL, it is imperative to begin treatment. Once invasive carcinoma is diagnosed only a small proportion of these women show a favorable clinical response to treatment- these being the women who are caught at the early stages of carcinoma and/or those with the least degree of immunocompromise (as indicated by the single most important marker-- helper/inducer [CD4] cell counts). Thus it is critical that clinicians be able to forecast the development of high-grade SILs based upon the most reliable data available. By identifying psychosocial factors associated with disease outcome, we might clarify specific targets for behavioral intervention in early, high-risk cases. In addition, knowledge of specific immunologic measures that predict the development of AIDS-related cancers such as cervical carcinoma could be useful in directing and monitoring response to treatment.

Immunosuppression is a pathophysiological consequence of HIV-1 infection and, as previously mentioned, has also been linked with HPV-associated CIN. Thus, one potential means of increasing the power to observe psychoneuroimmunological effects may be to study them in women with HPV-associated CIN and co-infection with HIV-1. Few studies have been conducted thus far on this group of women. While women with either HPV or HIV-1 infection alone are likely to be at increased risk for cervical cytological abnormalities, one recent study indicates that women with both infections are at highest risk.[90] In addition, more advanced HIV-1 infection was associated with greater risk, suggesting that the clinical changes in level of cervical (like anorectal) dysplasia parallel the degree of immunosuppression. Moreover, HPV-associated genital warts have been reported to regress with increased CD4 lymphocyte counts following zidovudine (AZT) therapy in HIV-1 infected women, suggesting a pathophysiological linkage - i.e., with decreased replication of HIV-1. In conclusion, HIV-1 infection is associated with both an increased incidence of HPV infection (in men and women) and with an increased incidence and rate of progression of CIN.

The frequent occurrence of cervical dysplasia among women at risk for HIV-1 infection may have implications for HIV-1 transmission as well as progression. HPV infection of cervical cells could facilitate HIV-1 transmission through disruption of normal mucosal integrity or decreased local immunosurveillance. Cervical squamous epithelium infected by HPV shows a decreased CD4/CD8 cell ratio, and a decreased response of cervical lymphocytes to mitogen has also been demonstrated in the setting of HPV infection.[92,93] Systemic changes are also found in CIN that parallel these local changes; CD4/CD8 cell ratio inversion and a decreased percentage of CD4 cells in the peripheral blood have been observed in 44% and 56%, respectively, of women with CIN.[94] Thus, the transmission as well as progression of HIV-1 infection may be facilitated by the presence of HPV infection. In addition, HPV-induced genital warts, CIN and invasive cervical carcinoma all may contribute further to the immunosuppression resulting from HIV-1 infection alone. It might be concluded, then, that HPV and HIV-1 infection may facilitate both the transmission and the progression of the other; if there is a true synergism (beyond an additive effect), then this co-infection may be of considerable clinical significance.

As mentioned in the introduction, women comprise over 11% of patients with AIDS in the United States; as many as 31% of HIV-1 infected women may have CIN, and 95% may show evidence suggestive of HPV infection.[95] Hence, cervical cancer is now considered an indication of AIDS as one of the specific criteria currently listed in the AIDS case definition.[96] Since women co-infected with HIV-1 and HPV are more likely to be minority group members of lower socioeconomic status, they might be expected to have less access to health care and to be less likely to have a history of regular Pap smear screening. Hence, the clinical implications of this co-infection are of special significance for minority group women and suggest the need for culturally sensitive interventions specifically targeted to minority group populations.[97]

The increased frequency of HPV infection among HIV-1 infected women could be due to: similar risk factors for exposure to both pathogens, facilitation of HIV-1 transmission among those who are HPV infected, or facilitation of risk for contracting HPV infection among those who are HIV-1 infected. Regardless, the long latency period of HIV-1 may allow cervical dysplasia or cervical cancer to occur prior to other complications of HIV-1. An even higher risk for cervical cancer in this population may be observed with the lengthening incubation period from infection with HIV-1 to AIDS,[98] which is at least partly due to advances in anti-retroviral drug treatment (e.g., the availability of ddI and ddC and combination drug treatments with zidovudine) as well as improved prophylaxis and treatment of common complications of HIV-1 infection (e.g., prophylaxis of *Pneumocystis carinii* pneumonia, the leading cause of mortality among AIDS patients).

HIV-1 infection itself has been clearly associated with decrements in CD4 cell counts and CD4/CD8 cell ratios[99] and also, though less strongly, with decrements in NKCC.[100] Psychosocial factors have been associated with changes in immunological measures known to be relevant to progression of HIV-1 infection (e.g., CD4 cell counts, CD4/CD8 ratio, mitogen response to PHA) and to cervical dysplasia (e.g., NK cell counts and cytotoxicity, CD4 cell counts, and CD4/CD8 cell ratios).[68,73,100-102] Therefore, it seems reasonable to expect that women with HPV-associated CIN and HIV-1 co-infection along with predisposing psychosocial and life style factors may be an especially high risk group for rapid clinical progression to cervical cancer. If psychosocial factors are associated with clinically significant immunological decrements, then HPV and HIV-1 co-infection might be expected to decrease survival time after diagnosis of AIDS in a similar fashion as has been demonstrated with HTLVI/II and HIV-1 co-infection.[103] This co-infection has been shown to increase mortality by a factor of three compared to those with HIV-1 infection alone among intravenous substance users.[103]

The present state of affairs in this country suggests that co-infection with HIV-1 and different HPVs- a reasonably commonly observed phenomenon in certain populations-- may be associated with the development of high-grade SILs, invasive cervical carcinoma, and a greater frequency of recurrence and shorter survival time following treatment. At the present time, however, it is impossible to predict, even in healthy people, which pre-cancerous lesions will progress, regress, or persist. In HIV+ women there is also evidence that once a woman advances to a high-grade SIL, it is imperative to begin treatment. Once invasive carcinoma is diagnosed only a small proportion of these women show a favorable clinical response to treatment- these being the women who are caught at the early stages of carcinoma and/or those with the least degree of immunocompromise (as indicated by CD4 cell counts).

THE ROLE OF PNI IN CERVICAL NEOPLASIA RESEARCH

The preliminary work completed to date argues for a multivariate model that incorporates morphologic, histologic, molecular virology, behavioral and psychosocial data to predict a woman's risk for developing high-grade SILs at the earliest possible point in time. Given the research on psychosocial factors in cervical dysplasia and PNI factors in HIV-1 infection reviewed in the previous sections, it may be expected that women with higher levels of major life stressors and lower levels of social support and active coping style may represent a risk group for more rapid progression of HPV-associated CIN among the HIV-1 infected. This might be especially true for women infected with HPV types independently associated with a greater likelihood of progression. Other factors associated with cervical cancer (e.g., cigarette smoking, substance use, and high levels of sexual activity) are independently associated with immune measures related to CIN progression -- lower NKCC and CD4/CD8 ratios. More recent work has suggested the importance of NKCC[104] and HLA-restricted T lymphocyte responses[105] in therapeutic strategies for cervical carcinoma. Hence, psychosocial factors (life stressors, social support, coping style) and virologic factors (HPV type; HIV-1 co-infection) may interact in a multi-factorial mechanism determining the likelihood of progression to cervical cancer and a common denominator of these factors may be their immunomodulatory effects.

Although we have found reasonably consistent evidence for a moderate association between certain psychosocial factors and the level of cervical neoplasia, it remains likely that a significantly large proportion of the variance in CIN and cervical carcinoma risk may be attributable to certain types of HPV infections that seem to precede the onset of this type of carcinoma. However, not all women with these viral infections go on to develop cervical neoplasia. We have reasoned that the relationships observed between psychosocial factors (such as uncontrollable negative life events and passive, repressive coping styles) and level of CIN may be mediated by changes in immune system status.[32] Such a mediational pathway is supported by the following: (a) virally-induced cancers are believed to be more immunogenic than other forms of neoplasia and thus their clinical course may be more dependent on the immunocompetency of the host; (b) pharmacologically immunosuppressed transplant patients and HIV-1 infected women show a several-fold increase in risk for CIN and cervical cancer over their sociodemographically-equivalent immunocompetent counterparts; and (c) uncontrollable stressful life events and passive or repressive coping styles have been related to both impaired immune functioning and poor tumor outcomes.

Much remains to be demonstrated regarding the biological factors responsible for the initiation and progression of CIN. Prospective studies of the rates of progression among various HPV subtypes with repeated measures on larger samples are warranted. Use of emerging laboratory techniques like the polymerase chain reaction may improve subtype assay sensitivity and specificity. Also, documentation of other potential infectious co-factors including cytomegalovirus, chlamydia, gonorrhea, syphilis and even retro-viruses must be done to isolate the impact of HPV infection. The stepwise, well-defined progression of CIN renders it amenable to research investigating the concomitant factors (i.e., immunologic status) associated with the promotion of HIV-1 infection. Lymphomas and Kaposi's sarcoma represent traditional opportunistic neoplasias in AIDS. Cervical neoplasias are part of the list of such malignancies which identify those HIV-1 infected individuals as having AIDS and as candidates for imminent progression to other AIDS-related conditions due to the documented immunosuppression and HPV infections (as potential co-factors of HIV progression) which are probable criteria for the manifestation of CIN.

This further suggests that a comprehensive, biopsychosocial model may be most powerful in predicting clinical outcomes among women with HPV and HIV-1 co-infection. We are currently examining the cross-sectional association between CIN level and immune functioning (NKCC) on the one hand and behavioral (cigarette smoking, substance use, sexual behaviors) and psychosocial factors (uncontrollable stressors, coping strategies and social support) on the other hand in women co-infected with HIV-1 and different types of HPV (e.g., 16/18/31/33/35 vs. 6/11) as part of an NCI-funded study. This work will allow us to determine the prevalence of varying levels of CIN in co-infected women as a function of psychosocial and immunologic status. To the degree that the HIV-infected woman's immune system is at the threshold necessary for conferring surveillance of particularly high-risk HPV infections (especially types 16 and 18), then even small behavioral and psychosocially-associated immunomodulatory changes may contribute to manifest cervical neoplasia. Future work, however, must test the *predictive power* of data on HPV subtype, behavioral and psychosocial factors, and immunologic status at regular time intervals in order to predict the likelihood of the development of treatable high-grade SILs and cervical carcinoma in HIV-infected women. This information could also enhance our understanding of the factors that predict the development of cervical carcinoma in the general population.

ACKNOWLEDGMENTS

Portions of the ideas and text of this chapter are adapted from those presented in **Antoni, M.H., Esterling, B. and Goodkin, K.** (in press) Interactions between HIV-1 and other sexually transmitted viruses: Herpesviruses and papillomaviruses; in Johnson, E. Amaro, H., Antoni, M.H. and Jemmott, J. Eds., *AIDS in African Americans and Hispanics: The role of behavioral and psychosocial factors*. Praeger Press; **Goodkin, K., Antoni, M.H., Helder, L., and Sevin, B.**, Psychoneuroimmunological aspects of disease progression among women with human papillomavirus-associated cervical dysplasia and human immunodeficiency virus type 1 co-infection, *Int. J. Psychiatry in Med.*, 23 (2), 119, 1993; and from **Goodkin, K., Antoni, M.H., Sevin, B. and Fox, B. H.**, A partially testable, predictive model of psychosocial factors in the etiology of cervical cancer. II. Bioimmunological, psychoneuroimmunological, and socioimmunological aspects, critique and prospective integration, *Psycho-oncology*, 2, 99, 1993. The reader is directed to these works for further detailed reviews.

This work was supported from NIH grant #NCI5P30CA14395-20.

REFERENCES

1. **Larsen, P.M., Vetner, M., and Hansen, K.**, Future trends in cervical cancer, *Cancer Letters*, 41,123, 1988.
2. **Elliott, P.M., Tattersall, M.H.N., Coppleson, M., Russell, P., Wong, T., and Coates, A.S.**, Changing character of cervical cancer in young women, *Brit. Med. J.*, 298,288, 1989.
3. **Munoz, N., Bosch, X., and Kaldor, J.M.**, Does human papillomavirus cause cervical cancer? The state of the epidemiological evidence, *Br. J. Cancer*, 57,1, 1988.
4. **Rapp, F.**, Sexually transmitted viruses, *The Yale Journal of Biology and Medicine*, 62,173, 1989.
5. **Nuovo, G. and Pedemonte, B.**, Human papillomavirus types and recurrent cervical warts, *JAMA*, 263,1223, 1990.
6. **Azocar, J., Abad, S. M.J., and Acosta, H.**, Prevalence of cervical dysplasia and HPV infection according to sexual behavior, *Int. J. Cancer*, 45, 622, 1990.
7. **Herrero, R., Brinton, L.A., and Reeves, W.C.**, Injectable contraceptives and risk of invasive cervical cancer: evidence of association, *Internatl. J. Cancer*, 46,5, 1990.
8. **Zunzunegui, M.V., King, M.C., Coria C.F., and Charlet, T.**, Male influences on cervical cancer risk, *Amer. J. of Epidemiol.*, 123,302, 1986.
9. **Brinton, L.A., Reeves, W.C., and Brenes, M.M.**, The male factor in the etiology of cervical cancer among sexually monogamous women, *Internatl. J. Cancer*, 44,199, 1989.
10. **Barrasso, R., deBrux, J., Croissant, O., and Orth, G.**, High prevalence of papillomavirus-associated penile intraepithelial neoplasia in sexual partners of women with cervical intraepithelial neoplasia, *New England J. Med.*, 317,916, 1987.
11. **Schneider, A., Kirchmayr, R., De Villiers, E.M., and Gissmann, L.**, Subclinical human papillomavirus infections in male sexual partners of female carriers, *J. Urol.*, 140,1431, 1988.
12. **van den Bruele, A.J.**, *New detection methods for human papillomavirus genotypes and possible implications for cervical cancer screening*, University of Amsterdam, 1991.
13. **Beaudenon, S., Kremsdorf, D., Groissant, O.** *et al.*, A novel type of human papillomavirus associated with genital neoplasias, *Nature*, 321,246, 1986.
14. **Durst, M., Gissman, L., Ikenberg, H., and zurHausen, H.**, A papillomavirus DNA from a cervical carcinoma and its prevalence in cancer biopsy samples from different geographic regions, *Proc. Natl. Academy of Sci.*, USA, 80,3812, 1983.

15. **Claas, E., Quint, W., Pieters, W., Burger, M., Ooterhuis, W. and Lindeman, J.,** Human papillomavirus and the three group metaphase figure as markers of increased risk for the development of cervical carcinoma, *Amer. J. Pathol.*, 140(2),497, 1992.

16. **Lorincz, A., Reid, R., Jenson, B., Greenberg, M., Lancaster, W., and Kurman, R,** Human papillomavirus infection of the cervix: relative risk associations of 15 common anogenital types, *Obstet. & Gynecol.*, 79(3),328, 1992.

17. **Tidy, J., Vousden, K.H., Mason, P.,** *et al.,* A novel deletion within the upstream regulatory region of episomal human papillomavirus type 16, *J. Gen. Virol.*, 70,999, 1989.

18. **Krchnak, V., Vagner, J., Suchankova, A., Krcmar, M., Ritterova, L., and Vonka, V.,** Synthetic peptides derived from E7 region of human papillomavirus type 16 used as antigens in ELISA, *J. Gen. Virol.*, 71,2719, 1990.

19. **Stoler, M., Rhodes, C., Whitbeck, A., Wolinsky, S., Chow, L., and Broker, T.,** Human papillomavirus type 16 and 18 gene expression in cervical neoplasias, *Human Path.*, 23, 117, 1992.

20. **Munger, K., Werness, B.A., Dyson, N., Phelps, W.C., Harlow, E., and Howley, P.M.,** Complex formation of human papillomavirus E7 proteins with the retinoblastoma tumor suppressor gene product, *EMBO J.*, 8,4099, 1989.

21. **Mann, V., de Lao, S., Brenes, M., Brinton, L., Rawls, J., Green, M., Reeves, W., and Rawls, W.,** Occurrence of IgA and IgG antibodies to select peptides representing human papillomavirus type 16 among cervical cancer cases and controls, *Cancer Res.*, 50,7815, 1990.

22. **Kochel, H., Monazahian, M., Hohne, M., Thomssen, C., Teichmann, A., Arendt, P., and Thomssen, R.,** Occurence of antibodies to L1, L2, E4 and E7 gene products of human papillomavirus types 6b, 16 and 18 among cervical cancer patients and controls, *Int. J. Cancer*, 48,682, 1991.

23. **Ambros, R.A. and Kurman, R., J.,** Current concepts in the relationship of human papillomavirus infection to the pathogenesis and classification of precancerous squamous lesions of the uterine cervix, *Seminars in Diagnostic Pathol.*, 7,158, 1990.

24. **Antoni, M., Schneiderman, N., LaPerriere, A.,** *et al.,* Mothers with AIDS, in *Living and dying with AIDS*, Ahmed, P., Ed., Plenum, NY, 1990.

25. **Goodkin, K., Antoni, M. H., and Blaney, P. H.,** Stress and hopelessness in the promotion of cervical intraepithelial neoplasia to invasive squamous cell carcinoma of the cervix, *J. Psychosomatic Res.*, 30 (1),67, 1986.

26. **Antoni, M. H., and Goodkin, K.,** Host moderator variables in the promotion of cervical neoplasia-I. Personality facets, *J. Psychosomatic Res.*, 32,327, 1988.

27. **Antoni, M. H. and Goodkin, K.,** Host moderator variables in the promotion of cervical neoplasia-II. Dimensions of life stress, *J. Psychosomatic Res.*, 33,457, 1989.

28. **Helder, L., Antoni, M.H., Goodkin, K., Donato, D., and Sevin, B.,** Life stress, situational coping responses, and cervical neoplasia, Unpublished manuscript, University of Miami.

29. **Antoni, M.H.,** Life stress, moderator variables, and the promotion and persistence of cervical intraepithelial neoplasia to invasive cervical cancer, *Dissertation Abstracts*, Ann Arbor, MI, 1986.

30. **Folkman, S., and Lazarus, R. S.,** An analysis of coping in a middle-aged community sample, *J. Hlth. and Soc. Behav.*, 21,219, 1980.

31.   **Goodkin, K., Antoni, M. H., Helder, L., and Sevin, B.,** Psychosocial factors in cervical intraepitheilial neoplasia--results of three studies and implications for HIV-infected women (letter to the Editor), *J. Psychosomatic Res.*, 37 (3),1, 1993.

32.   **Goodkin, K., Antoni, M. H., Sevin, B., and Fox, B.H.,** A partially testable model of psychosocial factors in the etiology of cervical cancer. II. Psychoneuroimmunological aspects, critique, and prospective integration, *Psycho-Oncology*, 2 (2),99, 1993.

33.   **Winkelstein, W.,** Cigarette smoking and cancer of the uterine cervix, *Banbury Rep.*, 23,329, 1986.

34.   **Reid, R. and Campion, M.J.,** The biology and significance of human papillomavirus infections in the genital tract, *Yale J. Biol. Med.*, 61,307, 1988.

35.   **Zur Hausen, H.,** Human genital cancer: synergism between two virus injections or synergism between a virus infection and initiating events, *Lancet*, II, 1370, 1982.

36.   **Jenson, B., Kurman, R., and Lancaster, W.,** Tissue effects of and host response to human papillomavirus infection, *Obstet. & Gynecol. Clinics of N. Amer.*, 14,397, 1987.

37.   **Schneider, A., Kay, S., and Lee, H.M.,** Immunosuppression as a high risk factor in the development of condyloma acuminatum and squamous neoplasia of the cervix, *Acta Cytol.*, 27,220, 1983.

38.   **Sood, A.,** Cigarette smoking and cervical cancer: Meta-analysis and critical review of recent studies, *Amer. J. Prev. Med.*, 7 (4),208, 1991.

39.   **Sasson, I. M., Haley, N. S., Hoffman, D., Wynder, E. L., Hellberg, D., and Nilsson, S.,** Cigarette smoking and neoplasia of the uterine cervix: smoke constituents in cervical mucus, *N. Engl. J. Med.*, 312,315, 1985.

40.   **Slatterly, M., Robison, L., Schuman, K.,** *et al.*, Cigarette smoking and exposure to passive smoke are risk factors for cervical cancer, *JAMA*, 261,1593, 1989.

41.   **Buckley, J., Harris, R., Doll, R., Vessey, M., and Willims, P.,** Case-control study of the husbands of women with dysplasia or carcinoma of the cervix uteri, *Lancet*, 2,1010, 1981.

42.   **Penn, I.,** Depressed immunity and the development of cancer, *Clin. Exp. Immunol.*, 46,59, 1981.

43.   **Penn, I.,** Cancers of the anogenital region in renal transplant recipients, *Cancer*, 58,611, 1986.

44.   **Reif, A.,** *Immunity and cancer in man: An introduction*, Reid, A., Ed., Dekker, New York, 1975.

45.   **Shokri-Tabibzadeh, S., Koss, L., Molnar, J., and Romney, S.,** Association of human papillomavirus with neoplastic processes in the genital tract of four women with impaired immunity, *Gynecol. Oncol.*, 12,S129, 1981.

46.   **Castello, G., Esposito, G., Stellato, G., Mora, L.D., Abate, G., and Germano, A.,** Immunological abnormalities in patients with cervical carcinoma, *Gynecol. Oncol.*, 25,61, 1986.

47.   **Tay, S., Jenkins, D., and Singer, A.,** Natural killer cells in cervical intraepithelial neoplasia and human papillomavirus, *Br. J. Obstet. Gynecol.*, 94,901, 1987.

48.   **Pillai, M., Balaram, P., Abraham, T.,** *et al.*, Natural cytotoxicity and serum blocking in cervical neoplasia, *Am. J. Reproductive Immunol. and Microbiol.*, 16,159, 1988.

49. **Satam, M., Suraiya, J., Nadkarni, T.,** *et al.*, Natural killer and antibody dependent cellular cytotoxicity in cervical carcinoma patients, *Cancer Immunol. Immunother.*, 23,56, 1986.

50. **Herberman, R.B.,** Possible role of natural killer cells and other effector cells in immune surveillance against cancer, *The J. Investigative Dermatol.*, 83 (Suppl. 1),137s, 1984.

51. **Deeley, T.,** Cancer of the cervix uteri: An epidemiologic survey, *Clin. Radiol.*, 27,43, 1976.

52. **Graham, S., Snell, L., Graham, J.,** *et al.*, Social trauma in the epidemiology of cancer of the cervix, *J. Chronic Dis.*, 24,711, 1971.

53. **Grail, A., and Norval, M.,** Significance of smoking and detection of serum antibodies to cytomegalovirus in cervical dysplasia, *Br. J. Obstet. Gynecol.*, 95,1103, 1988.

54. **Peters, R., Thomas, D., Hagan, D.,** *et al.*, Risk factors for invasive cervical cancer among Latinas and non-Latinas in Los Angeles county, *J. Natl. Canc. Inst.*, 77,1063, 1986.

55. **Kiecolt-Glaser, J. and Glaser, R.,** Methodologic issues in behavioral immunology research with humans, *Brain, Behavior and Immunity*, 2,67, 1988.

56. **Glaser, R., Thorn, B., Tarr, K., Kiecolt-Glaser, J., and d'Ambrosio, S.,** Effects of stress on methyltransferase synthesis: An important DNA repair enzyme, *Hlth. Psychol.*, 4 (5), 403, 1985.

57. **Goodkin, K., Antoni, M.H., Helder, L., and Sevin, B.,** Psychoneuroimmunologic aspects of disease progression among women with human papillomavirus-associated cervical dysplasia and human immunodeficiency virus type 1 co-infection., *Int. J. Psychiatry in Med.*, 23 (2),119, 1993.

58. **Fox, B.,** Psychosocial factors and the immune system in human cancer, in *Psychoneuro-immunol.*, Ader, R., Ed., Academic Press, New York, 1981.

59. **Fox, B. H.,** Current theory of psychogenic effects on cancer incidence and prognosis, *J. Psychosocial Oncol.*, 1,17, 1983.

60. **Esterling, B., Antoni, M.H., Kumar, M., and Schneiderman, N.,** Emotional repression, stress disclosure responses and Epstein Barr viral capsid antigen titers, *Psychosomatic Med.*, 52,397, 1990.

61. **Kiecolt-Glaser, J., Garner, W., Speicher, C., Penn, G.M., Holliday, J., and Glaser, R.,** Psychosocial modifiers of immunocompetence in medical students, *Psychosomatic Med.*, 46,7, 1984.

62. **Glaser, R., Thorn, B., Tarr, K., Kiecolt-Glaser, J., and d'Ambrosio, S.,** Effects of stress on methyltransferase synthesis: An important DNA repair enzyme, *Hlth. Psychol.*, 4 (5), 403, 1985.

63. **Glaser, R., Rice, J., Sheridan, J., Fertel, R., Stout, J., Speicher, C., Pinsky, D., Kotur, M., Post, A., Beck, M., and Kiecolt-Glaser, J.,** Stress-related immune suppression: Health implications, *Brain, Behav. Immun.*, 1(19),7, 1987.

64. **Glaser, R., Pearson, G.R., Jones, J.F., Hillhouse, J., Kennedy, S., Mao, H. Y., and Kiecolt- Glaser, J.K.,** Stress-related activation of Epstein-Barr virus, *Brain, Behav. Immun.*, 5,219, 1991.

65. **Laudenslager, M., Ryan, S., Drugan, R., Hyson, R., and Maier, S.,** Coping and immunosuppression: Inescapable but not escapable shock suppresses lymphocyte proliferation, *Science*, 221,568, 1983.

66. **Visintainer, M., Volpicelli, J., and Seligman, M.,** Tumor rejection in rats after inescapable or escapable shock, *Science*, 216(4544),437, 1982.

67. **Levy, S., Herberman, R., Maluish, A., Schlein, B., and Lippman, M.,** Prognostic risk assessment in primary breast cancer by behavioral and immunological parameters, *Hlth. Psychol.*, 4,99, 1985.

68. **Levy, S., Herberman, R., Lippman, M., and d'Angelo, T.,** Correlation of stress factors with sustained depression of natural killer cell activity and predicted prognosis in patients with breast cancer, *J. Clin. Oncol.*, 5,348, 1987.

69. **Goldstein, D., and Antoni, M. H.,** The distribution of repressive coping styles among non-metastatic and matastatic breast cancer patients as compared to non-cancer patients, *Psychol. & Hlth: An Internatl. J.*, 3,245, 1989.

70. **Temoshok, L., Heller, B., Sagebiel, R., Blois, M., Sweet, D., DiClemente R., and Gold, M.,** The relationship of psychosocial factors to prognostic indicators in cutaneous malignant melanoma, *J. Psychosomatic Res.*, 29(2),139, 1985.

71. **Greer, S., Morris, T., and Pettingale, K.W.,** Psychological response to breast cancer: Effect on outcome, *Lancet*, 2,785, 1971.

72. **Antoni, M.H., Goodkin, K., Goldstein, D., Ironson, G., LaPerriere, A., Schneiderman, N. and Fletcher, M.A,** Coping responses to HIV-1 serostatus notification predict short and longer-term affective distress, *Psychosomatic Med.*, 53,227, 1991 (abstract).

73. **Goodkin, K., Blaney, N.T., Feaster, D., Fletcher, M.A., Mantero-Atienza, E. Klimas, N.G., Morgan, R.O., Millon, C., Szapocznik, J., and Eisdorfer, C.,** Active coping style is associated with natural killer cell cytotoxicity in asymptomatic HIV-seropositive homosexual men, *J. Psychosomatic Res.*, 36,635, 1993.

74. **Ironson, G., Friedman, A., Klimas, N., Antoni, M.H., Fletcher, M.A., LaPerriere, A., Simoneau, J. and Schneiderman, N.,** Distress, denial, and low adherence to behavioral interventions predict faster disease progression in HIV-1 infected gay men, *Internatl. J. Behav. Med.*, 1,90, 1994.

75. **Rabkin, J.G., Willimas, J.B.W., Remien, R.H., Goetz, R,. Kertzner, R., and Gorman, J.M.,** Depression, distress, lymphocyte subsets, and human immunodeficiency virus symptoms on two occasions in HIV- positive homosexual men, *Arch. Gen. Psychiatry*, 48, 111, 1991.

76. **Esterling, B., Antoni, MH., Schneiderman, N., Carver, C.S., LaPerriere, A., Ironson, G., Klimas, N, and Fletcher, M.A.,** Psychosocial modulation of antibody to Epstein-Barr viral capsid antigen and human herpesviruses type-6 in HIV-1 infected and at-risk gay men, *Psychosomatic Med.*, 54,354, 1992.

77. **Esterling, B., Antoni, M.H., Fletcher, M.A., Marguilles, S.and Schneiderman, N.,** Emotional disclosure through writing or speaking modulates latent Epstein-Barr virus reactivation, *J. Consult. and Clin. Psychol.*, 62 (1),130, 1994.

78. **Chu, S., Buehler J, and Berkelman, R.** Impact of the human immunodeficiency virus epidemic on morality in women of reproductive age, United States, *J. of Amer. Med. Assoc.*, 264,225, 1990.

79. *Centers for Disease Control: HIV/AIDS Surveillance*, Centers for Disease Control, April, 1992.

80. Guinan, M., and Hardy, A., Epidemiology of AIDS in women in the United States: 1981-1986, *J. Amer. Med. Assoc.*, 257(15),2039, 1987.

81. Ellerbrock, T.V., Bush, T.J., Chamberland, M.E., and Oxtoby, M.J., Epidemiology of women with AIDS in the United States, 1981 through 1990, *J. Amer. Med. Assoc.*, 265 (22),2971, 1991.

82. Fischl, M., Dickenson, G., Scott, G., Klimas, N., Fletcher, M., and Parks, W., Evaluation of heterosexual partners, children, and household contacts of adults with AIDS, *JAMA*, 257,640, 1987.

83. Maiman, M., Fruchter, R., Guy, L., Cuthill, S., Levine, P., and Serur, K. (in press) HIV Infection and invasive cervical carcinoma, *Cancer*, 71(2),402, 1993.

84. Monfardi, S., Vaccher, E., Pizzocaro, G., *et al.*, Unusual malignant tumors in 49 patients with HIV infection, *AIDS*, 3,449, 1989.

85. Schwartz, L.B., Carcangiu, M.L., Bradham, L., and Schwartz, P.E., Case Report: Rapidly progressive squamous cell carcinoma of the cervix coexisting with human immunodeficiency virus infection: Clinical opinion, *Gynecol. Oncol.*, 41,255, 1991.

86. Byrne, M., Taylor-Robinson, D., Munday, P., *et al.*, The common occurrence of human papillomavirus infection and intraepithelial neoplasia in women infected by HIV, *AIDS*, 3,379, 1989.

87. Schrager, L., Friedland, G., Maude, D., Schreiber, K., Adachi, A., Pizzuti, D., Koss, L., and Klein, R., Cervical and vaginal squamous cell abnormalities in women infected with human immunodeficiency virus, *J. AIDS*, 2,570, 1989.

88. Vermund, S., Kelley, K., Klein, R., Feingold, A., Schreiber, K., Munk, G., and Burk, R., High risk of human papillomavirus and cervical squamous intraepithelial lesions among women with human immunodeficiency virus infection, *Am. J. of Obstet. and Gynecol.*, 165,392, 1991.

89. Schafer, A., Friedman, W., Mielke, M., Schwartlander, B., and Koch, M., The increased frequency of cervical dysplasia-neoplasia in women infected with the human immumodeficiency virus is related to degree of immunosupression, *Am. J. Obstet. & Gynec.*, 164,593, 1991.

90. Henry, M., Stanley, M.W., Cruikshank, S., *et al.*, Association of human immunodeficiency virus-induced immunosuppression with human papillomavirus infection and cervical intraepithelial neoplasia, *Am. J. Obstet. Gynecol.*, 160,352, 1989.

91. Petersen, O., Spontaneous course of cervical precancerous conditions, *Am. J. Obstet. Gynec.*, 72,1063, 1956.

92. Creasman, W., and Clarke-Pearson, D., Abnormal cervical cytology: Spotting it, treating it, *Contemporary OB/GYN*, 21,53, 1983.

93. Tay, S.K., Jenkins, D., Maddox, P., and Singer, A., Lymphocyte phenotypes in cervical intraepithelial neoplasia and human papillomavirus infection, *Brit. J. Obstet. and Gynecol.*, 94, 16, 1987.

94. Turner, M.J., Ford, M.R., Barrett, M., White, J.O. and Soutter, W.P., T lymphocytes and cervical intraepithelial neoplasia, *Irish J. of Med.*, 157,184, 1986.

95. Byrne, M.A., Taylor-Robinson, D., Munday, P.E., and Harris, R.W., The common occurrence of human papillomavirus infection and intraepithelial neoplasia in women infected by HIV, *AIDS*, 3,379, 1990.

96. *American Medical Association, Leads from the MMWR.* Revision of the CDC Surveillance case definition for the acquired immunodeficiency syndrome, *J. Am. Med. Assoc.*, 258,1143, 1153, 1987.
97. **Mays, V.M. and Cochran, S.D.**, Issues in the perception of AIDS risk and risk reduction activities by Black and Hispanic/Latin women, *Amer. Psychol.*, 43,949, 1988.
98. **Taylor, J.M.G., Kuo, J.M., and Detels, R.**, Is the incubation period of AIDS lengthening? *J. Acq. Immunodef. Syndr.*, 4,69, 1991.
99. **Miesels, A. and Morin, C.**, Morphology of lesions of the uterine cervix related to human papillomavirus (HPV), *J. Exp. and Clin. Cancer Res.*, 9:1,L94, 1990.
100. **Goodkin, K., Fuchs, I., Feaster, D., Leeka, J. and Dickon-Rishel, D.**, Life stress and coping style are associated with immune measures in HIV-1 infection. A preliminary report, *Internatl. J. of Psychiatry in Med.*, 22,155, 1992.
101. **Kiecolt-Glaser, J.K., Glaser, R., Strain, E.C., Stout, J.C.**, *et al.*, Modulation of cellular immunity in medical students, *J. Behav. Med.*, 9,5, 1986.
102. **Irwin, M., Daniels, M., Bloom, E., Smith, T.L., and Weiner, H.**, Life events, depressive symptoms, and immune function, *Amer. J.of Psychiatry*, 144,437, 1987.
103. **Page, J.B., Lai, S.H., Chitwood, D.D., Smith, P.C., and Fletcher, M.A.**, HTLV-I/II seropositivity and death from AIDS, *Lancet*, 335,1439, 1990.
104. **Garzetti, G., Ciavattini, A., Provinciali, M., Valensise, H., Romanini, C., and Fabris, N.**, Influence of neoadjuvant polychemotherapy on natural killer cell activity in patients with locally advanced cervical squamous carcinoma, *Gynecol. Oncol.*, 52 (1),39, 1994.
105. **Hilders, C., Houbiers, J., vanRavenswaay Claasen, H., Veldhuizen, R. and Fleuren, G.**, Association between HLA-expression and infiltration of immune cells in cervical carcinoma, *Lab. Invest.*, 69 (6),651, 1993.

Chapter 14

PSYCHOIMMUNOLOGY AND PSYCHOBIOLOGY OF PARASITIC INFESTATION

Béla Bohus
Department of Animal Physiology
Center for Behavioral and Cognitive Neurosciences,
University of Groningen
Haren, The Netherlands

Jaap M. Koolhaas
Department of Animal Physiology
Center for Behavioral and Cognitive Neurosciences,
University of Groningen
Haren, The Netherlands

INTRODUCTION

The recent war at the Persian Gulf represented the most high-tech activity of mankind during hostilities. The technical and health care preparations were supposed to preserve human life including protection against chemical and biological weapons. Despite these measures the appearance of diverse symptoms, including fatigue, muscle and joint pain, migraine, gastrointestinal complaints, amnesia, depression, dermal reactions, etc., of mysterious origin affected tens of thousands of American and British veterans. Posttraumatic Stress Disorder, known for many Vietnam veterans, was one of the candidates as cause of the syndromes. Parasite infection - a possible means of biological warfare - was also mentioned as a causal factor, but later rejected by both military and health authorities. Surprisingly, an interaction between subclinical traumatic stress and subclinical parasite infections as a possible cause of the "mysterious" disease picture was not openly considered according to our best knowledge. Our aim is not to suggest a solution for this group of chronic sufferers. It is rather the coincidence of thoughts by reviewing the psychoimmunology of parasitic infections. A wealth of data suggests the ecological and evolutionary and psychobiological consequences of the interactions between psychosocial and biological environment, behavior, and immunity. The evidence on the mechanisms of alterations is, however, mostly circumstantial, and often not devoid of speculations. This review is an attempt to cover findings in humans and other animals under diverse, natural field and laboratory conditions. We have to realize that, in contrast to rather extensive literature on psychosocial factors and immunity, the knowledge is rather scarce in relation to parasitic infection.

SOCIAL STRESS AND PARASITE INFECTION IN RODENTS: ECOLOGICAL IMPORTANCE

A general biological requirement for survival is the ability to adapt to changes in the environment. The majority of animals, like humans, live in complex social structures. Under natural life conditions survival, that is healthy, means functioning in a social environment. The importance of the relation between stress and survival was demonstrated under natural conditions in a species of small marsupials.[1,2] In natural populations the male brown antechinus *(Antechinus stuartii)*, a small insectivorous dasyurid marsupial common in the forests of eastern Australia, does not survive more than 11.5 months. There is an abrupt increase in total mortality of males after the beginning of a single, intensive mating period in the late winter. Some females survive to rear their young and may reproduce again in a second season. The late winter period is characterized by adverse climatic conditions with low temperatures, rain, and occasional snow. Extreme intra species aggressiveness and invasion of others' territory occur also during this physically adverse period of mating.[3] Animals captured at about this time are heavily infested by endo- and ectoparasites, and an increased degree of tissue pathology is associated with the site of infestation.[2] A loss of weight,[4] enlargement of adrenal glands, an increase in plasma corticosteroid and androgen concentrationists are characteristic for this period. In addition, animals found moribund or dying shortly after capture show ulceration of the gastric or duodenal mucosa associated with hemorrhage into the lumen of the gut.[2] Males captured before mating and isolated in the laboratory survive beyond the time of natural mortality and may live double their normal life span.[5,6]

It was suggested that the state of stress, induced by aggressive interactions at the time of mating and exacerbated by the fall of corticosteroid-binding globulin concentration with a consequent rise in plasma free corticosteroid concentration, results in suppression of the immune and inflammatory system of the male antechini.[5] This suggestion can be judged from very low levels of antibody titers following sheep red blood cell (SRBC) challenge and low serum immunoglobulin (IgA and IgG) concentrations. This state of stress then causes death from gastrointestinal hemorrhage as well as invasion by parasites and microorganisms.[5] Thus, social aggressive behavior can control population dynamics not only by affecting reproduction, but also has consequences for health, disease, and ultimately life span. The natural condition seems to be an extreme one. However, it is a very suggestive example of the biological significance of the social and physical environment to the reaction of the body in terms of behavioral, endocrinological, and immunological responses. In addition, parasite infestation is playing an important role in affecting survival of the individual.

PARASITE INFESTATION AND MATE PREFERENCE: AN EVOLUTIONARY ISSUE

Charles Darwin[7] was the first to introduce the hypothesis that female mate preference is a selective force of the development of extravagant male ornamentation in many vertebrates, an expression of secondary sexual characteristics. More than a century later in a controversial theory it was proposed that parasites play an important role in the driving force behind the evolution of female choice.[8] The male secondary sexual characters function in a way to facilitate female appraisal of a potential mate's ability to be resistant against the detrimental effects of parasite infestation.

That the expression of secondary sexual characters depends upon the health state of the individuals and that ability was heritable were the basic assumptions of this hypothesis. In addition, a further assumption was that females would select for parasite resistance in their future offspring. More recently an immunocompetence-endocrine hypothesis was proposed that emphasized the role of parasites in a multiple way.[9] The main aspect of the hypothesis that the male androgen testosterone is actually a 'double-edged sword' that creates a 'physiological dilemma' for the individual. Whereas an individual benefits from high testosterone levels both for the development of secondary sexual characteristics and the actual mating success, it causes immunosuppression and thereby increases the incidence of parasites infestation. The ecological and evolutionary aspect of this hypothesis is beyond the aim of this review. However, its importance is in calling attention to the role of parasites in a number of steps of complex interactions between the environment and the behavioral, endocrine and immunological features of an individual. We know now the way parasites influence immune function and how the immune state affects parasite dynamism.[10,11] Empirical evidence suggests parasites' effects on the endocrine system and vice versa,[12,13] on behavioral aspects like dominance, stimulus function, defense mechanisms, etc.[14,15] In addition, neuroendocrine effects on the immune system are well documented.[16,17]

PARASITE INFESTATION AND HOST BEHAVIOR

As mentioned before, important psychoimmunological aspects of parasite-host interactions are the modified behavior of the hosts and its consequence for the behavior of individuals encountering parasitized conspecifics. Changes in host behavior involve altered response to predators and increase in the host's vulnerability to predators.[18,19,20] Laboratory studies with a variety of behaviors reinforce the notion of the vulnerability of parasitized individuals.[18,21] Vulnerability may be related to decreased wariness or fearfulness of the infected individual; and against predation threat.[19] The healthy individuals' response to the threat of predation is defense including flight, immobilization, freezing, risk assessment, increased wariness and fearfulness.[19,22,23] Stimuli associated with predation appear to activate antinociceptive (analgesic) mechanisms.[15,24,25] The advantage of analgesia in predation is considered to prevent disruption of behavioral actions to noxious stimuli. It is suggested that the analgesic mechanisms play a fundamental role in the organization and expression of species specific defense.[26,27] Subclinical parasite infection with *Eimeria vermiformis*, a naturally occurring enteric sporozoan parasite, induces hypalgesia or analgesia to noxious thermal stimuli and increased opioid activity in the mice.[28,29,30] The same parasite infection, however, attenuates analgesia induced by a natural predator like the cat.[31] These controversial findings can be explained by the fact that, in contrast to thermal analgesia, predator induced analgesic responses are non-opioid dependent and related to serotoninergic mechanisms in the brain.[32,33] The attenuated non-opioid analgetic behavior of the infected mice fits the suggestion of the vulnerability to predation.[18,19,20,21] The exact mechanisms remain to be determined. One may not exclude the possibility that the infection directly influences serotoninergic mechanisms in the brain. Experimental *Trypasonoma gambiensis* infection is known to affect serotonin function in the brain.[34,35]

INFLUENCE OF STRESS ON PARASITIC INFESTATION

Although a wealth of evidence support the view that stress, including psychosocial factors, profoundly and differentially affect the onset and course of bacterial and viral infections,[36,37] surprisingly less is known about parasite infestation and stress. Few of the studies use solid immunological measures and some of the reports represent simply 'case studies'. A general view of the few available studies is presented in Table 1. The studies by Leite de Moraes *et al.*[38] represent the most detailed immunological analysis of the consequences of parasite infestation and its relation to stress hormones, particularly elevated plasma corticosterone level. They studied the changes in thymic T-cell subsets in mice acutely infected with *Trypanosoma cruzi*. The infestation results in significant decline in thymocyte number, a drastic decrease in CD4+CD8+ cell number, and an increase in CD3 cell frequency. This indicates that the infestation affects the cortical thymus, whereas medullary T-cells are spared. Infections are known to increase circulating plasma corticosterone level[39] inducing thereby an internal stress signal. Infected adrenalectomized mice display the same thymic alterations as their intact counterparts. This suggests that the thymic alterations are not linked to the high levels of circulating corticosterone that show up in relation to infestation.[38] Barnard *et al.*[40] housed unrelated and initially unfamiliar male mice in groups of 6, and infected them with *Babesi microti*. Both the rate of clearance and the time taken to reach peak parasitemia was related to the animals' preinfection agressive behavior. Mice displaying more aggression were

Table 1

Parasite Infestations and Stress in Animals and Humans

Infestation	Stress	Species	Measure	Effect	Reference
Protozoa					
Trypanosoma cruzi	Unknown	Mice	Thymocyte	Involutie	38
Babesia microti	Social	Mice	Immune	Enhance	40
Plasmodium berghei	Housing	Mice	Mortality	Enhancement	41
Babesiidae	Trauma	Macaca	Antibody	Present	42
Babesiidae	Polo game	Horse	Blood cells	Reduction	43
Toxoplasma gondii	Emotional	Man	Uvea	Uveitis	44
Helminths					
Globo-cephaloides	Cold	Kangaroo	Parasites	Increased	45
Trichinella spiralis	Exercise	Rats	Parasites	Increased	46
'Parasite'	Captivity	Half ape	Death rate	Increased	47
Giardia lambia	Family stress	Man	Immune	Enhance	48

slower to clear infection, and males with more aggression reached the peak of parasitemia sooner. Stress-induced immunodepression as judged from plasma IgG and corticosterone levels showed a relation to aggression: more aggressive mice with higher corticosterone showed a reduced level of IgG. Accordingly, increased susceptibility to disease may be the consequence of aggressively maintaining high social status. Plaut *et al.*[41] examined the mortality rate of malaria *(Plasmodium berghei)*-infected mice in relation to population size. They concluded that the effect of housing is dependent on population size rather than density. The resistance was independent of sex, maturity, preinfection conditions, and of the number of infected mice in a group. Accordingly, the social factor of size and not infection factors were the essential requisites of differential resistance to death. A case story was reported by Emerson *et al.*[42] on two long-tailed macaques suffering from the stress of severe trauma in one case and type D retrovirus infection in the other. Parasites were found in the red blood cells of these two animals that were identified as *Entopolypoides macaci* (Babesiidae). In a study of 40 Argentinian horses subjected to stress of a polo game, 67.5% appeared to develop *Babesia parasitemia*.[43] In humans, uveitis, a disease in which tissue injury by diverse factors like *Toxoplasma gondii* is essential for initiation or recurrence, stress is an important modulator.[44] An epidemiological study was performed by Arundel *et al.*[45] of the nematoda parasites of the eastern gray kangaroo *(Macropus giganteus)*. *Globocephaloides trifidospicularis* caused considerable mortality in juvenile kangaroos. There was a strong relationship between the number of parasites found, falling plasma proteins, haemoglobin concentration and haemoglobin values. This nematoda can cause heavy mortality in juvenile young kangaroos in enclosed populations when the small animals with no fat reserves experience maximal cold stress. The effect of exercise on a parasitic infection was studied on *Trichinella spiralis* infested Fischer inbred rats.[46] They were trained for 2 months by running in a drum daily for 60 or 90 min. Neither the invasive and chronic nor the recovery phase of infestation was affected by exercise. It was concluded that the rat *Trichinella spiralis* system is relatively insensitive to stress. Massive parasitic infestation occurs as in the half ape *Microcebus murinus* kept in captivity for years. This and other changes like chronic nephrosis with nephritis and myocardial necrosis may be ascribed to dysbalance of adrenal cortical and medullary hormone balance by stress factors occurring in captivity, the most important of which would be social stress.[47] Finally, the association between stress, humoral and cellular immune response and intestinal parasites was evaluated in patients with respiratory allergic disease (RAD) and in normal individuals.[48] Higher levels of IgE associated with RAD were observed in patients with family dysfunctions. Other immune parameters such as total T and B lymphocytes, helper (CD4) and suppressor/cytotoxic (CD8) T lymphocytes, IgG and IgA levels were similar in patients and normal humans. *Giardia lamblia* infestation occurred in 22.5% of the investigated population, 86% of whom were suffering from RAD. Patients with high levels of IgE and eosinophilia had *G. lamblia* infection. It was suggested that a higher frequency of *lambliasis* occurred in RAD, and family stress was significantly associated with the allergic condition.

CONCLUDING REMARKS

Parasite infestation is a complex issue in both animals and humans, and the biomedical aspect represents only one of many important factors. The immunological mechanism is already of a complex nature. Parasite infections typically stimulate more than one immunological defense mechanism, and the response, humoral or cellular that predominates, depends upon the kind of parasite involved.[11] The immunological complexity may explain why knowledge about the psychoimmunology of parasite infestation is less understood than the relationship between brain, behavior and immunity during microbial or viral infection.[36,37] Another aspect, i.e., the social-economical consequence of parasite infestation - in the Western world with highly developed immunological research is not as dramatic than in the less developed or underdeveloped countries. Besides, in an undeveloped biomedical research environment, the climatic, hygienic, and health care issues are not optimal for extensive experimental studies. In summary, ecological, evolutionary, stimulus function, and stress aspects of parasite infestation were reviewed here. The collected knowledge in most cases only allows guesses and speculations rather than solid hypotheses.

Studies during the last decade and a half indicate that a bidirectional connection exists between the brain and the immune system. Brain-immune system interactions involve neuroendocrine mechanisms and the autonomic nervous system. Brain functions and behavior are affected by lymphokines produced by the activated immune system.[16,17,49] There is empirical evidence suggesting parasite effects on the endocrine system, and endocrine effects on parasite infestations.[12,13] Stress effects on the immune system, especially in the case of psychosocial stressors, are differential and depend on the nature of the environment, personality factors, the immune processes and its location within different immuno-compartments, etc.[50,51] Despite this complexity, the neuroendocrine system may play an important role in the mediation of stress effects on immune function,[36] even in the case of parasite infestation. Corticosteroids, classically considered stress hormones and suppressors of immune functions, may be important mediators of stress effects. Earlier studies using the injection of exogenous corticosteroids showed that cortisol in the mouse decreases parasitemia reticulocyte counts and increases survival time following infestation with the malaria *Plasmodium berghei*.[52] Cortisol injection decreases nematode clearance and cellular infiltration together with lymphocytopenia and eosinopenia in mice infected with *Trichinella spiralis*.[54,55] That endogenous corticosteroids, or other immunological hormones of the hypothalamus-pituitary-adrenal axis, are involved remains to be shown.

Extensive knowledge is available on cellular and molecular biology of the action of Corticotropin Releasing Hormone, ACTH, and corticosteroids on the function of diverse components of the immune defense systems.[36,56] In addition, recent findings open novel ways of understanding parasite-host interactions. It was reported that the human parasite *Schistosoma mansoni* contains and releases immunoactive proopiomelanocertin (ACTH, β-endorphin). These peptide molecules either after conversion or without conversion serve as signal molecules and interfere with the immune response of the host.[57] This mechanism represents an example of molecular mimicry by which parasites use phylogenetically conserved molecules to interfere with the host immuno-defense. Additionally to opiomelanocortins, an opioid enkephalinergic system of *Schistosoma mansoni* may also participate in parasite-host signaling.[58] This parasite signaling may profoundly influence the behavior of the host, including analgesia[28,29,51] and other adaptive behavioral changes.[59]

ACKNOWLEDGMENT

The authors wish to thank Mrs. Joke Poelstra for editorial assistance.

REFERENCES

1. **Barker, I.K., Beveridge, I., Bradley, A.J., and Lee, A.K.**, Observations on spontaneous stress-related mortality among males of the dasyurid marsupial *Antechinus stuartii* Macleay, *Austr. J. Zool.*, 26, 435, 178.

2. **Barnett, J.L.**, A stress response in *Antechinus stuartii* (Madeay), *Austr. J. Zool.*, 21, 501, 1973.

3. **Braithwaite, R.W.**, Behavioural changes associated with the population cycle of *Antechinus stuartii*, *Austr. J. Zool*, 22, 45, 1974.

4. **Woollard, P.**, Differential mortality of *Antechinus stuartii* (Macleay): Nitrogen balance and somatic changes, *Austr. J. Zool.*, 19, 347, 1971.

5. **Bradley, A.J., McDonald, I.R., and Lee, A.K.**, Stress and mortality in a small marsupial (*Antechinus stuartii* Macleay), *Gen. Comp. Endocrinol.*, 40, 188, 1980.

6. **Woolley, P.**, Reproduction in Antechinus spp. and other dasyurid marsupials, in *Comparative Biology of Reproduction in Mammals* (15th Symposium of the Zoology Society), Acad. Press, New York, 1966, p.281.

7. **Darwin, C.**, *The Descent of Man and Selection in Relation to Sex*, J. Murray, London, 1871.

8. **Hamilton, W.D., and Zuk, M.**, Heritable true fitness and bright birds: a role for parasites? *Science*, 218,384, 1982.

9. **Folstad, I., and Karter, A.J.**, Parasites, bright males, and the immunocompetence handicap, *Amer. Naturalist*, 139,603, 1992.

10. **Grossman, C.J.**, Regulation of the immune system by sex steroids, *Endocrine Rev.*, 5,435, 1984.

11. **Roitt, I.M., Brostoff, J., and Male, D.K.**, *Immunology*, Churchill Livingstone, Edinburgh, 1985.

12. **Huber, S.A., Job, L.P., and Auld, K.R.**, Influence of sex hormones on coxackie B-3 virus infection in Balb/c mice, *Cellular Immunol.*, 67,173, 1982.

13. **Spindler, K.D.**, Parasites and hormones, in *Parasitology in Focus: Facts and Trends*, Mehlborn, H., Ed., Springer, Heidelberg, 1988, p.465.

14. **Rau, M.E., and Putter, L.** Running responses of *Trichinella spiralis*-infected CD-1 mice, *Parasitology*, 89, 579, 1984.

15. **Kavalier, M., and Colwell, D.D.**, Parasitism, opioid systems and host behavior, *Adv. Neuroimmunol*, 2,287, 1992.

16. **Reichlin, S.**, Neuroendocrine-immune interactions, *New Engl. J. Med.*, 329,1246, 1993.

17. **Black, P.H.**, Central nervous system-immune system interactions: Psychoneuroendocrinology of stress and its immune consequences, *Antimicrob. Agents & Chemother.*, 39,1, 1994.

18. **Dobson, A.P.**, The population biology of parasite-induced changes in host behavior, *Quart. Rev. Biol.*, 63,139, 1988.

19. Milinsky, M., Parasites and host decision-making, in *Parasitism and Host Behaviour*, Barnard, C.J., and Behnke, J.M., Eds., Taylor and Francis, London, 1991, p.95.

20. **Temple, S.A.**, Do predators always capture substandard individuals disproportionately from prey populations? *Ecology*, 68,669, 1987.

21. **Brassard, P., Ray, M.E., and Curtix, M.A.**, Parasite-induced susceptibility to predation in diplostomiasis, *Parasitology*, 85,495, 1981.

22. **Blanchard, D.C., Blanchard, R.J., Rodgers, R.J., and Weiss, S.M.**, The characterization and modeling of antipredator defensive behavior, *Neurosci. Biobehav. Rev.*, 14,463, 1990.

23. **Endler, J.A.,** Defense against predators, in *Predator-Prey Relationships,* Feder, M.E., and Lauder, J.M., Eds., University of Chicago Press, Chicago, 1988, p. 109.

24. **Hendrie, C.A.,** The calls of murine predators activate endogenous analgesia mechanisms in laboratory mice, *Physiol. Behav.,* 49,569, 1991.

25. **Lester, L.S., and Fanselow, F.S.,** Exposure to a cat produces opioid analgesia in rats, *Behav. Neurosci.,* 99,756, 1985.

26. **Bolles, R.C., and Fanselow, M.S.,** A perceptual-recuperative model of fear and pain. *Behav. Brain Sci.,* 3,291, 1980.

27. **Rodgers, R.J., and Randall, J.I.,** Defensive analgesia in rats and mice, *Psychol. Rec.,* 37, 335, 1987.

28. **Colwell, D.D., and Kavaliers, M.,** Altered nociceptive responses of mice infected with *Eimeria vermiformis*: Evidence for involvement of endogenous opioid systems, *J. Parasitol.,* 79,751, 1993.

29. **Kavaliers, M., and Colwell, D.D.,** Exposure to the scent of male mice infected with the protozoan parasite, *Eimeria vermiformis,* induces opioid and nonopioid mediated analgesia in female mice, *Physiol. Behav.,* 52,373, 1992.

30. **Kavaliers, M., and Colwell, D.D.,** Aversive response of female mice to the odors of parasitized males: Neuromodulatory mechanisms and implications for mate choice, *Ethology,* 95,202, 1993.

31. **Kavaliers, M., and Colwell, D.D.,** Parasite infection attenuates nonopioid mediated predato-induced analgesia in mice, *Physiol. Behav.,* 559,505, 1994.

32. **Blanchard, D.C., Shepherd, J.K, Rodgers, R.J., Magee, L., and Blanchard, C. D.,** Attenuation of antipredator defensive behavior in rats following chronic treatment with imipramine, *Psychopharmacology,* 110,245, 1993.

33. **Kavaliers, M.,** Responsiveness of deer mice to a predator, the short-tailed weasel: Population differences and neuromodulatory mechanisms, *Physiol. Zool.,* 63,388, 1990.

34. **Stibbs, H.H.,** Neurochemical and activity changes in rats infected with *Trypanosoma brucei gambiense, J. Parasitol.,* 70,428, 1984.

35. **Stibbs, H.H., and Curtis, D.A.,** Neurochemical changes in experimental African trypanosomiasis in voles and mice, *Ann. Trop. Med. Parasitol.,* 81,673, 1987.

36. **Sheridan, J.F., Dobbs, C., Brown, D., and Zwilling, B.,** Psychoneuroimmunology: Stress effects on pathogenesis and immunity during infection, *Clin. Microbiol. Rev.,* 7,200, 1994.

37. **Solomon, G.F., Kemeny, M.E., and Temoshok, L.,** Psychoneuroimmunologic aspect of human immunodeficiency virus infection, in *Psychoneuroimmunology,* Ader, R., Felten, D. and Cohen, N., Eds., Acad. Press, San Diego, 1991, p.1081.

38. **Leire de Moraes, M.C., Hontebeyrei-Joskowicz, M., Leboulenger, F., Savino, W., Dardenne, M., and Lepault, F.,** Studies on the thymus in Chagas' disease. II. Thymocyte subset fluctuations in Trypanosoma cruzi-infected mice, *Scand. J. Immunol.,* 133,267, 1991.

39. **Besedowsky, H.G., and Del Rey, A.,** Physiological implications of the immune-neuroendocrine network, in *Psychoneuroimmunology,* Ader, R., Felten, J., and Cohen, N., Eds., Acad. Press, San Diego, 1991, p.589.

40. **Barnard, C.J., Behnke, J. M., and Sewell, J.,** Social behavior, stress and susceptibility to infection in house mice (*Mus musculus*): Effects of duration of grouping and aggressive behaviour prior to infection on susceptibility to *Babesia microti, Parasitology,* 107,183, 1993.

41. **Plaut, S.M., Ader, R., Friedman, S.B., and Ritterson, A.L.,** Social factors and resistance to malaria in the mouse: Effects of group vs individual housing on resistance to *Plasmodium berghei* infection, *Psychosom. Med.,* 31,536, 1969.

42. **Emerson, C., L'Tsai C.-C., Holland, C.J., Ralston, P., and Diluzio, M.E.,** Recrudescence of *entopolypoides-macaci* Mayer 1933 Babesiidae infection secondary to stress in long tailed macaques *Macaca fascicularis, Lab. Anim. Sci.,* 40,169, 1990.

43. **Oladosu, L.A., and Olufemi, B.E.,** Haematological parameters of Argentinian Polo horses with acute Babesiosis, *Trop. Vet.,* 8,163, 1990.

44. **O'Connor, G.R.,** Factors related to the initiation and recurrence of uveitis. *Am. J. Ophthalmol.,* 96,577, 1983.

45. **Arundel, J.H., Dempster, K.J., Harrigan, K. E., and Black, R.,** Epidemiological observations on the helminth parasites of *Macropus giganteus Shaw* in Victoria, Australia, *Austr. Wildl. Res.,* 17,39, 1990.

46. **Ahmad, R.A., and Harpur, R.P.,** Trichinellosis in the exercised rat, *Int. J. Parasitol.,* 12,59, 1982.

47. **Perret, M.,** Stress effects in *Microcebus murinus, Folia Primatol.,* 39,63, 1982.

48. **Herrera, J.A., De Herrera, M.A., Hurtado, H., and Herrera, S.,** Stress and lambliasis in patients with respiratory allergic disease, *Stress Med.,* 8,105, 1992.

49. **Ader, R., Felten, D.L., and Cohen, N.,** *Psychoneuroimnunology,* Acad. Press, San Diego, 1991.

50. **Bohus, B., Koolhaas, J.M., de Ruiter, A.J.H., and Heijnen, C.J.,** Stress and differential alterations in immune system functions: conclusions from social stress studies in animals, *Ned. J. Med.,* 39,306, 1991.

51. **Bohus, B., Koolhaas, J.M., Heijnen, C.J., and de Boer, O.,** Immunological responses to social stress: dependence on social environment and coping abilities, *Neuropsychobiology,* 29,95, 1993.

52. **Singer, I.,** The effect of cortisone on infections with *Plasmodium berghi* in the white mouse, *J. Infect. Dis.,* 94,164, 1954.

53. **Jackson, G.J.,** The effect of cortisone on *Plasmodium berghi* infections in the white rat, *J. Infect. Dis.,* 97,152, 1995.

54. **Coker, C.M.,** Cellular factors in acquired immunity to cortisone treatment of mice, *J. Infect. Dis.,* 98,187, 1956.

55. **Coker, C.M.,** Some effects of cortisone in mice with a spiralis, *J. Infect. Dis.,* 98,39, 1956.

56. **Munck, A., and Guyre, P.M.,** Glucocorticoids and immune function, in *Psychoneuroimmunology,* Ader, R., Felten, D.L., and Cohen, N., Eds., Academic Press, San Diego, 1991, p.447.

57. **Duvaux-Miret, O., Stefano, G.B., Smith, E.M., Dissous, C., and Capron, A.,** Immunosuppression in the definitive and intermediate hosts of the human parasite *Schistosoma mansoni* by release of immunoactive neuropeptides, *Proc. Natl. Acad. Sci. USA.,* 89, 778, 1992.

58. **Duvaux-Miret, O., Leung, M.K., Capron, A., and Stefano, G.B.,** *Schistosoma mansoni*: an enkephalinergic system that may participate in internal and host-parasite signaling, *Exp. Parasitol.,* 76,76, 1993.

59. **Bohus, B.,** Opiomelanocortins and behavioral adaptation, *Pharmac. Ther.,* 26,417, 1984.

INDEX